Decision Support Tools for Water Quality Management

Decision Support Tools for Water Quality Management

Editors

Nigel W. T. Quinn
Ariel Dinar
Iddo Kan
Vamsi Krishna Sridharan

MDPI • Basel • Beijing • Wuhan • Barcelona • Belgrade • Manchester • Tokyo • Cluj • Tianjin

Editors
Nigel W. T. Quinn
Berkeley National Laboratory
USA

Ariel Dinar
University of California
USA

Iddo Kan
Hebrew University of Jerusalem
Israel

Vamsi Krishna Sridharan
University of California
USA

Editorial Office
MDPI
St. Alban-Anlage 66
4052 Basel, Switzerland

This is a reprint of articles from the Special Issue published online in the open access journal *Water* (ISSN 2073-4441) (available at: https://www.mdpi.com/journal/water/special_issues/decision_support_tools).

For citation purposes, cite each article independently as indicated on the article page online and as indicated below:

LastName, A.A.; LastName, B.B.; LastName, C.C. Article Title. *Journal Name* **Year**, *Volume Number*, Page Range.

ISBN 978-3-0365-7794-4 (Hbk)
ISBN 978-3-0365-7795-1 (PDF)

Cover image courtesy of Nigel W. T. Quinn

Contents

water MDPI

Editorial

Decision Support Tools for Water Quality Management

Nigel W. T. Quinn [1,*], Ariel Dinar [2] and Vamsi Sridharan [3]

1 Lawrence Berkeley National Laboratory, 1 Cyclotron Road, MS 64-209, Berkeley, CA 94720, USA
2 School of Public Policy, University of California, Riverside, 900 University Ave., Riverside, CA 92521, USA
3 Tetra Tech, 10306 Eaton Place, Fairfax, VA 22030, USA
* Correspondence: nwquinn@lbl.gov

1. Introduction

The sustainability of inland water resources worldwide is becoming increasingly endangered as climate change contributes to the human-induced problems of water supply scarcity and maldistribution. Environmental problems associated with water quality have been receiving some research attention; however, the litany of natural disasters that have accompanied changes faced by water-reliant ecosystems has created a current-day crisis. Multisectoral stressors imposed on water-related ecosystems exacerbate environmental problems. Environmental challenges associated with agriculture faced by the modern world include aquifer depletion [1–6], land subsidence [7–9], the seasonal drying of river flows [10,11], waterlogging [12–14], salinization of river water and aquifers [15,16], and human health impacts from excessive use of fertilizers and pesticides [17–19] as well as the use of a wide range of household chemicals. These problems have a water quality component that requires a radical re-thinking of resource management policy and new tools to help analysts and regulators craft novel solutions. Likewise, municipal and industrial sectors that rely on a high-quality drinking water supply are cognizant of the challenges associated with curtailing pollution, while minimizing the costs of treatment and pollutant disposal (e.g., References [20–22]). As a consequence, urban areas are increasingly looking to holistic [23] and nature-based pollution-abatement strategies [24,25].

While there is a general consensus among policy, scientific, practice, regulatory and management communities that science-based decision support is necessary to manage and mitigate the deleterious effects of water pollution under climate change (Figure 1), how these decision support tools (DSTs) are designed and implemented for different applications remains an open-ended question. Over the past four decades, with the advent and rapid progress in modeling capacity and computational technology, watershed models have increasingly become effective tools for tackling a wide range of issues regarding water resources and environmental management and supporting regulatory compliance. Statistical and machine learning methods are being used to support and even supplant more traditional simulation models to improve the estimation of temporal dynamics and patterns of variability in pollutant concentrations and loads. With the advancements in data-driven analyses and modeling approaches for water quality, there are also rapid developments in such model-based DSTs for water quality management.

These DSTs are playing central roles, in the following aspects: (i) driving socio-economic decision making by helping multi-sectoral participants make better operational decisions; (ii) informing scientific policy and funding investments and guiding research by revealing data and knowledge gaps; (iii) allowing regulatory agencies to track progress towards achieving water quality goals and facilitating policy guidance; (iv) aiding managers and practitioners to make evidence-based water management decisions; and (v) serving as a conduit to the public, providing a means for leveraging citizen science initiatives (Figure 1).

Citation: Quinn, N.W.T.; Dinar, A.; Sridharan, V. Decision Support Tools for Water Quality Management. Water 2022, 14, 3644. https://doi.org/10.3390/w14223644

Received: 2 November 2022
Accepted: 3 November 2022
Published: 11 November 2022

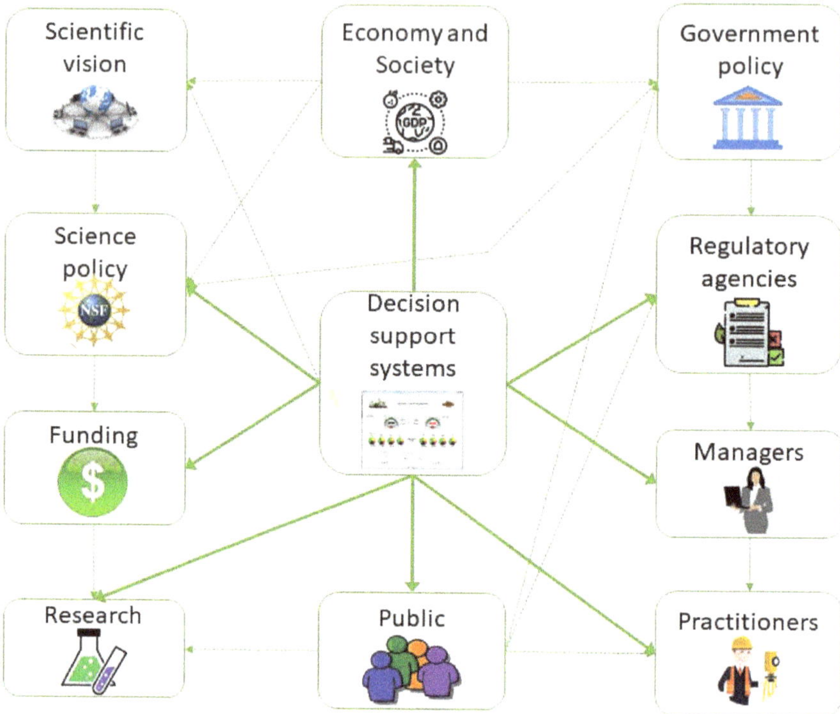

Figure 1. Central role of decision support systems in water quality management. (All icons used in this figure are available freely for public use under creative commons licensing).

The objectives of this Special Issue are to demonstrate the usefulness of decision support tools applied to different types of water quality management issues and to showcase select examples of these issues where contemporary science and technology are used to overcome associated challenges. The aim of this Special Issue within the scientific community is to drive research on emerging tools in water quality management from large-scale, programmatic scopes to small-scale, localized applications. At the same time, it is crucial to highlight the critical role played by stakeholders in supporting programmatic and implementation initiatives, and the need for stakeholder buy-in to ensure the success of water quality management programs. Thus, this Special Issue also highlights how decision support tools can aid in stakeholder participation and engagement.

2. Invitation to Submit to This Special Issue

The call for papers for this Special Issue on "decision tools for water quality management" sought contributions that describe innovative decision support approaches from around the world and across sectors that can be applied by stakeholders, government entities and regulators to reduce environmental pollution, and that can be cost-effective and sustainable.

Submissions from agriculture, municipal and industrial sectors, and environmental ecosystems were encouraged. The selected papers address broad aspects of environmental DSTs, including:

- An overview of water quality sustainability challenges and opportunities.
- Novel or successful techniques to measure and monitor water quantity and quality to achieve sustainable management.
- Sensor and remote sensing technologies that can be integrated with other and more traditional approaches to develop sustainable water quality enhancement strategies.

- Computer-based simulation modeling and other analytical techniques that enhance our understanding of the water quality issues and help formulate solution strategies.
- Benefit–cost analyses that demonstrate economic benefits and costs associated with the development and application of the decision support tool in question.
- Use of various decision support systems for the optimal management of quality aspects of water resources in a regional context and welfare consequences of water quality regulations.

https://www.mdpi.com/journal/water/special_issues/decision_support_tools (accessed on 15 June 2021).

3. Topics Covered by Papers in This Special Issue

This Special Issue consists of eleven papers covering a number of applications and approaches to the provision of decision support to tackle water quality issues. The first paper in the collection provides a policy overview of the importance of Decision Support Tool (DST) development for addressing municipal water quality problems, which is also relevant in other sectors. The applications of DSTs to other sectors is addressed by the other ten papers in this collection. Trends in the use of remote sensing for the development of DSTs used for watershed water quality regulation is explored next, followed by two papers that focus on salinity regulation as it pertains to drainage return flows from irrigated agriculture. The first of these papers is policy-focused and provides an economic perspective on salinity regulation, whereas the second paper describes and assesses the relative performance of two DSTs used in forecasting salt load assimilative capacity in a California river basin. This is followed by three papers that demonstrate various aspects of machine learning and artificial intelligence (AI) in the development of DSTs for water quality regulation at watershed and river basin scales. All three papers deal with forecasting and the interpretation of water quality data using various modeling frameworks. Next, two papers from Israel with an agricultural focus are presented, addressing the blending of saline water supply for optimal crop production and the rehabilitation and preservation of ecosystems, respectively. The next study presents a novel application in the pharmaceutical industry that details the development of a DST to regulate wastewater pollution from the household consumption of personal care products. The DST in this case is a simple ecological impact calculator based on the personal consumption of various products. A final methods paper describes the development of a DST based on tree-ring modeling data that indirectly use the effects of salinity on ecosystem damages. This ordering of papers was chosen in an attempt to provide a coherent thread of topics that starts with the description of various DSTs used for water quality forecasting, management and regulation, including the application of novel modeling approaches. More detailed summaries of each paper follow.

The first paper in this Special Issue—"The Municipal Water Quality Investigations Program: A Retrospective Overview of the Program's First Three Decades" [26] (https://www.mdpi.com/2073-4441/14/21/3426, accessed on 30 October 2022) by Paul Hutton, Sujoy Roy, Stuart Krasner and Leslie Palencia—is a policy paper that presents the history and evolution of the Department of Water Resources' Municipal Water Quality Investigations (MWQI) Program in California, USA. This paper will be of interest to readers involved in developing science-based decision-support capability related to the management of water quality. This paper focuses on regulating water quality in the Sacramento–San Joaquin Delta (Delta), which supplies nearly two-thirds of the population of California with drinking water. Some features of the Program is its ability to provide an early warning of changing conditions in source water quality, as well as data- and knowledge-based support for State Water Project (SWP) operational decision making for a wide variety of urban water users. In their paper, the authors follow Program's formation and its evolution in response to changing regulations and technological advances in water treatment and field monitoring. The paper particularly notes the development of federal drinking water quality regulations, such as the Disinfection By-Products Rule impacted the Program. The MWQI Program is the first drinking water supply program in the United States to conduct a continuous, real-time monitoring of organic carbon, bromide, and anions and to report these data on the internet. Future programs are likely to be guided by factors that trigger changes in treatment plant processes and operations, such as emerging contaminants, changes in land and water management practices, permanent Delta island flooding, sea level rise and climate change.

The second paper—"Can Remote Sensing Fill the United States' Monitoring Gap for Watershed Management?" [27] (https://www.mdpi.com/2073-4441/14/13/1985, accessed on 30 June 2022) by Vamsi Sridharan, Saurav Kumar, and Swetha Kumar—addresses the utility of various remote sensing tools for improving watershed water quality management using regulatory policy vehicles such as total maximum daily loads (TMDLs). An example of remote sensing discussed in the paper is the use of maps with different cost–payoff relationships to help stakeholders plan and incentivize remote-sensing-based water quality monitoring campaigns. One cogent application is the use of cloud cover as a proxy for the likelihood of acquiring remote scenes. The shortest time of travel to population centers was also discussed as a proxy for access to ground-truth imagery for water quality monitoring. Combining spatial indices of population, water demand, ecosystem services, pollution risk, and monitoring coverage deficits in remote-sensing-based in maps can help guide environmental management and the future use of remote sensing products. The authors found that remote sensing applications were most cost-effective for watershed monitoring in the southwestern United States and the central plains regions.

The third paper in this Special Issue—"Developing a Decision Support System for Regional Agricultural Nonpoint Salinity Pollution Management: Application to the San Joaquin River, California" [28] (https://www.mdpi.com/2073-4441/14/15/2384 (accessed on 2 August 2022)) by Ariel Dinar and Nigel Quinn—is one of a number of papers that focus on pressing water quality issues in California. The authors describe a novel stakeholder-centric approach that is being used to manage salinity in the San Joaquin River, which drains some of the most productive agricultural land in the nation and discharges into the Sacramento–San Joaquin River Delta. The authors argue that environmental problems such as salinity, the degradation of receiving waters, and groundwater resource contamination associated with irrigated agriculture require a paradigm shift in resource-management policy and a suite of new decision support tools to create sustainable solutions. The concept of real-time water quality management with a regulatory schema for sharing and allocating cost is described in the paper, as well as the application of this schema to a 20-year time series of flow and salinity data of the San Joaquin River Basin. The paper describes the simulation models and other decision support tools being applied by regulators and stakeholders.

The fourth paper—"Comparison of Deterministic and Statistical Models for Water Quality Compliance Forecasting in the San Joaquin River Basin, California" [29] (https://www.mdpi.com/2073-4441/13/19/2661 (accessed on 30 September 2021)) by Nigel Quinn, Michael Tansey and James Lu—examines two modeling approaches for the forecasting of water flow and quality in the San Joaquin River of California. This San Joaquin River Basin application complements the first paper of this Special Issue, reviewing the models used to implement real-time salinity management (RTSM). Web-based information portals have been developed to share model input data, salt-assimilative-capacity forecasts and provide stakeholder decision support in the River Basin. The paper describes two modeling approaches. The first approach is a statistical model that relies on the relationship between flow and salt concentration at three compliance monitoring sites and the use of these regression relationships for forecasting. The second approach relies on a comprehensive, data-driven computer simulation model of the Basin. The Watershed Analysis Risk Management Framework (WARMF) operates on a daily timestep and estimates daily river salt assimilative capacity along each major river reach. Although the daily regression-based forecasting model provided a marginally better performance, this was partly because the model was updated daily and was able to correct itself, whereas the WARMF model was run weekly and did not have the ability to self-adjust. The physical-based WARMF model has more utility for providing decision support to stakeholders who need to schedule salt loads from contributing watersheds.

The fifth paper—"A Hybrid Model for Water Quality Prediction Based on an Artificial Neural Network, Wavelet Transform, and Long Short-Term Memory" [30] (https://www.mdpi.com/2073-4441/14/4/610 (accessed on 20 February 2022)) by Junhao Wu and Zhaocai Wang—is a machine learning application for water quality assessments in China. This study focuses on water quality prediction in the Jinjiang River using a combination of artificial neural network (ANN), discrete wavelet transform (DWT), and long short-term memory (LSTM) models. Water quality predictions are compared to results from other models, and the results of the study show the superiority of the ANN-WT-LSTM model and suggest its potential as a decision support tool for water quality prediction.

The sixth paper—"Classification and Prediction of Fecal Coliform in Stream Waters Using Decision Trees (DTs) for Upper Green River Watershed, Kentucky, USA" [31] (https://www.mdpi.com/2073-4441/13/19/2790 (accessed on 30 October 2021)) by Abdul Hannan and Jagadeesh Anmala—is another machine leaning application for water quality simulation modeling.

The paper focuses on the classification of stream waters using the parameter of fecal coliform count for instances where there is body contact, such as in recreation, fishing and boating, and domestic utilization. The machine learning techniques involving decision trees can shed light on the structure of input variables, such as climate and land use for stream water quality prediction. The evaluated techniques include the classification and regression tree (CART), iterative dichotomiser (ID3), random forest (RF), and ensemble methods such as bagging and boosting. Input variables are used in the classification of the unknown stream water quality behavior. Of the techniques tested for the classification of fecal coliforms in the upper Green River watershed, Kentucky, USA, DTs with adaptive boosting and bagging were found to be the most accurate.

The seventh paper—"Meeting the Moment: Leveraging Temporal Inequality for Temporal Targeting to Achieve Water-Quality Load-Reduction Goals" [32] (https://www.mdpi.com/2073-4441/14/7/1003 (accessed on 30 March 2022)) by Nicole Opalinski, Daniel Schultz, TamieVeith, Matt Royer, and Heather Preisendanz—focuses on the phenomenon of hydrologic and water quality variation, often expressed as "hot moments" or episodes of unexpectedly high pollutant loading. In their study, the authors developed a Lorenz inequality decision-making framework using Lorenz curves and Gini coefficients to quantify temporal inequality using eight impaired catchments in the Chesapeake Bay watershed. The framework helps to guide the development of site-specific, cost-effective tools for contaminant load reduction and compliance with water quality objectives.

The eighth paper—"Blending Irrigation Water Sources with Different Salinities and the Economic Damage of Salinity: The Case of Israel" [33] (https://www.mdpi.com/2073-4441/14/6/917 (accessed on 20 March 2022)) by Yehuda Slater, Ami Reznik, Israel Finkelshtain, and Iddo Kan—is one of two international papers that focuses on agricultural salinity decision support. Israel has long been recognized as a leader in innovative water conservation and irrigation management technologies. Two strategies for blending water sources with different salinities as an irrigation water supply are compared using a dynamic mathematical programming model that captures this interdependence of hydrology and production economics. The two strategies compare field blending, which enables farmers to assign water with a specific salinity to each crop, with blending at a regional scale. Elevated crop salinity was observed for the regional-scale strategy, and the model results show the largest yield reductions for salt-sensitive crops.

The ninth paper—"Environmental Decision Support Systems as a Service: Demonstration on CE-QUAL-W2 Model" [34] (https://www.mdpi.com/2073-4441/14/6/885 (accessed on 20 March 2022)) by Yoav Bornstein, Ben Dayan, Amir Cahn, Scott Wells, and Mashor Housh—is another contribution from Israel that describes the application of an environmental decision support system (EDSS) for the rehabilitation and preservation of ecosystems. The EDSS is built upon the popular CE-QUAL-W2 model platform with enhancements that are configured to leverage new open-source technologies in software development (i.e., Docker, Kubernetes, and Helm) with cloud computing to significantly reduce implementation costs. For Python programmers familiar with the GitHub repository, new algorithms and executable code can be accessed by the EDSS from GitHub and employed in river basin water quality simulations. A case study is described in the paper from the Yarquon River Authority that combines agriculture and urban stakeholders and a variety of water sources with water quality concerns.

The tenth paper—"Development and Demonstration of an Endocrine-Disrupting Compound Footprint Calculator" [35] (https://www.mdpi.com/2073-4441/14/10/1587 (accessed on 20 May 2022)) by Rachel Taylor, Kathryn Hayden, Marc Gluberman, Laura Garcia, Serap Gorucu, Bryan Swistock, and Heather Preisendanz"—addresses an important topic in municipal water supply decision support. Few wastewater treatment plants were designed to remove chemicals in commonly used personal care products; hence, many of these products and their metabolites persist in plant effluent. Many of these chemicals are potential endocrine-disrupting compounds (EDCs) that cause adverse impacts to aquatic organisms at trace concentrations. The authors developed a public-domain EDC footprint calculator that prompts users to input the number of products they own into categories of health and beauty, laundry, and cleaning and estimate a user's EDC footprint (mass) together with the ranked importance of each product. A case study is presented involving 39 citizen scientists in the northeastern United States, which found the average household EDC footprint to be around 150 g. This decision-making tool can help reduce household footprint impact on future water supply by substituting certain products with greener alternatives.

The eleventh paper in this Special Issue—"Supporting Restoration Decisions through Integration of Tree-Ring and Modeling Data: Reconstructing Flow and Salinity in the San Francisco Estuary over the Past Millennium" [36] (https://www.mdpi.com/2073-4441/13/15/2139 (accessed on 20 August 2022)) by Paul Hutton, David Meko and Sujoy Roy—also focuses on salinity, although this application is further downstream in the Sacramento–San Joaquin River Delta. This paper presents updated reconstructions of watershed runoff into the Estuary from tree-ring data coupled with models that associate rainfall runoff with freshwater flow to the estuary and salinity intrusion. The authors' aim is to better understand the long-term magnitude and seasonality of changes in the Estuary, and thus provide decision support to agency engineers and regulators charged with sustaining adequate freshwater flow to the Estuary and protecting the important anadromous fishery resource. This paper confirms a dramatic decadal-scale hydrologic shift in the watershed from very wet to very dry conditions, which occurred during the late 19th and early 20th centuries, causing an increase in salinity intrusion in the first three decades of the 20th century. Population growth and extensive watershed modification during this period exacerbated this underlying hydrologic shift. Understanding the anthropogenic drivers behind this process is important for setting realistic salinity targets for estuarine restoration.

4. Conclusions

This Special Issue provided a number of modern-day examples of decision support capabilities directed toward improving environmental water quality. The collection of papers in this Special Issue suggest a wide range of issues and regulatory means that can make good use of current state-of-the-art achievements in DSTs and DSSs to improve the policy regulation of deteriorated water quality.

These examples embrace the use of state-of-the-art computer-aided technologies, in particular the use of remote sensing and machine learning, both of which dominate research applications in this field of endeavor. The DSTs derived from the applications of these technologies provide direct and actionable links between scientific research and practice, and this Special Issue highlights how such linkages can be achieved in programmatic, thematic and operational ways. However, despite the excitement generated from these activities, we should not lose sight of the important need to support and carry out basic data collection and data quality assurance activities, as well as the need to strive toward greater dissemination and our ability to share these data. Assessment and interpretation techniques count for nothing if they are based on flawed and inadequate background data. These activities need to march forward in lockstep with advances in decision support capabilities.

Author Contributions: Conceptualization, writing—review and editing, N.W.T.Q., A.D. and V.S. All authors have read and agreed to the published version of the manuscript.

Funding: This research received no external funding.

Acknowledgments: The co-guest editors would like to thank all of the authors, reviewers, and editors for their contributions. We are also grateful to the *Water* journal staff for their help and guidance in this Special Issue on "Decision Support Tools for Water Quality Management".

Conflicts of Interest: The authors declare no conflict of interest.

References

1. Feng, W.; Zhong, M.; Lemoine, J.M.; Biancale, R.; Hsu, H.T.; Xia, J. Evaluation of groundwater depletion in North China using the Gravity Recovery and Climate Experiment (GRACE) data and ground-based measurements. *Water Resour. Res.* **2013**, *49*, 2110–2118. [CrossRef]
2. Famiglietti, J. The global groundwater crisis. *Nat. Clim. Chang.* **2014**, *4*, 945–948. [CrossRef]
3. Foster, S.; Custodio, E. Groundwater resources and intensive agriculture in Europe–can regulatory agencies cope with the threat to sustainability? *Water Resour. Manag.* **2019**, *33*, 2139–2151. [CrossRef]
4. Perrone, D.; Jasechko, S. Deeper well drilling an unsustainable stopgap to groundwater depletion. *Nat. Sustain.* **2019**, *2*, 773–782. [CrossRef]
5. Frappart, F. Groundwater storage changes in the major North African transboundary aquifer systems during the GRACE era (2003–2016). *Water* **2020**, *12*, 2669. [CrossRef]
6. Jain, M.; Fishman, R.; Mondal, P.; Galford, G.L.; Bhattarai, N.; Naeem, S.; Balwinder-Singh, U.L.; DeFries, R.S. Groundwater depletion will reduce cropping intensity in India. *Sci. Adv.* **2021**, *7*, eabd2849. [CrossRef]
7. Erban, L.E.; Gorelick, S.M.; Zebker, H.A.; Fendorf, S. Release of arsenic to deep groundwater in the Mekong Delta, Vietnam, linked to pumping-induced land subsidence. *Proc. Natl. Acad. Sci. USA* **2013**, *110*, 13751–13756. [CrossRef]
8. Becker, M.; Papa, F.; Karpytchev, M.; Delebecque, C.; Krien, Y.; Khan, J.U.; Ballu, V.; Durand, F.; Le Cozannet, G.; Islam, A.S.; et al. Water level changes, subsidence, and sea level rise in the Ganges–Brahmaputra–Meghna delta. *Proc. Natl. Acad. Sci. USA* **2020**, *117*, 1867–1876. [CrossRef]
9. Perkins, S. Often driven by human activity, subsidence is a problem worldwide. *Proc. Natl. Acad. Sci. USA* **2021**, *118*, e2107251118. [CrossRef]
10. Mukherjee, A.; Bhanja, S.N.; Wada, Y. Groundwater depletion causing reduction of baseflow triggering Ganges river summer drying. *Sci. Rep.* **2018**, *8*, 12049. [CrossRef]
11. Olson, K.R.; Lang, J.M. Rio Grande an International Boundary River Is Drying up and in Need of Restoration. *Open J. Soil Sci.* **2021**, *11*, 587–610. [CrossRef]
12. Kang, M.; Jackson, R.B. Salinity of deep groundwater in California: Water quantity, quality, and protection. *Proc. Natl. Acad. Sci. USA* **2016**, *113*, 7768–7773. [CrossRef]
13. Tauro, F. River basins on the edge of change. *Science* **2021**, *372*, 680–681. [CrossRef]
14. Shaw, R.E.; Meyer, W.S.; McNeill, A.; Tyerman, S.D. Waterlogging in Australian agricultural landscapes: A review of plant responses and crop models. *Crop Pasture Sci.* **2013**, *64*, 549–562. [CrossRef]
15. Rajmohan, N.; Prathapar, S.A.; Jayaprakash, M.; Nagarajan, R. Vertical distribution of heavy metals in soil profile in a seasonally waterlogging agriculture field in Eastern Ganges Basin. *Environ. Monit. Assess.* **2014**, *186*, 5411–5427. [CrossRef]
16. Singh, A. Soil salinization and waterlogging: A threat to environment and agricultural sustainability. *Ecol. Indic.* **2015**, *57*, 128–130. [CrossRef]
17. Ward, M.H. Too much of a good thing? Nitrate from nitrogen fertilizers and cancer. *Rev. Environ. Health* **2009**, *24*, 357–363. [CrossRef]
18. Carvalho, F.P. Pesticides, environment, and food safety. *Food Energy Secur.* **2017**, *6*, 48–60. [CrossRef]
19. Hurtado-Barroso, S.; Tresserra-Rimbau, A.; Vallverdú-Queralt, A.; Lamuela-Raventós, R.M. Organic food and the impact on human health. *Crit. Rev. Food Sci. Nutr.* **2019**, *59*, 704–714. [CrossRef]
20. Dasgupta, S.; Huq, M.; Wheeler, D.; Zhang, C. Water pollution abatement by Chinese industry: Cost estimates and policy implications. *Appl. Econ.* **2001**, *33*, 547–557. [CrossRef]
21. Zhou, Y.; Khu, S.T.; Xi, B.; Su, J.; Hao, F.; Wu, J.; Huo, S. Status and challenges of water pollution problems in China: Learning from the European experience. *Environ. Earth Sci.* **2014**, *72*, 1243–1254. [CrossRef]
22. Belayutham, S.; Gonzalez, V.A.; Yiu, T.W. A cleaner production-pollution prevention based framework for construction site induced water pollution. *J. Clean. Prod.* **2016**, *135*, 1363–1378. [CrossRef]

23. Wang, M.; Janssen, A.B.; Bazin, J.; Strokal, M.; Ma, L.; Kroeze, C. Accounting for interactions between Sustainable Development Goals is essential for water pollution control in China. *Nat. Commun.* **2022**, *13*, 730. [CrossRef] [PubMed]
24. Ramachandra, T.V.; Va, S.; Asulabha, K.S. Efficacy of rejuvenation of lakes in Bengaluru, India. *Green Chem. Technol. Lett.* **2020**, *6*, 14–26. [CrossRef]
25. Chung, M.G.; Frank, K.A.; Pokhrel, Y.; Dietz, T.; Liu, J. Natural infrastructure in sustaining global urban freshwater ecosystem services. *Nat. Sustain.* **2021**, *4*, 1068–1075. [CrossRef]
26. Hutton, P.; Roy, S.; Kransner, S.; Palencia, L. The Municipal Water Quality Investigations Program: A Retrospective Overview of the Program's First Three Decades. *Water* **2022**, *14*, 3426. [CrossRef]
27. Sridharan, V.K.; Kumar, S.; Kumar, S.M. Can Remote Sensing Fill the United States' Monitoring Gap for Watershed Management? *Water* **2022**, *14*, 1985. [CrossRef]
28. Dinar, A.; Quinn, N.W.T. Developing a Decision Support System for Regional Agricultural Nonpoint Salinity Pollution Management: Application to the San Joaquin River, California. *Water* **2022**, *14*, 2384. [CrossRef]
29. Quinn, N.W.T.; Tansey, M.K.; Lu, J. Comparison of Deterministic and Statistical Models for Water Quality Compliance Forecasting in the San Joaquin River Basin, California. *Water* **2021**, *13*, 2661. [CrossRef]
30. Wu, J.; Wang, Z. A Hybrid Model for Water Quality Prediction Based on an Artificial Neural Network, Wavelet Transform, and Long Short-Term Memory. *Water* **2022**, *14*, 610. [CrossRef]
31. Hannan, A.; Anmala, J. Classification and Prediction of Fecal Coliform in Stream Waters Using Decision Trees (DTs) for Upper Green River Watershed, Kentucky, USA. *Water* **2021**, *13*, 2790. [CrossRef]
32. Opalinski, N.; Schultz, D.; Veith, T.L.; Royer, M.; Preisendanz, H.E. Meeting the Moment: Leveraging Temporal Inequality for Temporal Targeting to Achieve Water-Quality Load-Reduction Goals. *Water* **2022**, *14*, 1003. [CrossRef]
33. Slater, Y.; Reznik, A.; Finkelshtain, I.; Kan, I. Blending Irrigation Water Sources with Different Salinities and the Economic Damage of Salinity: The Case of Israel. *Water* **2022**, *14*, 917. [CrossRef]
34. Bornstein, Y.; Dayan, B.; Cahn, A.; Wells, S.; Housh, M. Environmental Decision Support Systems as a Service: Demonstration on CE-QUAL-W2 Model. *Water* **2022**, *14*, 885. [CrossRef]
35. Taylor, R.; Hayden, K.; Gluberman, M.; Garcia, L.; Gorucu, S.; Swistock, B.; Preisendanz, H. Development and Demonstration of an Endocrine-Disrupting Compound Footprint Calculator. *Water* **2022**, *14*, 1587. [CrossRef]
36. Hutton, P.H.; Meko, D.M.; Roy, S.B. Supporting Restoration Decisions through Integration of Tree-Ring and Modeling Data: Reconstructing Flow and Salinity in the San Francisco Estuary over the Past Millennium. *Water* **2021**, *13*, 2139. [CrossRef]

water

MDPI

Article

Developing a Decision Support System for Regional Agricultural Nonpoint Salinity Pollution Management: Application to the San Joaquin River, California

Ariel Dinar [1,*] and Nigel W. T. Quinn [2]

[1] School of Public Policy, University of California, Riverside, CA 92521, USA
[2] HydroEcological Engineering Advanced Decision Support Research Group, Lawrence Berkeley National Laboratory, Berkeley, CA 94720, USA; nwquinn@lbl.gov
* Correspondence: adinar@ucr.edu

Abstract: Environmental problems and production losses associated with irrigated agriculture, such as salinity, degradation of receiving waters, such as rivers, and deep percolation of saline water to aquifers, highlight water-quality concerns that require a paradigm shift in resource-management policy. New tools are needed to assist environmental managers in developing sustainable solutions to these problems, given the nonpoint source nature of salt loads to surface water and groundwater from irrigated agriculture. Equity issues arise in distributing responsibility and costs to the generators of this source of pollution. This paper describes an alternative approach to salt regulation and control using the concept of "Real-Time Water Quality management". The approach relies on a continually updateable WARMF (Watershed Analysis Risk Management Framework) forecasting model to provide daily estimates of salt load assimilative capacity in the San Joaquin River and assessments of compliance with salinity concentration objectives at key monitoring sites on the river. The results of the study showed that the policy combination of well-crafted river salinity objectives by the regulator and the application of an easy-to use and maintain decision support tool by stakeholders have succeeded in minimizing water quality (salinity) exceedances over a 20-year study period.

Keywords: San Joaquin River; salinity; decision support system; policy; assimilative capacity; real-time management economics

Citation: Dinar, A.; Quinn, N.W.T. Developing a Decision Support System for Regional Agricultural Nonpoint Salinity Pollution Management: Application to the San Joaquin River, California. *Water* **2022**, *14*, 2384. https://doi.org/10.3390/w14152384

Academic Editor: Athanasios Loukas

Received: 24 April 2022
Accepted: 28 July 2022
Published: 1 August 2022

1. Introduction

Irrigated agriculture in semi-arid regions typically produces drainage return flows with high salinity content. Tanji and Kielen [1], review the conditions of low precipitation and high evaporation in semi-arid regions that lead to a high level of salinity in the drainage return flows. They provide examples from several locations such as the Nile Delta in Egypt, the Aral Sea Basin, and the San Joaquin Valley of California. An additional important region affected by salinity and the salinity pollution of water ways is the Murry Darling River Basin in Australia (Hart et al. [2]). The authors indicate that the clearance of deep-rooted native vegetation for the development of dryland agriculture and the development of irrigation systems in the basin have resulted in more water now entering the groundwater systems, resulting in mobilization of salt to the land surface and to rivers. In these examples, the return flows are discharged to water bodies (the Nile River, the Aral Sea, the San Joaquin River, and the Murry Darling River) that could benefit from regulation aimed to minimize negative externalities in the form of damage to crops and the environment. When these negative externalities exceed certain thresholds, the regulator can respond by assessing fines or other means of encouraging compliance with water quality objectives. We have seen different types of regulators, such as a river basin authority, in charge of managing the quantity and quality of water in the basin. A river basin authority could be a state or federal entity with the authority to impose fines or restrict water allocations in the case of nonpoint source pollution. Baccour et al. [3] model the case of the Ebro River Basin in Spain, where regulations to control nonpoint source pollution of nitrates from livestock production are addressed, among other policy interventions, by restriction of water supply, imposed by the Ebro Basin Authority. Quinn [4], compares the performance of real-time, basin-scale salinity regulation in the San Joaquin River in California with those of the Hunter River Basin authority, Australia.

Some numerical simulation models can be configured to act as decision support tools, which provide alerts of potential violations of water quality objectives, and can assist in the development of schemas for creating incentives or assessing fines to encourage compliance [5,6]. Obropta et al. [5] developed a model to address hot spots for a water quality trading program intended to implement the total maximum daily load (TMDL) for phosphorus in the Non-Tidal Passaic River Basin in New Jersey. Zhang et al. [6] develop a model-based decision support system for supporting water quality management under multiple uncertainties. Such tools can have the added benefit of allowing equitable imposition of proposed incentives or fines on those polluters who bear the primary responsibility for the load exceedances. Similar decision support tools have been developed in other sectors and contexts. Ioannou et al. [7] designed a DSS to help managers in the process of decision making, in handling areas that have been burnt by forest fires, by running hypothetical (what-if) scenarios in order to achieve the best form of intervention in fire-affected regions of Greece. Makropoulos et al. [8] demonstrate the development and use of a DSS to facilitate the selection of bundles of water-saving strategies and technologies to support the delivery of integrated, sustainable water management in the UK. Rose et al. [9] identify factors affecting the selection and use of decision support tools by farmers and farm advisers in the UK for agricultural planning purposes.

Our paper presents an approach using a regional framework that enhances the utility of existing modeling tools (Watershed Analysis Risk Management Framework—WARMF), which are currently in use by practitioners in the San Joaquin Valley (Systech Water Resources Inc. [10]), to make forecasts of the salt load assimilative capacity of a major river. The river receives salt loads from catchments in the form of irrigation return flows that often exceed the river's salt load assimilative capacity. Uses of the modeling framework WARMF by practitioners are described by Quinn et al. [11], Fu et al. [12], and Quinn et al. [13], where WARMF features and performances are compared with other tools currently in use by practitioners.

The nonpoint source nature of agricultural salinity pollution poses a dual challenge for regulators by making it difficult to identify primary polluters, and to quantify pollution loads on a continuous basis. Not all drainage outlets can be monitored; therefore, calibrated simulation models play an important role in predicting pollutant loads under various permutations of hydrological and water quality inputs. Models allow alternative regulatory approaches, including schemes such as voluntary agreements and cap-and-trade in pollution permits to be evaluated, provided they can be adequately calibrated. Examples for such schemes are discussed and explained below [14–23].

Published literature on economic and regulatory aspects of nonpoint source pollution in irrigated agriculture highlights a variety of socio-political issues. These include the role of asymmetric information, value of information, effectiveness of policy interventions, and adoption of pollution reduction production practices. In an early work Griffin and Bromley [14], established a conceptual model for analyzing agricultural nonpoint pollution. An important aspect of pollution quantification is the representation of the biophysical processes linking production decisions to emission loads. Production decisions are reflected in the type and quantity of inputs in management practices and in local biophysical conditions.

An extension of the analysis in [14], was proposed in Shortle and Dunn [15], which included stochastic components in the pollution functions that arose from random natural processes as a means of addressing the lack of information about key biophysical processes. A review of various nonpoint source pollution control regulations (either incentives, taxes, or quotas) on inputs was also provided in Shortle and Dunn [15]. These are second-best interventions in the absence of direct measurements of polluter discharges. Shortle and Dunn [15], identified a reduction in the cost-effectiveness of these pollution control measures when applied uniformly across diverse agriculturally dominated subareas, which are heterogeneous in terms of water management practices and landscape characteristics, that can lead to different receiving water impact functions. To address such shortcomings, a cost-effective approach was developed by Shortle at al. [16] and was used to determine the best single-input tax policy for nonpoint source pollution in agriculture. The authors examined the question of reducing nitrate leaching from lettuce fields in California. Larson et al. [17] argue that under the certain circumstances applied, irrigation water can be the easiest single input to regulate since nitrate loading to groundwater is directly related to soil leaching rates. However, for other contaminants such as salinity, salt loads in subsurface drainage return flows may not be well correlated with surface water applications since most of the salt captured by the sub-surface drains may originate from deeper layers in the aquifer rather than from infiltrating water. Considerations of transaction costs and other political, legal, or informational constraints for dealing with nonpoint source pollution regulation were presented in Ribaudo et al. [18]. Such considerations could be applied to achieve specific environmental goals in a cost-effective manner. The authors discussed the economic characteristics of five instruments that could be used to reduce agricultural nonpoint source pollution (economic incentives, standards, education, liability, and research).

Several authors [9–23], considered regulation that had a spatial component in the presence of heterogeneity instead of regionally uniform instruments. In these works, authors demonstrated that spatially uniform policies resulted in economic efficiency losses and reduction in welfare. Kolstad [19], modeled a two-pollutant economy and showed that when marginal costs and benefits become steeper, the inefficiency associated with undifferentiated regulation increases. Wu and Babcock [20], demonstrate, among other things, that a uniform tax on polluter farmers may result in some farmers not using the chemical, and a uniform standard may have no effect on low-input land. Doole [21], finds that because of the disparity in the slopes of abatement–cost curves across dairy farms in New Zealand, a differentiated policy is more cost-effective at the levels of regulation required to achieve key societal goals for improved water quality. Doole and Pannell [22], find that due to variation in nitrate baseline emissions and the slopes of abatement–cost curves among polluting dairy farms renders a differentiated policy which is less costly than a uniform standard in the Waikato Region of New Zealand. Finally, the work by Esteban and Albiac [23], demonstrated and quantified the welfare loss from a spatially uniform regulatory policy to reduce salinity pollution and the efficiency gains from different policy measures based on the same spatial characteristics, applied in the Ebro River Basin of Spain.

Very few studies consider joint management of the nonpoint source pollution in a regional setting, using cooperative arrangements and trade, including trade-in water rights/quotas and trade-in pollution disposal permits in a regional setting. Several examples from actual cases exist. The Murray Darling Basin Authority [24], initiated a basin-wide agreement, a joint work program designed for setting salt disposal permits based on historical loads, including a revised cost-sharing formula and salinity credit allocation shares for Victoria, New South Wales, South Australia, and the Commonwealth [25]. In the Hunter Basin of New South Wales, Australia [25,26], a scheme of salt permit discharges has been put into work. The main idea of this scheme was to permit discharge of salt loads only when there was available salt load assimilative capacity in the Hunter River that drains the Hunter Basin. Quinn [4], reviews how salt load discharges to the river were scheduled by quantity, time, and location based on stakeholder need and calculations of salt load assimilative capacity using a simple spreadsheet mass–balance model.

Increasing use of high-salinity water as an irrigation source could be a serious problem. Yaron et al. [27] analyzed the economic potential to address such a problem by cooperative settlements in Israel and calculated income distribution schemes for three farms, using cooperative game theory (GT) algorithms. Work by Dinar et al. [28] also applied cooperative GT to the regional use of irrigation water under scarcity and salinity. Their model addressed inter-farm cooperation in water use for irrigation and determined the optimal water quantity and quality mix for each water user in the region.

Several additional works that represent various efforts and methods include Nicholson et al. [29] who conducted a comprehensive assessment of decision support tools used by farmers, advisors, water managers, and policy makers across the European Union as an aid to meeting the EU Common Agricultural Policy objectives and targets. Development and use of a GIS-based decision support framework was suggested by Chowdary et al. [30], integrating field scale models of nonpoint source pollution processes for assessment of nonpoint source pollution measures of groundwater-irrigated areas in India. A GIS was used to represent the spatial variation in input data over the project area and to produce a map that displayed the output from the recharge and nitrogen balance models. Different strategies for water and fertilizer were evaluated using this framework to foster long-term sustainability of productive agriculture in large irrigation projects.

The work by Quinn [4], which uses monitoring, modeling, and information dissemination for salt management in the Hunter River Basin in Australia, was compared to a more model-intensive approach deployed in the San Joaquin River Basin in California. Decision support systems for these river basins were developed to achieve environmental compliance and to sustain irrigated agriculture in an equitable and socially and politically acceptable manner. In both basins, web-based stakeholder information dissemination was a key for the achievement of a high level of stakeholder involvement and the formulation of effective decision support salinity management tools. The paper also compared the opportunities and constraints of governing salinity management in the two basins as well as the integrated use of monitoring, modeling, and information technology to achieve project objectives.

In the present paper, we provide a scalable water quality simulation model and decision support tool for a regional water quality (salinity) management problem that incorporates water/irrigation regions, each serving several individual farmers. The model operates at the subarea level where each subarea has distinct features that include political and hydrologic boundaries and which recognize different accesses to water supply and drainage resources. These subareas have been recognized by the State of California water regulatory agency with jurisdiction over the project area. We highlight the role of top-down regulations as well as market-based arrangements that might form a basis for cap and trade in pollution permits. We compare and discuss the physical as well as the welfare consequences of various policy interventions. The combination of monitoring system networks and decision support frameworks are scalable—hence, the final work product can be applied at the individual stakeholder level or aggregated at the water district level. Institutional and managerial components of the schema would need to be separately developed. Regional cooperation in the form of a market for tradeable salinity pollution permits [16], would be a significant outcome of a future study and one that is facilitated by the unique application of the simulation modeling tool.

The paper develops as follows: First, we present the analytical model aimed to evaluate the various options for pollution control at the subarea (regional) levels, responses of individual dischargers to introduced regulations, and the allocation of joint costs and benefits among the salt-discharging regions. Next, we introduce a proposed empirical framework to be applied to the San Joaquin River in California, given existing model resources in use by regulatory agencies. Then, we define a subset of seven subareas within the region as the basis for the empirical application aimed to test the analytical model. Finally, we evaluate the results to expand the method to incorporate future cooperative strategies.

2. Theoretical Model Building

2.1. Theoretical Perspective of the Analytical Model

We refer to a region that is composed of N subareas ($n = 1, 2, \ldots, N$). Each subarea n could comprise water/irrigation districts that incorporate several agricultural producers regulated by individual water district mandates. Each subarea n includes K_n (k_n = 1, 2, ..., K_n) agricultural producers that are considered nonpoint source polluters of a given pollutant, or of a set of several pollutants (for simplicity we will refer to salinity as the pollutant in question). Each agricultural producer applies water on agricultural crops to produce market products. A byproduct in the form of agricultural pollution is the irrigation return flow that may contain a regulated water quality pollutant, which we will specify to be salinity.

Each of the k producers in the n-th subarea may have different factors affecting agricultural production conditions (natural and technical) that can lead to different cropping patterns, crop yields, net revenue, and the salt concentration and salt load of the return flow. We define a production function of agricultural yield and return flow for producer k as (for simplicity we drop the indexes k and n):

$$\begin{cases} Y = f(W, C, T|\underline{X}) \\ Q = g(W,\ C,\ T|\underline{X}) \\ S = h(W,\ C,\ T|\underline{X}) \end{cases} \tag{1}$$

where Y is yield per acre of a given crop, W is water applied per acre, C is salinity level of applied water, T is irrigation technology used (expressed in integer values to indicate various irrigation technologies available to each agricultural grower within a designated subarea), Q is volume of return flow produced on that farm, S is the salt concentration of the return flows, and $\underline{X}$ is a vector of all fixed effects related to the location of that producer. We will discuss later the first and second order conditions of the production function derivatives, namely the shape of these three components of the production function.

Given Equation (1), agricultural producers within a designated subarea maximize their net revenue under constraints imposed by both natural and regulatory conditions:

$$\pi_{k_n} = \sum_{Crops} p \cdot Y \cdot L - w \cdot W - t \cdot T \tag{2}$$

s.t.:

$$\sum_{Crops} L \leq \overline{L} \tag{3}$$

$$\sum_{Crops} L \cdot W \leq \overline{W} \tag{4}$$

Additional constraints imposed on each agricultural producer within a designated subarea by subarea management, are summarized in (5), and discussed below

$$\left.\begin{matrix} Land\ fallowing\ constraints \\ Irrigation\ technology\ constraints \\ Specific\ crop\ constraints \end{matrix}\right\} \tag{5}$$

where for each agricultural producer within a designated subarea, p is the price per unit of crop, L is the area grown with that crop, w is the price of water, t is the per-acre cost of the technology, $\overline{L}$ is the total cultivable land of the agricultural producer, and $\overline{W}$ is the water quota imposed by the subarea on the agricultural producer. Net revenue is defined as the revenue from crop sales minus the variable costs of production and payments of fees for exceedance of pollution load.

The solution to (2)–(5) provides for each agricultural grower within a designated subarea: the area under production with each crop selected; the total amount of water applied; the technology selected for each crop; the total profit; the total volume of return flow from the designated subarea; and the salt concentration of the return flow that can be used to compute drainage salt (mass) loads. While we may predict the volume Q and salt loading with associated S for each subarea, such information is not available to either the subarea management or to the federal regulator.

The subarea managers have access to monitoring data that provides the total volume of Q from all agricultural producers and its quality, S, that can be used to estimate salt loading. Salt loading is the factor each subarea manager is obligated not to exceed on a monthly and annual basis by the regulator, as defined within the Total Maximum Daily Load (TMDL) allocation for each subarea. TMDLs are the policy vehicles that are used by the US Environmental Protection Agency to limit nonpoint source pollution to levels that do not exceed the assimilative capacity of the receiving water body. TMDLs are keyed into water quality standards or objectives at a compliance monitoring station for the pollutant in the receiving water, and are designed to be protective. The agricultural non-point source pollutant management problem is a typical principal–agent problem under circumstances of asymmetry of information. Hence, we need to introduce several simplifying assumptions. We start by drawing (Figure 1) a schematic regional setting, using four agriculturally dominated subareas located on the valley floor, and a water body in the form of a river (describing the actual situation in the region that we will empirically analyze). The remaining three subareas are tributary river watersheds where water flow is controlled by upstream dams and reservoirs and whose operation is largely independent of agricultural drainage decision making.

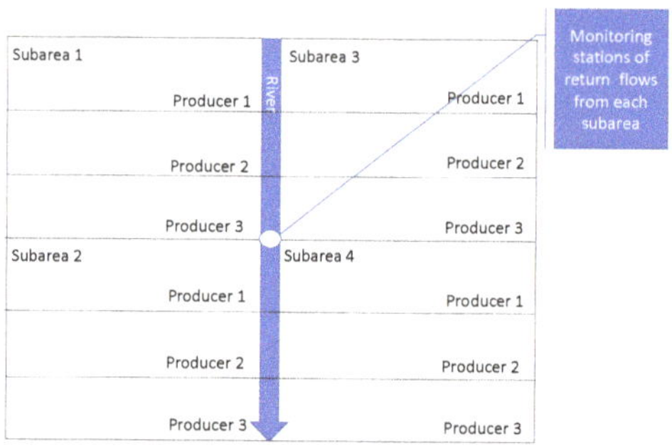

Figure 1. A schematic representation of the region with subareas, agricultural producers, and a regulated receiving water body (river).

2.2. Real-World Model Fitting

Water supply for the westside of the San Joaquin River Basin (SJRB) is provided by a water agency (e.g., United States Bureau of Reclamation) to the two westside subareas (Grasslands and the North-West Side subareas), according to water contracts negotiated between the water agency and individual water districts within each subarea. The individual water districts, in turn, allocate and distribute water supply according to agreements made with agricultural producers within each subarea. Water supply to subareas on the eastside of the SJRB derives largely from snowpack runoff from the Sierra Nevada Mountain Range, stored in downstream reservoirs along each major San Joaquin River (SJR) tributary. A state government water quality regulator (such as the California State Water Board), with the Regional Water Quality Control Board as enforcer, sets salt load objectives for the Basin in accordance with a Total Maximum Daily Load (TMDL) allocation developed by the Environmental Protection Agency for the basin. The load-based TMDL was further refined to develop subarea-level salt load allocations that take account of different water year hydrology. The conservative nature of the TMDL computation that utilizes the lowest 10% average low-flow condition resulted in allocations that were unattainable without major impact to the agricultural economy in each subarea. Hence, the initial TMDL allocations were replaced by concentration objectives, based on a 30-day running average electrical conductivity (EC), for the most downstream monitoring location on the SJR, Vernalis. A concentration objective allows agricultural producers and other salinity dischargers to utilize more of the available salt load assimilative capacity in the SJR. This initial compliance-monitoring objective has been supplemented with two additional upstream salinity objectives, ostensibly to protect the water quality of agricultural diversions made by westside agricultural producers. These additional salinity objectives are set at 1550 uS/cm year-round, as opposed to the 1000 uS/cm non-irrigation season, and 700 uS/cm objective set at Vernalis. The regulator suggested a number of approaches by which the original salinity load allocations, under the TMDL, might form the basis for salt load reduction strategies or cost allocation in situations where these various salt load objectives were violated.

The salinity load (mass of salt from all producers which is calculated by summing the product of drainage volume and salt concentration from each producer) produced on subarea n is the result of the return flows (drainage) from the agricultural activity of all producers, such that $\sum_{k_n} S_n Q_n \leq \overline{S}_n$. There is no practical way that the regulator could equitably assign salt pollution levels to the individual agricultural producers or enforce this regulation at a reasonable cost to individual agricultural producers. Therefore, the regulator has chosen to allow stakeholders to internally govern the strategies to attain and abide by river EC objectives, while encouraging stakeholders to consider the subarea as the organizing entity for stakeholder action. Stakeholder compliance is monitored by the Regional Board using data supplied by state and federal water agencies.

To maintain compliance the agricultural producers can dynamically allocate salt loads to each subarea given that available salt load assimilative capacity at each compliance station is the product of the total assimilative capacity (defined by the current flow multiplied by the EC objective) and the current salt load in the river.

The monthly salt load cap can be calculated for each subarea individually based on the calculated TMDL allocations and the current salt loading to the river from each subarea (measured in terms of tons of salt: $S_L = d$ [salt concentration, S; volume, Q]), where S_L is salt load (In the San Joaquin River the current TMDL criterion is a 30-day running average salt concentration that is multiplied by a monthly design flow to determine allowable salt loading). Using this stakeholder-maintained salinity load cap approach subareas would pay a fine (F) to the regulator, which could be a price per unit of salt load above the cap or some other equitable formula for dividing the fine amongst stakeholders. F can be specific to each subarea or similar for all subareas (see [23], for critique on uniform tax). F is then equitably distributed according to some formula (by land area, drainage volume, incremental salt load etc.) among all K_n agricultural producers in the different subareas (or by water user associations/districts in each subarea). We assume, for simplicity, that since we have a non-point source salinity management problem where the exact source of salt is not known, the most straight-forward and cost-effective initial approach to distribute F is to divide it equally per acre of land in production, or per acre–foot of irrigation water supply delivered. These initial approaches ignore the fact that some crops are associated with higher drainage return flow volumes and that subsurface drainage return flow salt loads may be poorly correlated with irrigation applications. Alternative allocation formulas may be relevant and will be considered in the empirical model. We assume that the (hypothetical) subarea manager (While there is no actual subarea manager, it is assumed that the model allocations are respected by the individual farmers and other decision makers at the water district level) has the authority or power to impose these allocations of river salt load assimilative capacity. We also do not want to set an optimal level for F, but rather take F as given in the empirical analysis. We will use several levels of F in a sensitivity analysis to evaluate the effect of F on the behavior of the agricultural decisionmakers at the subarea level.

An additional consideration is in the temporal administration of fines and fee schedules, which has a bearing on the design of a decision support system to aid the subarea manager to orchestrates stakeholder responses to potential violations of the river salinity objectives. An approach that attempts to respond to potential exceedance of salt load assimilative capacity at each compliance site in real-time would require model simulation tools that ran on a monthly timestep at a minimum. An optimization model would choose between available salt load reduction strategies, purchase of available water supply for dilution purposes, or payment of fines each month. Alternatively, accounting could be postponed until the end of each year and fines imposed retroactively. The latter strategy would rely on uncertainty and the fear of a potential exceedance to motivate compliance. However, the decision tool needed to support this strategy could be simplified to operate on an annual timestep.

2.3. Individual Responses to Water Quality Regulations

We expect that each stakeholder within each subarea will respond to F, depending on the level of F and the conditions (cropping patterns, physical conditions such as soil properties, etc.) in that subarea. In the empirical application, we will look at the effects of surface water and irrigation land and water limits and fees, as these are the main forms of regulation that can affect salinity load in the case of a nonpoint source pollution. In future empirical applications, we propose limits on the output of the model, specifically the salinity loading. Here, we outline the full analytical model.

Given F, each subarea faces the following two options:

(a) Maintain the current (status quo) level of salt loading if $F \leq$ the cost of abatement. The cost of abatement could include changing the crop mix and/or land use changes (e.g., fallowing land) (Changes in land use (crop mix or fallowed land) is an important component to maximize revenue and obtain maximum resource use efficiency. In the empirical model, changes in land use is incorporated at a later stage in the model development process), surface and/or subsurface drainage reuse, investing in more efficient irrigation technology, changing irrigation scheduling, and other options.

(b) Abate to a level of allowable salt loading $\leq$ than the cap. Each subarea will require abatement activities (detailed below) to the point where the marginal abatement cost equals F.

We consider the abatement in (a) and (b) to be "individual responses" to the salinity management regulation. That is, each subarea acts on its own responses, given its resources and local conditions (In the empirical application section, we also consider some of the individual responses, such as restrictions on water quotas, restrictions on cropping patterns, land fallowing, and investment in water-conserving irrigation technologies, as regulatory-imposed policies).

Each subarea is characterized by an aggregate revenue function (of all agricultural producers within each water district) minus fines on excess salt loading and minus abatement cost, such that

$$\Pi_n = \sum_{k_n=1}^{K_n} \pi_{k_n} - F_n - A_n \tag{6}$$

$$F_n = d(Q, S) \cdot \vartheta \tag{7}$$

$$A_n = \sum_{k_n=1}^{K_n} m_{k_n}(\Gamma_{k_n}\{1,\ 2,\ 3 \ldots N\}) \tag{8}$$

where F_n is a fine on excess salt loads, ϑ is pollution fine per unit of excess salt load, d is salt load, A_n is abatement cost, and $\Gamma_{k_n}\{1,\ 2,\ 3 \ldots N\}$ is a set of abatement options, such as changing cropping patterns, fallowing land, adopting more efficient irrigation technologies, and investing in monitoring drainage quantity and quality. Each subarea can select one of these abatement options or a subset of the abatement options.

2.4. Allocation of Joint Costs and Benefits in the Case of Individual Responses

In both individual and cooperative responses, we estimate the subarea net benefits as revenues minus variable operational costs and incremental costs. The incremental costs include costs associated with activities that polluters introduce to the agricultural production process in response to the regulatory objectives or constraints on input use imposed by the regulator for each subarea. In the case of a fine imposed on the entire region for exceeding the pollution EC objective, the subarea level of fine is allocated, based on several allocation principles, and the subarea amount of fine, Θ_j, is added to the incremental costs. In the case of cooperation among the subareas; we estimate first the regional net benefits to the entire region. The value of the regional benefits is obtained by running a regional optimization model, coined 'a social planner' model, which maximizes the entire regional welfare rather than looking at welfare of each subarea individually.

Economic theory [31], suggests that a social planner allocating regional benefits or costs among the agents involved maximizes the joint welfare of the region, subject to physical and institutional constraints relevant to the situation under study. Under a social planner optimization, the region is seen as one unit without political borders. An optimal social planner allocation is considered as first best and serves as reference (benchmark) to which other allocation schemes are compared. Deviations from the social planner outcomes represent inefficiency (welfare loss) of the alternative allocations.

Once a regional social planner allocation solution has been found, the regional gains (either welfare benefits or savings of joint costs—such as regulatory fines) must be distributed among the regional parties. We will consider a couple of schemes for the allocation of the joint benefits or the costs of pollution control, or regulatory fines, among resource management regions, namely, the subareas (and the individual farmers in each subarea). For example, allocation of benefits or fines could be based on annual drainage flows or based on irrigated area. The likelihood of subareas forming stable coalitions aimed to reduce salt loads can be measured by comparing the empirical attributes of the standard allocation schemes with game theoretic allocation schemes whose acceptability and stability can be measured. We introduced several allocation schemes, based on the subarea's contribution of pollution load and consistent with the strategy described previously [32].

2.4.1. Allocation of Regulatory Fines Based on Surface Water Applied

This allocation scheme simply suggests that each polluter (subarea) will be charged in proportion to the volume of surface water applied on that subarea. Therefore, the cost to subarea j is

$$\Theta_j = F\frac{SW_j}{\sum_{j\in N} SW_i} \tag{9}$$

where Θ_j is the regulatory fine allocated to subarea j; F is total regional regulatory fine. This scheme allocates all the regulatory fine among all N subareas. SW_j is the volume of surface water applied for irrigation in subarea j (a summation over all irrigated area). The disadvantage of this regulatory method is that it does not target those stakeholders who physically discharge to the SJR and not take into account the significant reuse that occurs in some areas that helps to curtail salt loading to the river. It is a blunt policy instrument that is nonetheless relatively easy to administer.

2.4.2. Allocation of Regulatory Fines Based on Total Irrigation Water Applied

This allocation scheme simply suggests that each polluter (subarea) will be charged in proportion to the total volume of water applied (surface water + groundwater + recycled wastewater) on that subarea. Therefore, the cost to subarea j is

$$\Theta_j = F\frac{W_j}{\sum_{j\in N} W_i} \tag{10}$$

The drawbacks to this policy are the same as those of the prior policy, although, it does account for groundwater use that can add significant salt to the total salt discharged from each region, since the EC of groundwater is typically more than double that of applied surface water on the westside of the Valley, and more than four times the average EC of eastside water applications.

2.4.3. Allocation of Regulatory Fines Based on Salt Load Generation

The allocation scheme in (11) simply suggests that each polluter (subarea) will be charged in proportion to the amount of salt load it generates. Therefore, the cost to subarea j is

$$\Theta_j = F\frac{S_j}{\sum_{j\in N} S_i} \tag{11}$$

where S_j is the quantity of salt load generated by subarea j.

This policy is the most equitable but also the most difficult to administer since current monitoring and modeling is insufficient to accurately measure or estimate the salt load export from each subarea. Current models are not capable of recognizing the amount off drainage reuse within each subarea.

2.4.4. Allocation of Regulatory Fines Based on Cultivated Area

The allocation based on cultivated area is built on a similar rule as in Equation (12), except that Q_j is cultivated land and not disposed drainage.

$$\Theta_j = F\frac{L_j}{\sum_{j\in N} L_i} \tag{12}$$

where L_j is the cultivated land area in subregion j.

3. Model Testing

3.1. Application to Water Quality Issues in the San Joaquin River

Salinity loads to the San Joaquin River (SJR), the receiving water body for agricultural drainage in the San Joaquin Basin, are regulated by the State of California through the Central Valley Regional (Water Quality Control) Board. A TMDL was developed that set load limits for each subarea [33,34]. The TMDL for each of the seven subareas (Figure 2) was largely based on basin hydrography, and existing water district and jurisdictional boundaries. Four of these subareas (Northwest Side, East Valley Floor, Grasslands, and San Joaquin River Above Salt Slough) are located on the valley floor, and drainage from these subareas is dominated by agricultural and managed wetland decision-makers. The other three subareas are watersheds serving three major east-side tributaries to the SJR, namely the Stanislaus, Tuolumne, and Merced rivers. Given the institutional history and management functions within the basin, these seven subareas are the most logical management units and any possible future trade in salinity load permits would initially occur between these entities.

The load allocations under the TMDL ended up being overly restrictive, following the typical TMDL development methodology and would have resulted in potential annual fines in the order of USD 300,000 per subarea based on a 9-year average of salt loads. Load allocations were based on a design flow hydrology representing the lowest 10% of monthly flows. An additional safety factor was applied to the allowable monthly salt loads which further reduced stakeholder ability to meet objectives. The Regional Board adopted a real-time concentration-based schema to substitute for the TMDL salt load based approach which allowed greater use of the river's assimilative capacity. Salinity management in the San Joaquin River Basin is complex, involving agricultural, wetlands, and municipal stakeholders within the basin. Being cognizant of this complexity and the difficulty of building coalitions among entities that had little history of working cooperatively, the Regional Water Board named the alternative approach to TMDL implementation "real-time salinity management". The current Basin Water Quality Control Plan for the San Joaquin Basin lays out the general requirements for a Board-approved Real-Time Salinity Management Program:

1. The program is a basin-wide program requiring all stakeholders discharging to the SJR to be signatories and active participants.

2. Real-time monitoring networks should be developed and maintained to allow the estimation of net salt loading to the San Joaquin River and residual salt load assimilative capacity at the Vernalis compliance monitoring station.
3. Provision should be made to allow for free and easy sharing of real-time flow and EC data that will be the basis for salt load assimilative capacity estimation.
4. A decision support tool or simulation model should be developed to allow reliable forecasting of salt load assimilative capacity to give stakeholders adequate time to make scheduling decisions to maintain compliance with River salinity objectives.

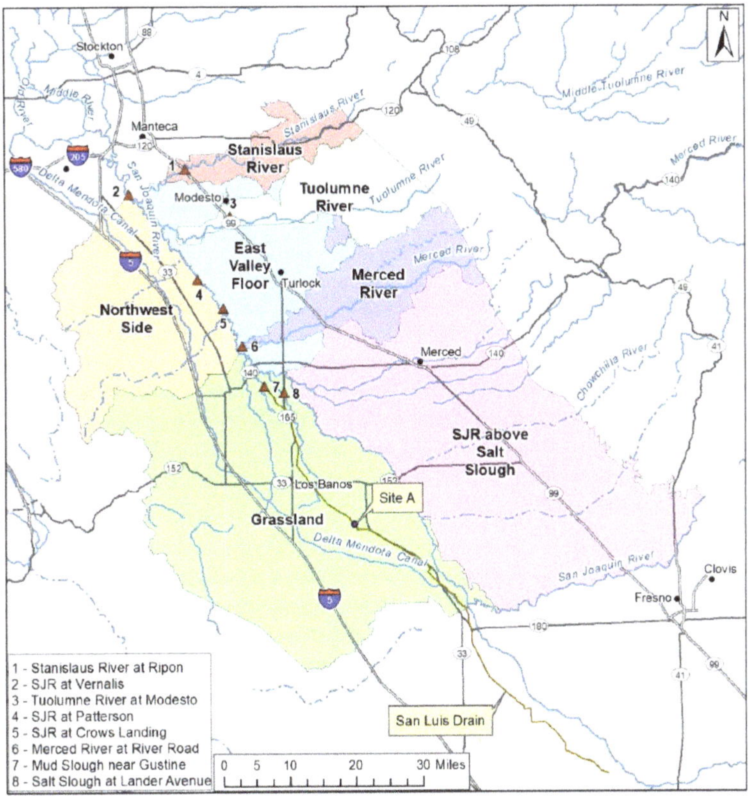

Figure 2. Map of the various San Joaquin River Basin contributing subareas as defined in the 2002 TMDL Regulation Plan: Northwest Side (NWS), Grasslands Agriculture (GRA), East Valley Floor (EVF), Merced River (MER), Stanislaus River (STL), Tuolumne River (TLU), and San Joaquin River above Salt Slough (SJR). Note: the eight red triangles in the Figure that are indicated in the legend by numbers 1–8 are the locations of the flow and salinity load monitoring stations. Source: [33,34].

Salinity concentration objectives for compliance monitoring stations at Crows Landing bridge, Maze Road bridge and Vernalis were set as 30-day running averages of EC. For Vernalis these salinity concentration objectives were a winter objective of 1000 uS/cm and a summer objective of 700 uS/cm. The summer irrigation season salinity objective was considered protective of irrigation agriculture and salt-sensitive crops. A year-round salinity objective of 1550 uS/cm was later set for compliance monitoring stations at Maze Road Bridge and Crows Landing Bridge, set to be protective riparian diversions used primarily on orchards in the river reach between Crows Landing Bridge and Maze Road Bridge.

Using a water quality simulation model as a real-time decision support tool, two-week forecasts of the 30-day running average salt load assimilative capacity were made routinely for the San Joaquin River at each compliance monitoring station. The model provides information on the salt loads from each of the seven subareas and computes the excess load that must be removed from the system by one or a number of salinity management options that can adjust the scheduling of salt load discharge into the river, including: temporary storage of these salt loads in ponds; shutting off discharge from agricultural drainage sump pumps into drainage conveyances for periods of time; recirculating or reusing return flows of high salinity; and/or providing dilution flows from eastside tributary reservoirs to lower ambient salt concentration in the river to meet salinity concentration objectives. To illustrate the potential use of assimilative capacity, the USBR calculated the available daily salt load assimilative capacity in 2008, a year when violations of the salinity objective were still common. During 2008, the SJR salt load assimilative capacity was available on 246 days of the year (conservatively estimated when the SJR salinity was less than 85 percent of the salinity objective) for a total of around 115,000 tons of salt (calculated on a daily basis). The salt load assimilative capacity of the SJR was exceeded for 119 days.

A "strawman" allocation policy was developed as part of an analysis by Regional Board staff in 2015, to demonstrate the potential fines that might have occurred under the published TMDL using a suggested daily fine of US$5000 per day for each overage of the EC objectives. The cultivated area in each subarea was the means by which the total fine was distributed among subareas and the stakeholders within each subarea. During periods when salt load assimilative capacity in the San Joaquin River was exceeded, stakeholders within each subarea were obliged to provide a collective response to salinity objective exceedances. Stakeholders in one subarea could engage in voluntary agreements with stakeholders in other subareas through their representatives to trade individual subarea salinity load exceedances for a number of management actions to be deployed in the other subregions. The schema that was developed (Table 1) made clear the merits of a collaborative and coordinated real-time water quality management program, even though the ability of stakeholders to develop the partnerships required to realize the benefits of the proposed alternative regulatory policy remained untested. One concession made by the Regional Board was that in critically dry years, when the option was typically foreclosed of releasing dilution flows from tributary reservoirs to help meet the three salinity concentration objectives in the San Joaquin River, was that all three salinity concentration compliance objectives would be waived.

Table 1. Potential Salt Discharge Load Exceedance Fees by subarea (2001–2012), ref. [35].

Subarea		NWS	GL	SJR	EVF
Days exceeded by period	Oct	0	0	0	0
	Nov	90	60	0	0
	Dec	124	248	0	0
	Jan	186	0	310	0
	Feb	28	196	0	0
	Mar	0	279	0	0
	Apr	28	56	42	14
	VAMP [a]	0	0	30	30
	May	0	0	51	17
	Jun	30	30	210	90
	Jul	0	0	248	91
	Aug	0	0	248	31
	Sep	0	0	0	0
Total days of exceedances		486	869	1139	273
Total penalties [US$5000 per day penalty]		US$2,430,000	US$4,345,000	US$5,695,000	US$1,365,000
Years calculated		8	10	10	3
Average penalty per year		US$303,750	US$434,500	US$569,500	US$455,000
Subarea acreage		118,000	353,000	187,000	201,000
Average penalty per acre		US$2.57	US$1.23	US$3.05	US$2.26

Note: [a] VAMP stands for Vernalis Adaptive Management Program. These were programmatic reservoir releases for fish migration that occurred each year from 15 April to 15 May. Values in the table are elaborated by the authors, based on data in [35].

3.2. Analysis of Exceedance Frequency of San Joaquin River Salinity Objectives

The first step in assessing salinity management options and developing policy for the trading of salt load credits was to look at the exceedance frequency at the three compliance monitoring sites. Given the changes in the San Joaquin Basin hydrology over the past fifty years, in part due to the hydrologic variability induced by a warming planet and climate change, only the past decade of the San Joaquin River hydrology was included in this analysis. A pattern is emerging in California of more persistent back-to-back droughts punctuated by years of abnormally wet weather (State of California [35]).

Table 1 presents a schedule of potential annual average penalties (2001–2012) that could have been assessed for exceedances of salinity objectives at Vernalis under the salinity TMDL, assuming a hypothetical penalty of USD 5000/day for each monthly overage [28]. Subareas include Northwest side (NWS), Grasslands (GL), Upstream San Joaquin River (SJR), and East Valley Floor (EVF).

In December 2018 the Environmental Protection Agency approved a Regional Water Board Basin Plan amendment that established upstream objectives at the Maze and Crows Landing monitoring locations on the San Joaquin River. These objectives were ostensibly to protect the water quality of riparian diversions to westside San Joaquin River irrigation districts. The objectives of 1550 uS/cm apply year-round at both monitoring stations and considered the field crops and orchards farmed adjacent to this reach of the river. Figures 3 and 4 show the 30-day running average EC at the Maze and Crows Landing compliance monitoring stations for the period 1 October 1995 through 1 September 2013. At Maze Road some of the highest EC values occurred in below-normal years as opposed to dry and critically dry years, although, all the data appear well below the 1550 uS/cm concentration objective. Maze Road is downstream of the Tuolumne River which appears to provide adequate dilution flow year-round. On the other hand, the Crows Landing site did show exceedances of the 1550 uS/cm objective during April for all three water year types (below normal, dry, and critically dry). Since the Vernalis compliance monitoring station is downstream of the Stanislaus River and was obligated to meet the winter 1000 uS/cm objective and 700 uS/cm objective through controlled releases by the USBR through their New Melones Reservoir operation, the incidence of exceedance was largely eliminated at this site [33,34].

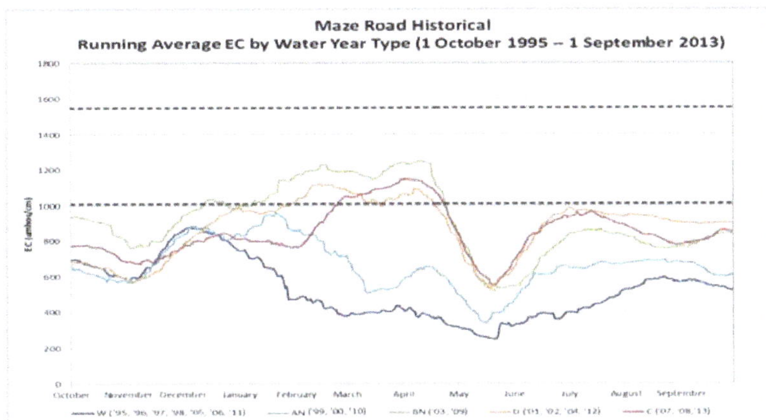

Figure 3. Maze Road Historical 30-day Running Average EC by Water Year Type [36].

An analysis of the 30-day running average EC was subsequently conducted for all three compliance monitoring stations using more decent data for a 21-year period between 2000 and 2021. The same logic was applied as in the prior Regional Water Board compliance penalty analysis [35], although in this instance a second policy was introduced that imposed a fine of USD 10 per ton of salt. This fine schedule was chosen ostensibly to be a round number that produced an outcome of similar magnitude to the first policy when the 30-day running average monthly objective was exceeded. Figure 5 is a bar chart showing the frequency (number of days) that the 30-day running average EC was exceeded at each of the compliance monitoring stations. In Tables 2 and 3 the potential fines were compared for the two policies for each month of the 21-year period. Note that this is a slight variation of the prior Regional Water Board policy, which imposed a penalty for all days of the month anytime the 30-day running average EC was exceeded in any month. With the use of the WARMF forecasting model and decision support system, calculation of the 30-day running average has become routine and provided a more equitable policy outcome. With better online real-time access to estimated salt load from each of the seven subareas the response time for remedial actions on the river was reduced. However, an institutional mechanism for deciding real-time actions has still to be ratified by designated resource managers for each of the seven subareas contributing salt load to the San Joaquin River.

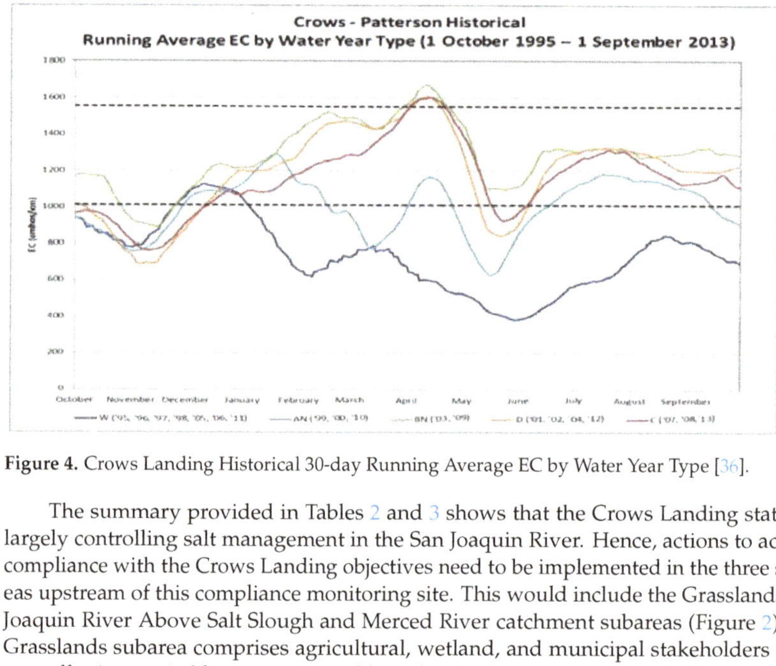

Figure 4. Crows Landing Historical 30-day Running Average EC by Water Year Type [36].

The summary provided in Tables 2 and 3 shows that the Crows Landing station is largely controlling salt management in the San Joaquin River. Hence, actions to achieve compliance with the Crows Landing objectives need to be implemented in the three subareas upstream of this compliance monitoring site. This would include the Grasslands, San Joaquin River Above Salt Slough and Merced River catchment subareas (Figure 2). The Grasslands subarea comprises agricultural, wetland, and municipal stakeholders and a cost-effective equitable response would need to be developed among these entities. The majority of exceedances at the Crows Landing compliance monitoring station occurred in the months of March–May (Figure 4), which coincided with the wetland drawdown period when irrigated agriculture in the subarea was also providing pre-irrigation to field crops and orchards. Although the last 20 years have seen a significant move to drip irrigation, more water conserving technologies and selenium management practices have eliminated a major source of salt load from the subarea. Subsurface tile drainage is now diverted to a dedicated 6000+ acres of reuse area, thus, any response would be expected to include actions from all three entities. Managed seasonal wetlands in the Grasslands subarea are constrained by the fact that any significant delay in wetland drainage from the 160 private duck clubs and state and federal refuge wetland complexes can change the germination success of high-value grasses potentially leading to lower value habitat for the overwintering waterfowl that rely on this resource [37,38]. Irrigated agriculture may need to invest in additional temporary storage ponds in areas free of selenium in shallow groundwater (selenium is a potential toxin when drainage concentrations exceed 5 ppb) or automation of sump pumps and new drainage conveyance plumbing to facilitate salt management through reuse and drainage recirculation. Municipal discharge of salts in the Grassland sub-basin is minimal with most ponded water being eliminated through seepage to groundwater. Municipal wastewater facilities may own storage ponds that might be usable and temporary storage facilities to reduce salt loading from the Grassland subarea. The potential for this cooperative and coordinated salt management strategy is unknown at this time. In the Merced River catchment subarea, the salt load contribution to the river is dominated by the reservoir releases from the McSwain and New Exchequer Dams and irrigation return flows into the Merced River. The Merced National Wildlife refuge contributes an insignificant return flow given its reliance on groundwater as the water supply source. The subarea south of Salt Slough is bottomland in the Valley trough and contributes to the San Joaquin River through groundwater seepage. Reservoir releases from Friant Dam can be diverted into the subarea through riparian pumping. Seepage may occur from the river into the subarea in response to local groundwater pumping.

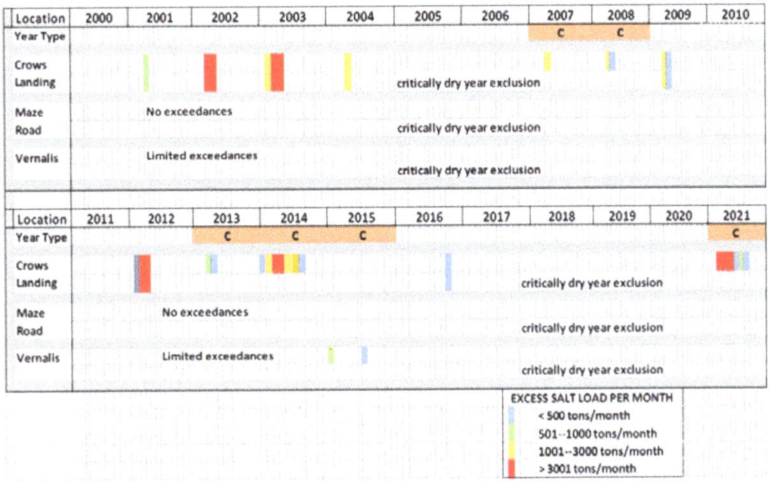

Figure 5. Exceedance frequency of 30-day running average EC at all three compliance monitoring stations of the San Joaquin River for 21 years: 2000–2021.

Table 2. Historical accounting on salt load exceedance, days of exceedance, monthly fees, and fine for salt load exceedance 2001–2021 at Crows Landing [36].

Month-Year	Excess Monthly Salt Load (ton)	Days of Exceedance	Monthly Fine (US$)	Fine (US$)
			US$5000/day	US$10/ton of Salt
Apr-2001	552	12	60,000	5518
Mar-2002	4967	16	80,000	49,669
Apr-2002	4510	23	115,000	45,099
Feb-2003	2238	18	90,000	22,377
Mar-2003	4403	31	155,000	44,034
Apr-2003	2245	18	90,000	22,453
Apr-2004	2034	18	90,000	20,344
Apr-2007	134	12	60,000	1340
Apr-2008	1807	21	105,000	18,066
May-2008	823	11	55,000	8233
Apr-2009	1559	25	125,000	15,593
May-2009	82	4	20,000	820
Feb-2012	66	4	20,000	658
Mar-2012	3374	31	155,000	33,743
Apr-2012	2898	29	145,000	28,979
Mar-2013	178	6	30,000	1777

Table 2. *Cont.*

Month-Year	Excess Monthly Salt Load (ton)	Days of Exceedance	Monthly Fine (US$)	Fine (US$)
			US$5000/day	US$10/ton of Salt
Apr-2013	1264	20	100,000	12,641
Jan-2014	147	6	30,000	1474
Feb-2014	2544	28	140,000	25,435
Mar-2014	7855	31	155,000	78,549
Apr-2014	7230	30	150,000	72,299
May-2014	1607	31	155,000	16,072
Jun-2014	1280	30	150,000	12,802
Jul-2014	1132	31	155,000	11,321
Aug-2014	82	25	125,000	823
Oct-2016	28	14	70,000	280
Mar-2021	77,514	27	135,000	775,145
Apr-2021	47,752	30	150,000	477,518
May-2021	30,491	31	155,000	304,910
Jun-2021	4719	8	40,000	47,193
Jul-2021	53	7	35,000	528
Aug-2021	547	31	155,000	5472
Sep-2021	14	5	25,000	40
		Total cost	US$3,320,000	US$2,161,305

Note: Total cost in nominal dollars. Values elaborated by authors, based on data in [36].

Table 3. Historical accounting on salt load exceedance, days of exceedance, monthly fees, and fine for salt load exceedance 2001–2021 at Maze Road [36].

Month-Year	Excess Monthly Salt Load (ton)	Days of Exceedance	Monthly Fine (US$) US$5000/day	Fine (US$) US$10/ton of Salt
Jan-2015	727	14	70,000	7276
Jul-2015	5	3	15,000	48
		Total cost	US$85,000	US$7315

Notes: (1) Exceedence during the analyzed period occurred only in January and July 2015. (2) Total cost in nominal US dollars. Values elaborated by authors, based on data in [36].

3.3. Economic Analysis of Selected Salt Management Strategies

The first part of this paper laid out the theoretical underpinning and principles of a collaborative strategy that was anticipated would be necessary by the state regulator and that is necessary to achieve cost-effective and equitable salt management strategies. The cost and institutional feasibility of these strategies should be juxtaposed against paying the fine that the Regional Water Board has formulated to provide an incentive for regional basin cooperation and coordination without creating an undue burden on stakeholders or incentivizing stakeholder litigation. The seven subareas were chosen, not only from a hydrological perspective but also from a jurisdictional and institutional viewpoint. This paper now provides an economic rationale for a couple of effective salt management strategies that were proposed for consideration in the region.

We start by comparing the strategy of building and maintaining temporary holding ponds to store salt loads until they can be safely discharged to the San Joaquin River without exceeding objectives at any of the three compliance monitoring sites. Other means of providing temporary storage are surface and subsurface drainage reuse or short-term storage of salt load in the shallow groundwater, achieved by temporarily switching off tile drainage sump pumps. The analyses in [39], and prior in the United States Department of the Interior, Bureau of Reclamation model report [40], provide realistic cost parameters for various temporary salt storage options relevant to real-time management of salt loading to the San Joaquin River.

The CDM Smith report [39], refers to two regions that can be relevant to two of our sub areas: Tulare Lake Bed (TLB) (Note that the Tulare Lake bed is outside our study area. It does not export salt to the San Joaquin River. However, the costs of ponds may be relevant to our study.) and San Joaquin River Quality Improvement Project (SJRIP). Table 4 summarizes the relevant results to our work from the report of [39].

We used the observed data for the region during 2001 through 2021 as the basis for calculating the excess annual salt load above the regional allowances that could either be disposed of to the river with an excess fine paid, or be temporarily stored and reduced through reuse, recirculation, or storage in the shallow groundwater system (shut off sump pumping). The cost values in Table 4 are based on several assumptions. First, it is assumed that the cost per ton of salt stored in the pond include variable, fixed, and the opportunity cost of the land assigned to the pond. We used the working assumption in [39], without changing them.

Table 4. Costs and effectiveness in removing salt of the evaporation pond technology in two locations in the Central Valley.

Region	Water Volume (AFY)	Salt Stored (Ton/Year)	Salt Accumulation (Ton)	Salt /Managed (%)	Cost/AF (US$)	Cost/Salt Stored (US$/Ton)
TLB	17,240	139,897	155,479	90	76.58	9.62
SJRIP	19,248	98,108	890,639	11	129.62	24.72

Source: Extracted from [39], Tables 2 and 3. Note: pond cost assumed to be equivalent to the cost of an evaporation pond. Selenium issues have pretty much nixed mixed evaporation ponds for the past 20 years. However, any selenium in drainage will warrant bird hazing and the cost of temporary holding ponds may not be that different from evaporation ponds. Values elaborated by authors, based on data in [39].

The pond technology in [39], was assessed in two regions in the Central Valley. However, in our analysis we used the cost and effectiveness values in each region as our higher and lower values within which the technology performed. One set of values suggested that it took USD 9.62 per ton of stored salt and with an effectiveness rate of 90%. The second set of values suggested a USD 24.72 per ton of temporary storage of salt with an effective rate of 11% (Table 4).

Salt load that exceeded the allowance for disposal was subject to a fine. Taxes were either by days of exceedance of salt concentration or by total tons of salt above the level of allowance (Tables 2 and 3). The analysis in this paper refers only to tons of salt observed on a monthly basis between 2001 and 2021. Two levels of fine rates have been arbitrarily used (USD 10/ton and USD 20/ton) to reflect incentive–provision fines on the part of the stakeholders in the subarea. Since the analysis spans over 20 years (2001–2021) we used a discount rate to allow comparisons of sums between 2001 and 2021. We used the directive for using a real discount rate of 7% [41]. That source requires the use also of a 3% discount rate, but for our purposes this was not necessary.

The WARMF model provided the salt load allocation for each of the seven subareas that was developed for the Real-Time Program by the Regional Water Quality Control Board based on the original salt load TMDL. These salt load allocations provide an initial basis for negotiation and allow the excess salt loads over and above the subarea allocation to be calculated. If treated, we applied the cost and effectiveness parameters to calculate the treatment cost. Since effectiveness is always less than 100%, the remainder of the salt to be disposed of was subject to the fine. Then, we compared the total annual cost of treating the salt plus paying the incremental fee to the total annual cost in the case that the salt load is not treated and is disposed of in its entirety.

4. Results and Discussion

4.1. Results

Table 5 presents the results of our analysis under the various assumptions regarding the parameters we presented earlier.

The results provide a clear distinction between the different scenarios: efficient-inexpensive-low fine ((90%; USD 9.62/ton); USD 10/ton); inefficient-expensive-low fine ((1%; USD 24.72/ton); USD 10/ton); efficient-inexpensive-high fine ((90%; USD 9.62/ton); USD 20/ton); and inefficient-expensive-high fine ((1%; USD 24.72/ton); USD 20/ton), which are reflected in Table 5. Clearly seen are the suggestions that, in three out of the four scenarios, the net present value of the differences between the fee payments for the total managed salt load and the storage or reuse cost plus the fee payment for the remainder of the unmanaged salt suggest that the subarea should not engage in managing the salt and should pay the cost of the fine. Only under the scenario of a high level of fee per ton and efficient and inexpensive treatment, the difference between the costs was positive, suggesting that treatment is justified.

Table 5. Cost of temporarily storing salt in surface ponds, reusing, or recirculating the salt or temporarily storing it in the shallow groundwater system under different assumptions of fee level, storage efficiency, and cost.

Year with Exceedance	Salt LOAD Excess (ton)	ANNUAL Fine Payment on Stored or Reused Quantity		Annual Storage or Reused Fee Cost on Remainder of Salt Retained and Managed (US$10/ton)		Annual Fee Cost on Remainder of Salt Not Temporarily Retained and Managed (US$20/ton)	
		US$10/ton	US$20/ton	(US$9.62/ton; 90%)	(US$24.72/ton; 1%)	(US$9.62/ton; 90%)	(US$24.72/ton; 1%)
2001	552	5520	11,040	5862	18,558	1159	23,471
2002	9477	94,770	189,540	100,645	318,616	19,901	402,962
2003	8886	88,860	177,720	94,369	298,747	18,660	377,832
2004	2034	20,340	40,680	21,601	68,383	4271	86,485
2007	134	1340	2680	1423	4505	281	5697
2008	2630	26,300	52,600	27,930	88,420	5523	111,827
2009	1641	16,410	16,410	17,427	55,170	3446	69,775
2012	6338	63,380	126,760	67,309	213,083	13,309	269,491
2013	1442	14,420	28,840	15,314	48,480	3028	61,313
2014	21,877	217,300	437,540	230,772	730,562	45,941	930,210
2015	732	7320	14,640	7773	24,609	1537	31,124
2016	28	280	560	297	941	59	1190
2021	161,090	1,610,900	3,221,800	1,710,775	5,415,845	338,289	6,849,547
Net Present Value of Difference Between Payment for Management of Entire Salt Load and Cost of treatment and payment for incremental storage or reuse (In 2021 dollars, 7% discount rate)				−172,574	−6,610,459	+5,030,130	−6,229,518

4.2. Discussion

As California battles another year of drought; records are being broken for climate extreme conditions impacting the sustainability of water resources and human-induced problems of water scarcity, quality, and misallocation. The State of California has taken aggressive action starting in 2014 with the passage of the Sustainable Groundwater Management Act and the formation of the stakeholder-led Central Valley salinity Coalition (CVSALTS) that together address problems of unsustainable groundwater pumping practices, problems of subsidence and water quality degradation of surface and groundwater resources, and the fertility of agricultural soils. Although salinity degradation of receiving waters, such as rivers and deep percolation of saline water to aquifers, have been studied for over 100 years and are well understood, there has been a reluctance of the state to commit to addressing this problem. The publication of the 2002 Total Maximum Daily Load (TMDL) by the federal EPA for salt and boron incentivized the search for effective and cost-effective policy tools to address salinity impairments in the San Joaquin River and to find a feasible and equitable schema that would be accepted by stakeholders and foreclose costly litigation that would result in a continuation of the status quo.

TMDL required the state government water quality regulator to set salt load objectives for the basin in 2002. However, the conservative nature of the TMDL computation and utilizing the lowest 10% average low-flow condition resulted in allocations that were unattainable without a major impact to the agricultural economy in each subarea. This created a need to replace the initial TMDL allocations with more flexible concentration objectives based on a 30-day running average electrical conductivity (EC) that will accommodate agricultural production and irrigation practices. A concentration objective allows agricultural producers and other salinity dischargers to utilize more of the available salt load assimilative capacity in the SJR. This initial compliance monitoring objective has been supplemented with two additional upstream salinity objectives, ostensibly, to protect the water quality of agricultural diversions made by westside agricultural producers.

The adjustments of the quality standards to seasons and to the different locations (subareas) and the real-time reporting to the stakeholders allow farmers more flexibility in adjusting their practices and responding to the standards in a creative way. In that respect, the approach suggested in our work is similar to Doole [21], and Doole and Pannell [22], which take into account the differences in abatement cost and ability to address quality standards by different dairy farms in New Zealand. Hence, a differential standard is suggested for more cost-effective results at the regional level.

Adjustment of the salinity standard to the conditions in the river assimilative capacity and in each subarea along the river also allows introduction of trade in salinity disposal permissions among subareas as another way of handling the salinity load to the river. In that respect, the experience of the Hunter Basin of New South Wales, Australia [25,26], is similar to the suggested scheme in the SJR as discussed in our paper. The main idea of the scheme in the Hunter was to permit discharge of salt loads only when there was available salt load assimilative capacity in the river that drains the Hunter Basin. Details in Quinn [4], explain how salt load discharges to the river were scheduled by quantity, time, and location, based on stakeholder need and calculations of salt load assimilative capacity using a simple spreadsheet mass–balance model. While our work has not explored regional cooperative arrangements, such as trade in salt load permissions, or side payments for improvements in on-farm salinity load disposal, our future work will focus on such options.

5. Conclusions and Policy Implication

This paper presents an alternative approach to salt regulation and control that follows first attempts to implement the 2002 TMDL, when it was realized that TMDL policy objectives could not be achieved without potential annual costs to stakeholders in the millions of dollars annually, using typical penalty schedules for daily exceedance of a 30-day running average EC objective at a single downstream compliance site. These costs would have potentially risen with the inclusion of two additional upstream compliance monitoring sites adopted to protect agricultural riparian diverters from high salt concentrations in irrigation applied water. The novel concept of "Real-Time Water Quality management" relies on a continually updated forecasting model to provide daily estimates of salt load assimilative capacity in the San Joaquin River and assessments of compliance with salinity concentration objectives at three monitoring sites on the river, based on the 30-day running average EC. A water quality forecasting model WARMF was developed as part of this alternative regulatory schema, which served both as a compliance forecasting tool and the means by which salt load allocations and salt exports from each of the seven contributing subareas could be estimated and compared. The trading of salt loads between subareas is now feasible as both the regulatory salt load allocation and actual salt load discharge to the river can be quantified. The results of the study have shown that the policy combination of well-crafted river salinity objectives by the regulator and the application of an easy-to use and maintain decision support tool by stakeholders have succeeded in minimizing water quality (salinity) exceedances over a 20-year study period. The WARMF model improvements, and consequent increase in stakeholder and agency confidence in this decision support tool, suggest its potential application in other river basins facing similar challenges. Our framework allows farmers and regulators to jointly understand and evaluate the meaning of various regulatory policy interventions on the emission of salinity and on the cost to be incurred by farmers at various locations along the river. The results of the paper support the development of close collaboration between farmers and regulators in the application of non-point source pollution policy. The paper also suggests significant benefit from better cooperation and coordination among and between farmers and other dischargers of salt load who rely on the river for drainage disposal and who are already organized into sensible subareas for salt management. This can provide a cost-effective pathway for agricultural sustainability.

Author Contributions: Conceptualization, A.D. and N.W.T.Q.; methodology, A.D. and N.W.T.Q.; validation, A.D. and N.W.T.Q.; formal analysis, A.D. and N.W.T.Q.; investigation, A.D. and N.W.T.Q.; resources, A.D.; data curation, A.D. and N.W.T.Q.; writing—original draft preparation, A.D.; writing—review and editing, A.D. and N.W.T.Q.; supervision, A.D. All authors have read and agreed to the published version of the manuscript.

Funding: This research was funded, in part, by the Gianni Foundation Mini-grant Program 2020 DAVIS CA, 95616, USA.

Institutional Review Board Statement: Not applicable.

Informed Consent Statement: Not applicable.

Data Availability Statement: All data used in this paper is reported in the various tables. Additional clarifications regarding the data can be obtained from the corresponding author upon request.

Acknowledgments: The work leading to this paper was funded by the Giannini Foundation of Agricultural Economics Mini-grant Program. Ariel Dinar would like to acknowledge support from the W4190 Multistate NIFA-USDA-funded project, "Management and Policy Challenges in a Water-Scarce World". Financial support for the development of the real-time salinity management concept has been provided by the US Bureau of Reclamation, Division of Planning, and the California Department of Water Resources through Proposition 84- funded grants.

Conflicts of Interest: The authors declare no conflict of interest.

References

1. Tanji, K.K.; Kielen, K.C. *Agricultural Drainage Water Management in Arid and Semi-Arid Areas*; FAO Irrigation and Drainage Paper 61; Food and Agriculture Organization of the United Nations: Rome, Italy, 2002.
2. Hart, B.; Walker, G.; Katupitiya, A.; Doolan, J. Salinity Management in the Murray–Darling Basin, Australia. *Water* **2020**, *12*, 1829. [CrossRef]
3. Baccour, S.; Albiac, J.; Kahil, T.; Esteban, E.; Crespo, D.; Dinar, A. Hydroeconomic modeling for assessing water scarcity and agricultural pollution abatement policies in the Ebro River Basin, Spain. *J. Clean. Prod.* **2021**, *327*, 129459. [CrossRef]
4. Quinn, N.W.T. Adaptive implementation of information technology for real-time, basin-scale salinity management in the San Joaquin Basin, USA and Hunter River Basin, Australia. *Agric. Water Manag.* **2011**, *98*, 930–940. [CrossRef]
5. Obropta, C.C.; Niazi, M.; Kardos, J.S. Application of an environmental decision support system to a water quality trading program affected by surface water diversions. *Environ. Manag.* **2008**, *42*, 946–956. [CrossRef] [PubMed]
6. Zhang, X.; Huang, G.H.; Nie, X.; Lin, Q. Model-based decision support system for water quality management under hybrid uncertainty. *Expert Syst. Appl.* **2011**, *38*, 2809–2816. [CrossRef]
7. Ioannou, K.; Lefakis, P.; Arabatzis, G. Development of a decision support system for the study of an area after the occurrence of forest fire. *Int. J. Sustain. Soc.* **2011**, *3*, 5–32. [CrossRef]
8. Makropoulos, C.K.; Natsis, K.; Liu, S.; Mittas, K.; Butler, D. Decision support for sustainable option selection in integrated urban water management. *Environ. Model. Softw.* **2008**, *23*, 1448–1460. [CrossRef]
9. Rose, D.C.; Sutherland, W.J.; Parker, C.; Lobley, M.; Winter, M.; Morris, C.; Twining, S.; Ffoulkes, C.; Amano, T.; Dicks, L.V. Decision support tools for agriculture: Towards effective design and delivery. *Agric. Syst.* **2016**, *149*, 165–174. [CrossRef]
10. Systech Water Resources, Inc. *Watershed Analysis Risk Management Framework (WARMF): Technical Model Documentation*; Systech Water Resources, Inc.: Walnut Creek, CA, USA, 2017.
11. Quinn, N.W.T.; Tansey, M.K.; Lu, J. Comparison of Deterministic and Statistical Models for Water Quality Compliance Forecasting in the San Joaquin River Basin, California. *Water* **2021**, *13*, 2661. [CrossRef]
12. Fu, B.; Horsburgh, J.S.; Jakeman, A.J.; Gualtieri, C.; Arnold, T.; Marshall, L.; Green, T.R.; Quinn, N.W.T.; Volk, M.; Hunt, R.J.; et al. Modeling water quality in watersheds: From here to the next generation. *Water Resour. Res.* **2020**, *56*, e2020WR027721. [CrossRef]
13. Quinn, N.W.T.; Hughes, B.; Osti, A.; Herr, J.; Wang, J. Real-time, web-based decision support for stakeholder implementation of basin-scale salinity management. In *Environmental Software Systems: Computer Science for Environmental Protection, Proceedings of the 12th IFIP WG 5.11 International Symposium, ISESS 2017, Zadar, Croatia, 10–12 May 2017*; Hrebicek, J., Denzer, R., Schimak, G., Pitner, T., Eds.; Springer: Cham, Switzerland, 2018.
14. Griffin, R.C.; Bromley, D.W. Agricultural runoff as a nonpoint externality: A theoretical development. *Am. J. Agric. Econ.* **1982**, *64*, 547–552. [CrossRef]
15. Shortle, J.S.; Dunn, J.W. The relative efficiency of agricultural source water pollution. *Am. J. Agric. Econ.* **1986**, *68*, 668–677. [CrossRef]

16. Shortle, J.S.; Abler, D.G.; Horan, R.D. Research issues in nonpoint pollution control. *Environ. Resour. Econ.* **1998**, *11*, 571–585. [CrossRef]
17. Larson, D.M.; Helfand, G.E.; House, B.W. Second best tax policies to reduce nonpoint source pollution. *Am. J. Agric. Econ.* **1996**, *78*, 1108–1117. [CrossRef]
18. Ribaudo, M.; Horan, R.D.; Smith, M.E. *Economics of Water Quality Protection from Nonpoint Sources: Theory and Practice*; Agricultural Economic Report No. 782; Resource Economics Division, Economic Research Service, U.S. Department of Agriculture: Washington, DC, USA, 1999.
19. Kolstad, C.D. Uniformity versus differentiation in regulating externalities. *J. Environ. Econ. Manag.* **1987**, *14*, 386–399. [CrossRef]
20. Wu, J.; Babcock, B.A. Babcock. Spatial heterogeneity and the choice of instruments to control nonpoint pollution. *Environ. Resour. Econ.* **2001**, *18*, 173–192. [CrossRef]
21. Doole, G.J. Evaluating input standards for nonpoint pollution control under firm heterogeneity. *J. Agric. Econ.* **2010**, *61*, 680–696. [CrossRef]
22. Doole, G.J.; Pannell, D.J. Empirical evaluation on nonpoint pollution policies under agent heterogeneity: Regulating intensive dairy production in the Waikato Region of New Zealand. *Aust. J. Agric. Resour. Econ.* **2011**, *56*, 82–101. [CrossRef]
23. Esteban, E.; Albiac, J. Salinity pollution control in the presence of farm heterogeneity: An empirical analysis. *Water Econ. Policy* **2016**, *2*, 1650017. [CrossRef]
24. Murray-Darling Basin Authority. Basin Salinity Management Strategy 2001–2015. Australian Government. 2001. Available online: https://www.mdba.gov.au/sites/default/files/pubs/BSMS-full.pdf (accessed on 31 July 2022).
25. Department of Environment and Conservation, New South Wales, Australia. Hunter River Salinity Trading Scheme: Working Together to Protect River Quality and Sustain Economic Development. 2003. Available online: https://www.cbd.int/financial/pes/australia-pesriver.pdf (accessed on 31 July 2022).
26. NSW Minerals Council. Review of Hunter River Salinity Trading Scheme. 2014. Available online: https://www.epa.nsw.gov.au/-/media/epa/corporate-site/resources/licensing/hrsts/nsw-minerals-council.pdf (accessed on 31 July 2022).
27. Yaron, D.; Ratner, A. Regional cooperation in the use of irrigation water: Efficiency and income distribution. *Agric. Econ.* **1990**, *4*, 45–58. [CrossRef]
28. Dinar, A.; Ratner, A.; Yaron, D. Evaluating cooperative game theory in water resources. *Theory Decis.* **1992**, *32*, 1–20. [CrossRef]
29. Nicholson, F.; Laursen, R.K.; Cassidy, R.; Farrow, L.; Tendler, L.; Williams, J.; Surdyk, N.; Velthof, G. How Can Decision Support Tools Help Reduce Nitrate and Pesticide Pollution from Agriculture? A Literature Review and Practical Insights from the EU FAIRWAY Project. *Water* **2020**, *12*, 768. [CrossRef]
30. Chowdary, V.; Rao, N.; Sarma, P. Decision support framework for assessment of non-point-source pollution of groundwater in large irrigation projects. *Agric. Water Manag.* **2005**, *75*, 194–225. [CrossRef]
31. Mas-Colell, A.; Whinston, M.D.; Green, J.R. Chapter 16: Equilibrium and Its Basic Welfare Properties. In *Microeconomic Theory*; Oxford University Press: Oxford, UK, 1995.
32. Dinar, A.; Howitt, R.E. Mechanisms for allocation of environmental cost: Empirical tests of acceptability and stability. *J. Environ. Manag.* **1997**, *49*, 183–203. [CrossRef]
33. (RWQCB) California Environmental Protection Agency, (CalEPA) Regional Water Quality Control Board, Central Valley Region. *Total Maximum Daily Load for Salinity and Boron in the Lower San Joaquin River*; Staff Report; Califorinia Environmental Protection Agency, Regional Water Quality Control Board, Central Valley Region: Sacramento, CA, USA, 2002.
34. (RWQCB) California Environmental Protection Agency, (CalEPA) Regional Water Quality Control Board, Central Valley Region. *Amendments to the Water Quality Control Plan for the Sacramento River and San Joaquin River Basins for the Control of Salt and Boron Discharges into the Lower San Joaquin River*; Final Staff Report; California Environmental Protection Agency, Regional Water Quality Control Board, Central Valley Region: Sacramento, CA, USA, 2004.
35. Brownell, J.; (Regional Water Quality Control Board, Central Valley Region, Rancho Cordova, CA 95670, USA). Personal communication, 2013.
36. (LWA) Larry Walker Associates, Inc.; Systech Water Resources; Carollo Engineers; PlanTierra. *Development of a Basin Plan Amendment for Salt and Boron in the Lower San Joaquin River (LSJR)—Task 4 Implementation Planning for Proposed Salinity Objectives*; San Joaquin Valley Drainage Authority: Los Banos, CA, USA, 2015.
37. Quinn, N.W.T.; Hanna, W.M. Real-Time Adaptive Management of Seasonal Wetlands to Improve Water Quality in the San Joaquin River. *Adv. Environ. Res.* **2002**, *5*, 309–317. [CrossRef]
38. Ortega, R. Swamp Timothy Production Response to a Modified Hydrology in Wetlands of the Grassland Ecological Area. Master's Thesis, Department of Avian Studies, UC Davis, Davis, CA, California, 2009.
39. CDM Smith. *Central Valley Salinity Alternatives for Long-Term Sustainability (CV-Salts). Strategic Salt Accumulation Land and Transportation Study (SSALTS), Draft Phase 3 Report–Evaluate Potential Salt Disposal Alternatives to Identify Acceptable Alternatives for Implementation*; Prepared for San Joaquin Valley Drainage Authority, October 2016; CDM Smith: Boston, MA, USA, 2016.
40. United States Department of the Interior, Bureau of Reclamation, San Luis Unit Drainage, Central Valley Project, California, Fresno, Merced, and Kings Counties. *Draft Environmental Impact Statement*; Report S621.M531 1991; United States. Bureau of Reclamation. Mid-Pacific Regional Office: Sacramento, CA, USA, 1991.
41. (OMB) US Office of Management and Budget. 2021 Discount Rates. 2021. Available online: https://www.energy.gov/sites/default/files/2021-04/2021discountrates.pdf (accessed on 17 April 2022).

water

MDPI

Article

Can Remote Sensing Fill the United States' Monitoring Gap for Watershed Management?

Vamsi Krishna Sridharan [1,*], Saurav Kumar [2,*] and Swetha Madhur Kumar [3]

1 Fisheries Collaborative Program, University of California, Affiliated with Fisheries Ecology Division, Southwest Fisheries Science Center, National Marine Fisheries Service, National Oceanic and Atmospheric Administration, Santa Cruz, CA 95060, USA
2 Texas A&M Agrilife and Department of Biological and Agricultural Engineering, Texas A&M Agrilife Research Center, El Paso, TX 79927, USA
3 Client Analytics, Evolent Health, Arlington, VA 22203, USA; mk.swetha89@gmail.com
* Correspondence: vamsi.sridharan@noaa.gov (V.K.S.); saurav@tamu.edu (S.K.)

Abstract: Remote sensing has been heralded as the silver bullet in water quality modeling and watershed management, and yet a quantitative mapping of where its applicability is likely and most useful has not been undertaken so far. Here, we combine geospatial models of cloud cover as a proxy for the likelihood of acquiring remote scenes and the shortest time of travel to population centers as a proxy for accessibility to ground-truth remote sensing data for water quality monitoring and produce maps of the potential of remote sensing in watershed management in the United States. We generate several maps with different cost-payoff relationships to help stakeholders plan and incentivize remote sensing-based monitoring campaigns. Additionally, we combine these remote sensing potential maps with spatial indices of population, water demand, ecosystem services, pollution risk, and monitoring coverage deficits to identify where remote sensing likely has the greatest role to play. We find that the Southwestern United States and the Central plains regions are generally suitable for remote sensing for watershed management even under the most stringent costing projections, but that the potential for using remote sensing can extend further North and East as constraints are relaxed. We also find large areas in the Southern United States and sporadic watersheds in the Northeast and Northwest seaboards and the Midwest would likely benefit most from using remote sensing for watershed monitoring. Although developed herein for watershed decision support in the United States, our approach is readily generalizable to other environmental domains and across the world.

Keywords: remote sensing; geographical information systems; watershed management; water quality; decision support; ambient monitoring; data collection

Citation: Sridharan, V.K.; Kumar, S.; Madhur Kumar, S. Can Remote Sensing Fill the United States' Monitoring Gap for Watershed Management?. *Water* **2022**, *14*, 1985. https://doi.org/10.3390/w14131985

Academic Editor: Thomas Meixner

Received: 1 May 2022
Accepted: 20 June 2022
Published: 21 June 2022

1. Introduction

Of the 37.6 million waterbodies including canals, stream segments, ponds, and lakes in the United States, fewer than three million are monitored in situ, with only about 60,000 monitoring sites providing information that can be compared with remotely sensed data (Figure 1a) [1,2]. Of the millions of waterbodies, only 430,893 have been assessed for water quality impairments as of 2022 (Figure 1b) [3,4]. Within the recent past, even for the assessed waterbodies, ambient monitoring of general state of the water quality and synoptic data collection of the load, transport, and fate processes associated with impairment happens sporadically in only about a third of the cases [5]. Remote sensing, both with airborne and space-borne platforms offers tremendous potential for bridging these massive data gaps. But there are significant data gaps to even be able to use remote sensing everywhere in the United States (Figure 1a). Our contribution herein is, albeit with several simplifying assumptions, to map out the potential for remote sensing to be a viable monitoring tool for all the subwatersheds—the smallest catchment class associated with waterbodies as defined by the United States Geological Survey (USGS)—within the conterminous United

States. As a corollary to these maps, we also present a geospatial estimate of where remote sensing is likely to have the greatest impact in the country by mapping the intersection of remote sensing potential and high risk of impairment and low data coverage. These maps will serve as decision support tools for researchers and practitioners to plan where to invest their energies and resources with respect to data collection for water quality management approaches.

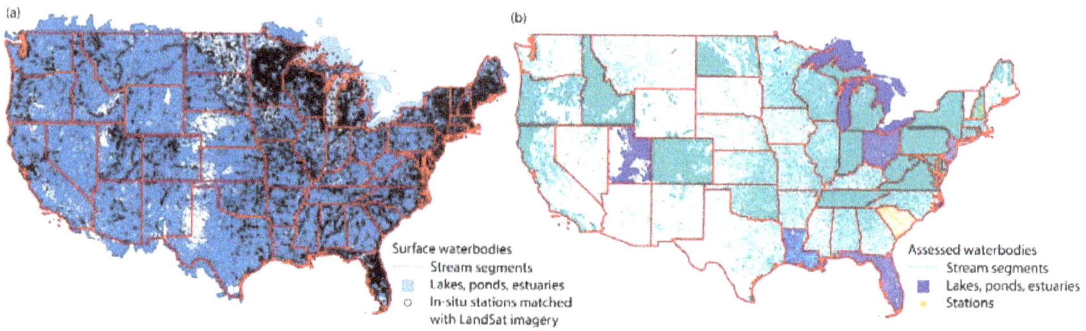

Figure 1. Assessment and in-situ monitoring gap for water quality in the United States: (**a**) 39.6 million stream segments, ponds, lakes and estuaries in the National Hydrography Dataset and 603,433 in-situ monitoring locations from the AquaSat database for Secchi disk depth, Chlorophyll-*a*, total suspended solids and dissolved organic carbon that have been matched to landSat scenes, and (**b**) the 430,893 assessed waterbodies in the United States obtained from the United states Environmental Protection Agency's Assessment and Total Maximum Daily Load Tracking and Implementation System (ATTAINS) database. Red lines indicate state borders.

Our focus in this paper is not to limit our treatment to a specific land surface process, water quality parameter, type of impairment or remote sensing data source. This is because water quality is a complex system response within the watershed to surface, subsurface and hydrologic processes that are modulated by human influence. Therefore, rather than focusing on a specific aspect of watershed processes, we develop a comprehensive scoping tool for adopting remote sensing in monitoring and data collection. Scoping tools such as the one presented here can be developed for specific land surface process monitoring, or water quality management of given impairments by using information relevant to certain remote sensing platforms and data collection methods.

Over the past two decades, a large number of many aerial and space-borne platforms and sensors that sample different parts of the electromagnetic spectrum have come online [6]. Platforms such as LandSat, GeoEye, WorldView, and technologies such as Light Detection and Ranging (LiDAR) are useful for water body delineation (e.g., [7]), while bathymetry of waterbodies can be obtained either from LiDaR, or by analyzing spectral band ratios of satellite and drone imagery [8,9]. Streamflow can be obtained using Radio Detection and Ranging (RaDAR) altimetry and rating curves from platforms such as Jason-3, Sentinel-3, and Saral/ALtika [10]. The terrestrial water budget can be quantified using gravimetric data, while floods can be characterized using optical sensors such as the Advanced Very-High-Resolution Radiometer (AVHRR), the Visible Infrared Imaging Radiometer Suite (VIIRS), the Moderate Resolution Imaging Spectroradiometer (MODIS), Sentinel, Landsat, Satellite Pour l'Observation de la Terre (SPOT), the Advanced Spaceborne Thermal Emission and Reflection Radiometer (ASTER), IKONOS, Worldview, RapidEye, Ziyuan 3 and Gaofen and synthetic aperture radar data [10]. Water quality parameters that are directly observable can be monitored using LandSat, Sentinel, MODIS, the MEdium Resolution Imaging Spectrometer (MERIS) and other specialized sensors such as Hyperion (chlorophyll-*a*), SPOT, AVHRR (Total suspended sediments), IKONOS (water clarity) and the Coastal Zone Color Scanner (CZCS) [colored dissolved organic matter] imagery [10].

Hyperspectral imagery [11], LiDAR and airborne laser scanning [12] can be used to infer crop characteristics and tree canopies in agrarian and forested watersheds. Within watersheds, indices developed by using reflectances from different spectral bands are useful for classifying land cover and land use types [13]. Also within watersheds, the state of best management practices such as low impact developments and green infrastructure can be monitored using visible and thermal imagery and spectral reflectance-based indices [14].

Authors have developed an array of methods ranging from the well-established to the experimental frontier have emerged to quantify watershed loading and receiving water quality pertaining to various types of impairment [15,16]. Broadly, for land surface processes such as crop cycles, land cover and land use changes, observations of surface reflectances and ratios of reflectance across multiple spectral bands are used with validation datasets such as land use classification maps or photographs are common (e.g., [17]). For water quality parameters that affect the inherent and apparent optical properties of the water column and spectral properties of the reflectances, such as total suspended solids, Chlorophyll-a, colored dissolved organic carbon, turbidity, and water surface temperature, statistical models linking observations of surface reflectances and ratios of reflectance across multiple spectral bands with the water quality parameters are popular [18]. For such processes, bio-optical and bio-geo-optical models that mechanistically link the irradiance-reflectance ratio across multiple spectral bands to the bulk water column properties are a reliable alternative [19]. For impairments such as harmful algal blooms and oxygen depletion, more sophisticated statistical (e.g., [20]) or mechanistic models (e.g., [21]) are used to relate watershed nutrient loads with impairment. For water quality parameters that do not directly affect the optical properties of the water column and spectral properties of the reflectances (e.g., mercury and heavy metals), statistical, machine learning and artificial intelligence models relating the quantities of interest with other water constituents (e.g., turbidity) are becoming increasingly common [15].

Two unifying requirements of all of these approaches are that for a given location, (i) it must be possible to obtain remotely sensed spectral imagery in the first place, and (ii) the land surface process such as fertilizer application, presence of a pollutant load attenuation best management practice like a detention pond or bioswale, and the receiving water quality must all be ground-truthed with in-situ observations [22–24]. This means that the vast potential of remote sensing can only truly be exploited in those areas where atmospheric conditions are generally conducive to remote sensing, and where some access to the watershed is possible. We leverage geospatial datasets containing information on atmospheric conditions and accessibility to rank subwatersheds according to their propensity to being amenable to remote sensing for water quality monitoring under a variety of cost-payoff scenarios.

Reasons for the gap between the number of this assessment and monitored waterbodies and the total number of waterbodies in the United states can include (i) applicability, (ii) accessibility, (iii) resource constrains and (iv) socioeconomics. First, in most cases, waterbodies might be located in remote, locations with minimal human development and limited beneficial uses to society apart from preserving pristine natural beauty and bounty. Second, there may be access issues due to rugged landforms, impassable land use features and limited mobility networks which make monitoring difficult or even impossible. Third, state, tribal and local agencies tasked with assessing and programmatically attaining designated uses for waterbodies may be hampered by resource constraints that make regular ambient monitoring and synoptic data collection difficult or even impossible in all watersheds. Fourth, there may be environmental justice issues with impairment being linked to socioeconomic and demographic conditions (e.g., [25]). In this paper, we only look at the first three factors, as we are only interested in the technical constraints of feasibility of using remote sensing for water quality monitoring and data collection here. The advantage of remote sensing over conventional monitoring programs is that for areas where the former is applicable, ground-truthing need not be performed very often, and in the overall life cycle of monitoring and data collection programs, can be significantly less expensive to implement, particularly if public domain data and tools are used.

The paper is organized as follows: in Section 2, we describe our classification approach, the datasets we used and the geostatistical methods we adopted. In Section 3, we showcase the remote sensing potential map. In Section 4, we discuss where in the country gaps need to be filled, given the potential for remote sensing by looking at the applicability and accessibility of the subwatersheds and the intersection of quantified risks of impairment, ecological vulnerability and lack of coverage. Finally, also in Section 4, we outline the caveats in our approach, and chart our future trajectory in making these classifications more robust and readily accessible to the watershed management community.

2. Materials and Methods

Our fundamental premise is that we can determine the potential of remote sensing to be useful for ambient monitoring and synoptic data collection in a given waterbody by mapping this potential at the USGS' Hydrologic Unit Code 12 (HUC-12) subwatershed scale. The HUC-12 subwatershed classification allows us to analyze watersheds at the 1:24,000 scale. At this scale, these subwatersheds typically contain four stream segments, except in a few larger subwatersheds in the Northeast and the Great Lakes. We feel that this spatial scale is sufficient to identify the potential for remote sensing, as the streams encompassed within these subwatersheds are usually higher order distributaries or small areal bodies with similar climate, land use and terrain characteristics. In total, there are 102,973 HUC-12 subwatersheds in the conterminous United States. We discuss our approach in Section 2.1 below.

We used numerous Geographical Information System (GIS) resources in this study to map access, acquisition, constraints and risks of impairment. All GIS operations and visualizations were performed in QGIS [26], while the model development was performed in MATLAB [27]. In determining the potential for remote sensing and its applications, we used existing accessibility and image acquisition conditions, and did not account for the effects of land use and land cover change or climate change. After determining the potential of remote sensing as a data source within each subwatershed, we subsequently subset subwatersheds on the basis of applicability and resource constraints as described in Section 2.2 below. We have listed all the datasets we used in these analyses in Table 2.

Table 1. Data sources used in this study.

Serial Number	Data	Use	Source	Location
1	Hydrologic Unit Class 12 subwatersheds	Smallest hydrologic feature for analysis	National Hydrography Dataset	https://www.usgs.gov/national-hydrography/national-hydrography-dataset (accessed on 20 April 2022)
2	Digital Elevation Model	Hillshade and basemaps	Advanced Spaceborne Thermal Emission and Reflection Radiometer (ASTER)	https://earthexplorer.usgs.gov (accessed on 20 April 2022)
3	Cloud cover database	Acquisition analysis	EarthEnv	https://www.earthenv.org/cloud (accessed on 20 April 2022)
4	Time to cities	Accessibility analysis	Malaria Atlas Project	https://malariaatlas.org/research-project/accessibility-to-cities/ (accessed on 20 April 2022)
5	United States administrative boundaries	Basemaps	Census cartographic boundary files	https://www.census.gov/geographies/mapping-files/time-series/geo/cartographic-boundary.html (accessed on 20 April 2022)

Table 2. *Cont.*

Serial Number	Data	Use	Source	Location
6	Assessed waterbodies	Evaluating assessment gaps	Assessment and Total Maximum Daily Load Tracking and Implementation System (ATTAINS)	https://www.epa.gov/waterdata/attains (accessed on 20 April 2022)
7	Population by HUC-12	Evaluating human footprint	EnviroAtlas	https://www.epa.gov/enviroatlas (accessed on 20 April 2022)
8	Agricultural, domestic, industrial and thermoelectric water demand by HUC-12 (MGD)	Evaluating water supply needs	EnviroAtlas	https://www.epa.gov/enviroatlas (accessed on 20 April 2022)
9	Protected areas under the International Union for Conservation of Nature and the United States protected areas by HUC-12 (% of land cover)	Evaluating conservation need	EnviroAtlas	https://www.epa.gov/enviroatlas (accessed on 20 April 2022)
10	Big game and bird hunting, fishing, and migratory bird watching demand by HUC-12 (days/year)	Evaluating recreational need	EnviroAtlas	https://www.epa.gov/enviroatlas (accessed on 20 April 2022)
11	Vulnerability index of native aquatic species by HUC-12	Evaluating biodiversity needs	EnviroAtlas	https://www.epa.gov/enviroatlas (accessed on 20 April 2022)
12	Wastewater discharge (MGD), total permitted discharge, daily agricultural runoff (mm)	Evaluating environmental needs	EnviroAtlas	https://www.epa.gov/enviroatlas (accessed on 20 April 2022)

Although our analysis pipeline extends to Alaska, Hawaii and other minor islands and United States protectorates, we have restricted all figures to only the conterminous United States for simplicity. To illustrate the role of topography on various watershed characteristics, in most figures, we have superimposed our analytical products on a hillshade relief we generated from the 30 m resolution ASTER digital elevation model obtained from the USGS Earth Explorer portal.

2.1. Potential for Remote Sensing as a Tool to Collect Data on the Subwatershed Scale

We used the mean annual fraction of cloudy days from the EarthEnv global cloud cover model dataset [28] as a surrogate to indicate how likely it is to acquire remote sensing imagery of a 1 Km^2 scene and how hard it is to process this imagery for use (Figure 2a). This dataset includes 1 Km resolution mean conditions between the period of 2001 and 2014. We used the global map of minimum travel times from a location to a population center (defined as a city or township with more than 50,000 people) produced by the Malaria Atlas Project [29] as a surrogate for accessibility to a given 1 Km^2 grid cell (Figure 2b). This raster map is the outcome of a geospatial travel time model that takes existing road and rail networks, land cover and 30 m resolution digital terrain models of the entire globe. The combination of these two datasets, respectively indicate how likely it is to acquire a scene remotely, and how easy it is to ground-truth reflectances to link the remotely sensed data to land cover characteristics and water quality impairments. The tacit assumption is that it will be possible for stakeholders tasked with managing the watershed to reach any major population center, and then travel from there to perform in-situ sampling work. To link these datasets to the hydrological units of interest, we used the USGS' National Hydrography Dataset's (NHDplus) HUC-12 layer [3].

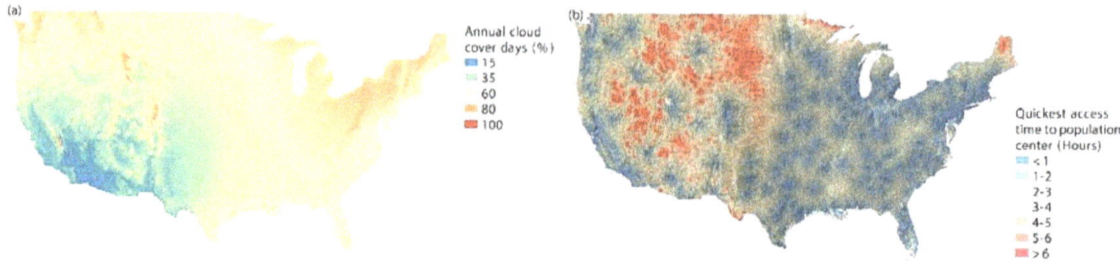

Figure 2. Datasets used to model potential for remote sensing in watershed monitoring: (**a**) mean annual fraction of cloudy days, and (**b**) maximum shortest travel time to population center within each HUC-12 subwatershed. These quantities are surrogates for acquisition and accessibility of remotely sensed scenes and in-situ data needed to ground-truth them.

We modeled the potential for using remote sensing for monitoring and data collection within each subwatershed as the product of various categorical levels of $\tau^{\text{Shortest}}_{\text{max}}$, the maximum of the shortest time to reach a population center from anywhere within the subwatershed, and various categorical levels of $\mu_{\text{Cloud Cover}}$, the mean annual fraction of cloudy days within the subwatershed. Within each subwatershed, we estimated $\tau^{\text{Shortest}}_{\text{max}}$ as the maximum over all pixels within the area of the shortest time to reach a population center. We estimated $\mu_{\text{Cloud Cover}}$ as the mean over all pixels within the area of mean annual fraction of cloudy days. Thus, for the ith subwatershed, we have

$$P_{\text{Remote Sensing},i} = \mathbb{C}\left(\tau^{\text{Shortest}}_{\text{max},i}\right) \times \mathbb{C}(\mu_{\text{Cloud Cover},i}) \quad (1)$$

Here, the function $\mathbb{C}(\cdot)$ assigns a category from 0 to 3 to both the acquisition ($\mu_{\text{Cloud Cover}}$) and access ($\tau^{\text{Shortest}}_{\text{max}}$) variables according to Table 3. 0 indicates a low potential for success, and 3 indicates the highest potential for success. The possible values of $P_{\text{Remote Sensing}}$ obtained in this way are binned into four categories for ease of application: 0 (unsuitable), 1–3 (low), 4–6 (good), and 9 (excellent). We note that the numeric values of $P_{\text{Remote Sensing}}$ themselves are only meaningful when they contribute to the different categories of "unsuitable," "low," "good" and "excellent."

Table 3. Cost-payoff matrix for modeling potential of remote sensing for watershed monitoring.

Category	Accessibility: Shortest Time to Population Center	Acquisition: Fraction Cloudy Days Per Year
Conservative		
0	>6 h	>75%
1	2–6 h	50–75%
2	1–2 h	25–50%
3	<1 h	<25%
Normal		
0	>24 h	>90%
1	6–24 h	75–90%
2	3–6 h	50–75%
3	<3 h	<50%
Optimistic		
0	>48 h	>95%
1	24–48 h	90–95%
2	6–24 h	75–90%
3	<6 h	<75%

As stakeholders in the watershed management domain may weigh the costs associated with the logistics of travel to remote locations for in-situ ground-truthing of remote sensing data, and the careful planning of periods when to collect in-situ data in conjunction with clear sky days, we represented these weights in cost-payoff matrices to set the categorical values of the acquisition and access variables (Table 3). These weights effectively lower the $\tau_{max}^{Shortest}$ and $\mu_{Cloud\ Cover}$ threshold values, respectively for the accessibility and acquisition criteria for the different categories going from conservative to normal to optimistic estimates of the cost-payoff relationships (Table 3).

For example, one regulatory agency may be willing to send an engineer on a six-hour long journey to collect water quality data. So, any value of $\tau_{max,i}^{Shortest}$ smaller than six hours for a subwatershed within the jurisdiction of this agency would get a $\mathbb{C}(\tau_{max}^{Shortest})$ value of 3. Another agency may decide that it does not want its engineers to spend more than one hour traveling to a remote site. In this case, the $\mathbb{C}(\tau_{max}^{Shortest})$ for any $\tau_{max}^{Shortest}$ larger than one hour would be 2 or lower. Similarly, a consultant may be willing to invest the effort into building robust processing pipelines to deal with largely cloudy scenes or invest time and resources in data fusion-based cloud removal algorithms that are becoming more popular across various platforms [30,31]. This consultant would perhaps be willing to assign a $\mathbb{C}(\mu_{Cloud\ Cover})$ value of 2 or higher for a subwatershed where the sky is likely to be cloudy for almost 75% of the days in the year. Another consultant who is perhaps geared to directly use analysis-ready products may assign a $\mathbb{C}(\mu_{Cloud\ Cover})$ value of 3 for their subwatershed only if the sky is likely to be clear for 75% of the days in the year. By presenting maps of the categorical levels of the potential across a range of $\mathbb{C}(\cdot)$ values, we allow decision-makers to choose the cost-payoff level they are comfortable with to plan their workflows accordingly.

2.2. *Quantifying Benefits of Remote Sensing in Watershed Monitoring*

We modeled the potential for remote sensing using Equation (1) and surrogates for accessibility and acquisition. However, these maps in themselves provide only one piece of the puzzle. It is only when the potential for watershed monitoring using remote sensing is combined with risk maps and lack of coverage by conventional monitoring that areas can be identified where remote sensing can play a crucial role. To do this, we developed maps of five metrics of risk using published GIS data from the EnviroAtlas [32] (Table 2): (i) the human footprint, (ii) anthropogenic water demand, (iii) ecosystem vulnerability, (iv) impairment, and (v) conventional monitoring and assessment coverage gaps. As the units of these metrics are different from one another, we simply overlaid these maps and determined the intersecting areas to determine where the remote sensing role would be most useful.

Human footprint: We collected the human population in each subwatershed. The conventional wisdom is that impairment and watershed management matter most where human settlements are prevalent. The subwatersheds populations were then ranked from 0 to 6 with decadal increase from less than 10 to more than one million people.

Anthropogenic water demand: We combined the domestic, agricultural, industrial and thermoelectric water demand in each subwatershed to estimate the total water demand from each subwatershed. Watersheds with greater demand are likely to experience larger ecological deficit flows and a higher propensity for critical conditions of impairment. The subwatershed demands were then ranked from 0 to 6 with decadal increase from less than 12,000 Gallons Per Day (GPD) [the per capital daily water demand is 1200 GPD estimated by dividing the total water demand by the total population].

Ecosystem vulnerability: We combined three key ecosystem services with equal weightage to each to estimate an overall vulnerability index. First, we summed the average number of days in a year that hunting, fishing, and recreational activities are typically engaged in within a subwatershed to obtain a fractional recreation time, $f_{Recreation}$. Large values of this number indicate greater pressure for recreational activities. Second, we combined the total fraction of land cover earmarked for conservation by both the International Union for Nature Conservation (IUCN) and the United States Government, $f_{Conservation}$. Large values of this number indicate that additional monitoring, assessment, and restoration effort must be expended in these subwatersheds. Third, we collected the native aquatic species vulnerability index, $f_{Vulnerability}$, a key measure of how many endangered, native species are threatened [33]. This index is also a proxy for ecosystem health in general. Higher values of this index imply generally poorer ecosystem health. All these numbers range from 0 to 1. The overall ecosystem vulnerability of the subwatershed is then

$$f = w\left(f_{\text{Recreation}}\mathbb{I}_{\text{Recreation}} + f_{\text{Conservation}}\mathbb{I}_{\text{Conservation}} + f_{\text{Vulnerability}}\mathbb{I}_{\text{Vulnerability}}\right) \quad (2)$$

where $\mathbb{I}_i$ is an indicator that can either be 0 or 1 depending on whether the service i, i.e., recreation, conservation or species vulnerability, is being provided by the subwatershed or not, and w is 1, 0.5 or 0.33 if there are one, two or all three of the services being provided by the subwatershed. The subwatersheds ecosystem vulnerability indicator was then ranked 0 for no vulnerability, and then from 1 to 4 in increments of 0.25.

Impairment risk: We developed subindex curves of the total wastewater discharge in Million Gallons per Day (MGD), the total permitted pollutant load in pounds per year, and the total agricultural overland, tile and non-tile subsurface runoff in mm as described in Walsh and Wheeler [34]. We then combined these subindices using a maximum subindex measure as the overall index of impairment within a subwatershed. This index was deemed to represent the impairment risk in the subwatershed. To develop these subindex curves the following approach was adopted: first, we summed the total load in each class, i.e., wastewater, permitted sources, and agricultural runoff, and divided it by the total surface area of waterbodies within the watershed to normalize loads across subwatersheds. Then we omitted zero values and obtained the 25th, 50th and 75th quantiles of these normalized loads across all the subwatersheds in the country and assigned rank values from 0 to 4 depending on whether there was no impairment from that loading class to whatever quantile range the normalized loads fell into. This effectively positioned the normalized loads in each load class onto a subindex curve. Then, as the nature of impairment is likely to vary depending on the dominant economic sector, human activity, and land surface processes in each subwatershed, we took the overall impairment risk to be the maximum of the three rank values. Thus, higher values of this index (ranging from 0 to 4) represents higher risk of impairment.

Conventional monitoring and assessment coverage gaps: For each subwatershed, we estimated the total length of assessed stream segments, L_{Stream}, the total area of assessed ponds, lakes, and estuaries, A_{Body}, and the total number of monitoring stations, N_{Station}, from the Assessment and Total Maximum Daily Load Tracking and Implementation System (ATTAINS) database [4]. A subwatershed was determined to have a monitoring gap if $L_{\text{Stream}} + A_{\text{Body}} + N_{\text{Station}} = 0$.

Finally, we also produced a remote sensing monitoring potential map using the normal cost-payoff estimates from Table 3 to indicate where remote sensing would be likely feasible. Subwatersheds where the role of remote sensing is likely to be most crucial were then classified as those that met the decision tree in Figure 3, that is, where all the layers after suitable thresholding to represent various risks intersect with at least good potential for remote sensing.

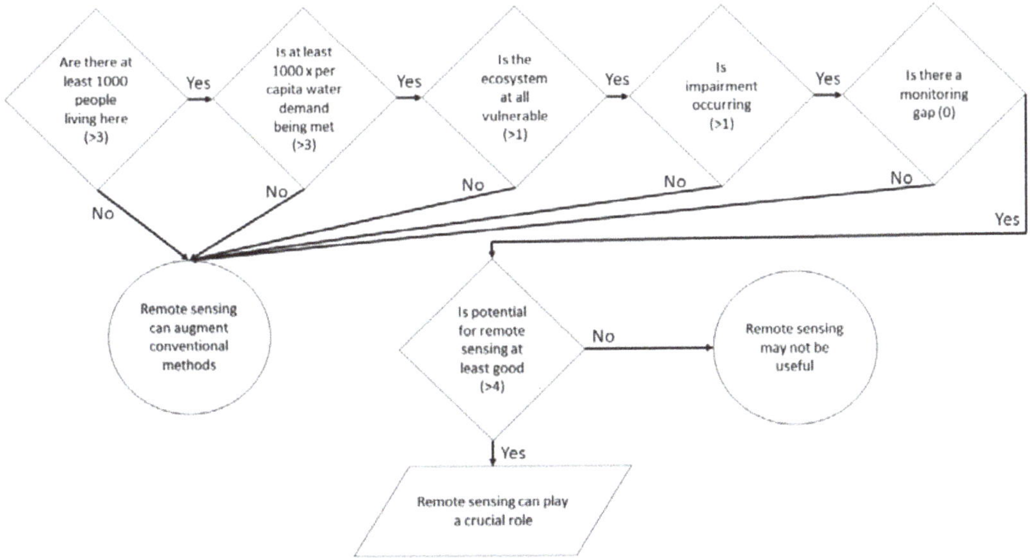

Figure 3. Flowchart for estimating where the role of remote sensing will be crucial in watershed management. The operation represents the intersection of geospatial layers. The numbers indicate threshold values of various index levels.

3. Results

We show the maps of potential for remote sensing in monitoring within watersheds for various cost-payoff scenarios in Figure 4. These potential rankings represent an interplay between the ambient cloud cover (Figure 2a) and the maximum of the shortest travel time to a population center (Figure 2b). The model predicts that when cloud cover is low, and travel times are short, the potential is maximized. Conversely, when cloud cover is high and travel times are long, the potential is inhibited. Typically, cloud cover is lowest in the Southwest, and increases generally through the Central Plains towards the East and is highest over the North and the Rockies and Appalachian mountains (Figure 2a). Travel times are generally relatively low (less than three hours) on the Eastern and Western seaboards, and throughout the Central plains and in the South, and increase to more than six hours in the mountainous Rockies and the remote Midwest. These patterns are reflected in the model.

Going from left to right in Figure 4, the expense of effort to build pipelines for occluded images and changing light conditions is budgeted increasingly generously. This reflects in the improved potential for remote sensing in more parts of the Mideast and East. Going from top to bottom in Figure 4, the effort required to collect in-situ data for ground-truthing is budgeted increasingly generously. This reflects in the improved potential for remote sensing in more parts of the Southwest and the Midwest, until in the bottom right of Figure 4, there is more or less uniformly high potential throughout the country. Nonetheless, most of the California Central Valley, Southern California, and the Central Plains have uniformly good to excellent potential under all cost-payoff scenarios, owing to generally clear skies and short travel times.

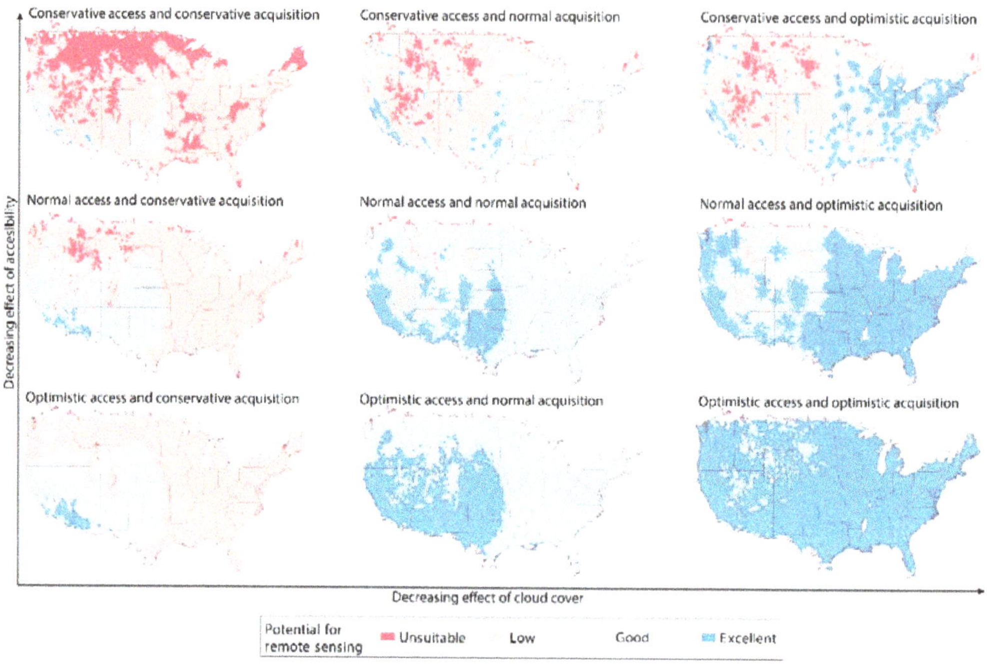

Figure 4. Modeled maps of the potential of remote sensing for data collection and monitoring of watersheds under various cost-payoff assumptions. As the effect of cloud cover decreases (more cloudy days are acceptable), more cloudy areas display higher potential. As the effect of access decreases (longer travel times are acceptable), more remote areas display higher potential. Thin red lines are state boundaries.

In Figure 5, we show the various risk metrics (a through e) and the associated remote sensing utility map (f) by applying the flowchart in Figure 3 to each subwatershed in these layers. Population centers, as expected, are generally concentrated near the major cities on the Eastern and Western seaboards and in the Mideast and the Gulf of Mexico coast (Figure 5a). The water demand generally tracks the population centers, except in California, the Midwest, and the Central Plains subwatersheds from where water is diverted to other places (Figure 5b). For instance, in California, water is gravity fed from reservoirs in the North to consumers in the South. Almost everywhere in the United States, ecosystems are vulnerable (Figure 5c). In most of the Midwest, this is likely due to large tracts of lands being designated as protected zones, and in the Sierra Nevada mountains in the West, due to ecological disasters by sustained drought and forest fires. Most of the population centers on the Eastern and Western seaboard are polluted by wastewater and permitted discharges, while the Central plains and Midwest impairments reflect of agricultural runoff into the Mississippi River tributaries (Figure 5d). The ATTAINS database indicates that most of the nation's subwatersheds have at least some waterbodies that have been assessed. But many subwatersheds in the United States-Mexico border region, the remote Midwest, and the Rockies remain unassessed (Figure 5e). Based on a combination of these layers in conjunction with the remote sensing potential model under the assumption of a normal cost-payoff scenario (middle panel in Figure 4 and Table 3), several hotspots emerge banded predominantly in the South and the North where remote sensing could be crucial, and often perhaps the only source of assessment of the condition of these subwatersheds (Figure 5f).

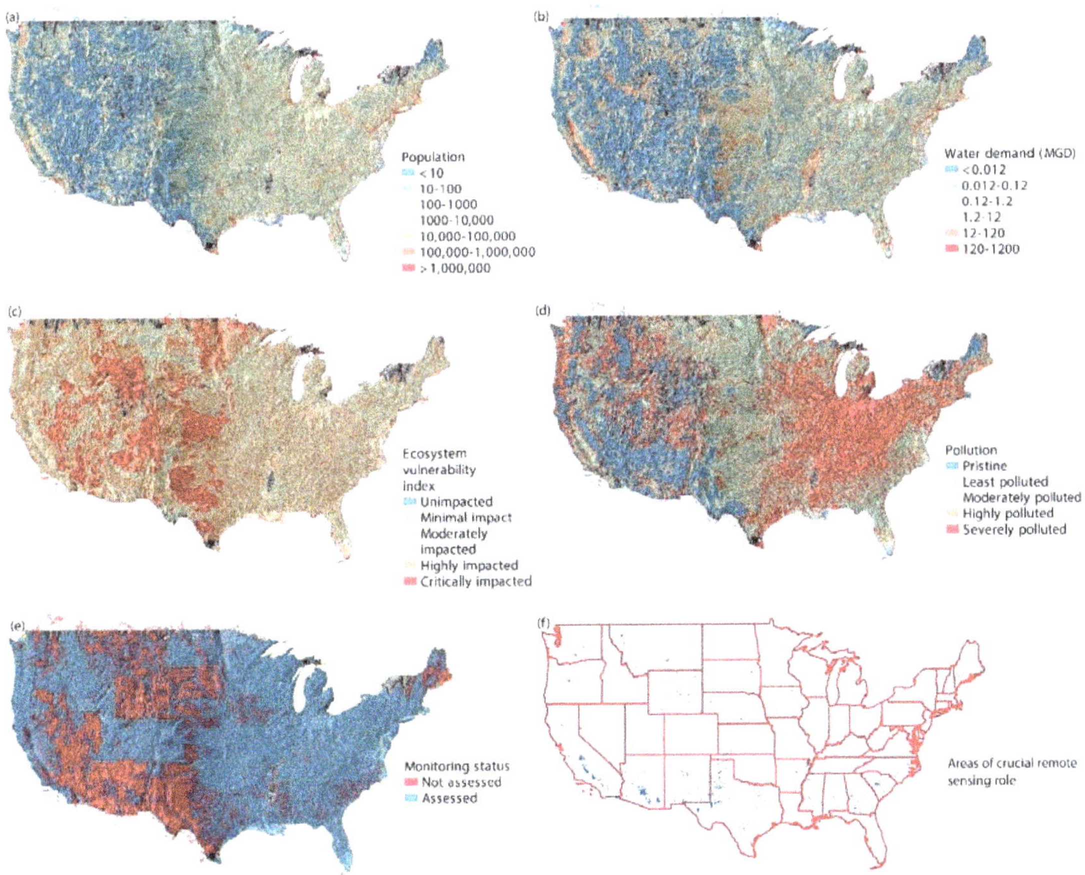

Figure 5. Predicted subwatersheds where the role of remote sensing could be crucial as the intersection of greater than good remote sensing potential, high impairment risk indicators, and low coverage. Across subwatersheds, (**a**) population, (**b**) total water demand, (**c**) ecosystem vulnerability index, (**d**) pollution entering streams and waterbodies, (**e**) water quality monitoring gaps, and (**f**) intersection of these layers with normal accessibility and normal cloud cover cost-payoff model predicted higher than good remote sensing potential subwatersheds (blue polygons).

It is evident from these remote sensing potential maps and the combination of risk metrics and remote sensing potential that there are specific areas within the country where watershed researchers and managers can beneficially leverage remote sensing data at various spatial scales ranging from stream segment-reach to the subwatershed scale. By allocating resources preferentially to occluded and poorly-lit scene imagery usage pipelines, the potential for using remote sensing can be tremendously maximized. This is borne out by rapidly improving potential going left to right in Figure 4, than by going from top to bottom in Figure 4.

4. Discussion and Conclusions

We have presented a set of maps that chart the potential of remote sensing to augment the monitoring and data collection in watersheds for land surface and water quality processes. These are available for open source download in a Git repo [35]. To enable easy access to, and visualization of these maps, the risk factors for impairment, and summary statistics of the potential of remote sensing under various assumptions within the cost-payoff matrix, we have also developed an app [36] using Google Earth Engine [37]. This app contains an interactive map within which the user can click on a specific subwatershed of interest to obtain additional information (Figure 6). A summarization of this paper is available as a guide to using the app in a separate panel. Once a subwatershed has been clicked, a summary visual report will be generated, and the relevant information on the impairment risk factors and the potential for remote sensing can be downloaded as comma-separated value files.

Figure 6. Google Earth Engine application to demonstrate the potential of remote sensing in monitoring watershed management in the United States. Attribution: Map Data © 2022 Google INEGI Imagery © 2022 NASA, TerraMetrics.

Additionally, we have shown how such maps can be modified within a cost-payoff landscape to be truly customizable by watershed decision-makers. The cost-payoff analysis also shows that improving the data processing pipelines for occluded and poorly lit scenes can tremendously benefit the power of remote sensing (going left to right in Figure 4). By combining the regions of high remote sensing potential with other risk factors describing human footprint, impairment risk, ecosystem vulnerability, and conventional monitoring coverage gaps, we can identify those subwatersheds where remote sensing is likely to have the highest benefit, and in fact, likely be an integral source of primary data. Large areas in the Southern United States, particularly in California, New Mexico, Texas, Mississippi, Georgia, and the Carolinas, and sporadic watersheds in the Northeast and Northwest seaboards (Washington and Maine) and the Midwest would likely benefit most from using remote sensing for watershed monitoring (Figure 5f).

Our approach represents a seminal and necessary step in aiding the decision-making process for resource-constrained regulatory agencies and contractors. By augmenting these maps with other socio-economic geospatial data, deeper insights can be gained to tackle challenges of environmental justice and equity. Rather than focusing on specific watershed processes, impairments, and remote sensing platforms, we have developed a method of assessing the general potential for remote sensing for whatever applications are envisioned by system managers in their monitoring and data collection workflows. This approach allows us to explore the benefits of remote sensing further, as (i) not all parts of the country are monitored regularly, and even when they are, in-situ monitoring costs can typically exceed several hundred thousands of dollars a year within a small watershed [38], and (ii) remote sensing can provide long-term cost savings. Remote sensing potential maps could be constructed for specific applications outlined in the introduction by suitably modifying the factors considered in the model presented in Section 2.1. For example, if a consultant proposes a data fusion pipeline using Sentinel 2, MODIS, and Planet imagery [39] to study watershed health, heavy metal concentrations in a nearby lake, and the role of watershed best management practices, then the principal costs of image acquisition will likely be the cost of procuring high-resolution Planet imagery, while in-situ monitoring costs will likely be driven by personnel costs rather than access time.

There are several caveats to the maps we have produced here. First, the assumptions built into the EarthEnv cloud cover and the Malaria Atlas travel time databases carry over to our maps. While these data sources are excellent for global coverage at the kilometer-scale, our pipeline may be refined with source DEMs and spectral imagery for specific watersheds to systematically downscale our national coverage maps. The shortest time to reach a population center, while somewhat ad-hoc, is an appropriate metric to characterize the difficulty of ground-truthing remote sensing imagery because typically waterbodies outside major population centers are not monitored. However, as we show in Figure 5d, such areas are also at high risk of impairment. Second, our remote sensing potential model is extremely simple and steady state, and yet captures the underlying tradeoff between scene acquisition and ground-truthing. Within a watershed, multiple process-scales exist, ranging from storms and coastal upwelling events that happen on hourly to daily timescales, land use and land cover changes that happen on annual to decadal timescales, and landform changes that happen over many decades to centuries [40]. Our maps do not factor in these different timescales. More sophisticated models, including spatial kriging approaches and Gaussian process models that incorporate seasonal variability in scene acquisition and travel time can be developed. Cloud cover, for instance, can vary seasonally so that even in areas with many cloudy days, there can be periods with clear skies during which period field campaigns could be planned. We do not capture such nuances here. For simple applications that require only a single remote sensing image such as a National Agricultural Image Program [41] or LandSat scene, cloud cover may not even be an issue if the acquisition window can be carefully determined. Although many non-dimensional metrics for land use and land cover classification for characterizing the environment can be derived from multi- and hyperspectral imagery in most locations within the United States, as we showed in Figure 1, there are still significant data gaps that require physical ground-truthing for which our maps will be useful. Third, we have operated at the HUC-12 subwatershed scale in this project for computational tractability. However, much finer reach-scale resolution of the impaired waterbodies can be incorporated within our pipeline by combining the NHDplus and ATTAINS datasets using cloud platforms such as Google's Earth Engine [37]. Fourth, our thresholds for cloud cover and minimum travel time to population centers for the various cost-payoff scenarios are somewhat arbitrary, but still indicate sensible and meaningful trends in the remote sensing potential maps. This is a reflection of the fact that we have sacrificed specificity to different remote sensing technologies for the generality of our guidance maps. These thresholds can be more rigorously defined for different types of remote sensing platforms by combining the atmospheric conditions and ground-truthing needs of specific technologies. Fifth, while we have chosen risk metrics that largely encompass most risk factors for watershed impairments, a more thorough exploration of available datasets, particularly socioeconomic data, is possible to identify other, transdisciplinary benefits of remote sensing within a watershed. It is likely that such exhaustive inclusion must be done on a local municipal, county, or state scale than on the national scale for disparate factors to sensibly be combined at meaningful physical scales. In subsequent iterations of these products, we will tackle these advances. Sixth, we only considered terrestrial subwatersheds in our current pipeline, and excluded all coastal and

Great Lakes impairments. In subsequent iterations of these maps, we will include these areas as well.

While we have developed the pipeline for the United States, a similar workflow could be developed anywhere in the world, or even for the entire planet. The only consideration is that while Earth Explorer, the United States Environmental Protection Agency, and the EnviroAtlas catalog a wide array of geospatial datasets for immediate use, such data may be difficult to find in the international landscape, and may have to be stitched together from disparate sources of information. Nonetheless, for any serious, reproducible, and scalable application of remote sensing for monitoring and data collection on watershed loads and impairments, this is a necessary first step. For the first time ever, a product is available that will allow researchers and managers to evaluate the potential of remote sensing to augment, or even fulfill ambient monitoring and data collection needs throughout the United States. We hope that our maps will be useful for the watershed managers within the United States, even as we continue to improve them with finer granularity and additional factors.

Author Contributions: Conceptualization, V.K.S., S.K. and S.M.K.; methodology, V.K.S. and S.K.; software, V.K.S. and S.K.; validation, V.K.S. and S.K.; formal analysis, V.K.S.; investigation, V.K.S.; resources, V.K.S., S.K. and S.M.K.; data curation, V.K.S.; writing—original draft preparation, V.K.S.; writing—review and editing, V.K.S. and S.K.; visualization, V.K.S.; supervision, V.K.S. and S.K.; project administration, V.K.S.; funding acquisition, Unfunded study. All authors have read and agreed to the published version of the manuscript.

Funding: This research received no external funding.

Institutional Review Board Statement: Institutional review and approval was not applicable for this study.

Informed Consent Statement: Informed consent was not applicable for this study as there were no human subjects.

Data Availability Statement: Data available in a publicly accessible repository that does not issue DOIs here: https://github.com/vamsiks2003/remoteSensingPotentialMapping (accessed on 20 April 2022). Publicly available datasets were analyzed in this study. This data can be found here: https://www.usgs.gov/national-hydrography/national-hydrography-dataset, https://earthexplorer.usgs.gov, https://www.earthenv.org/cloud, https://malariaatlas.org/research-project/accessibility-to-cities/, https://www.census.gov/geographies/mapping-files/time-series/geo/cartographic-boundary.html, https://www.epa.gov/waterdata/attains, https://www.epa.gov/enviroatlas (accessed on 20 April 2022). Data citation: Sridharan, V.K., Kumar, S.N., and Madhur Kumar, S. 2021. GitHub Repo remoteSensingPotentialMapping (https://github.com/vamsiks2003/remoteSensingPotentialMapping (accessed on 20 April 2022)).

Acknowledgments: Discussions with Rocky Talchabadel and Santosh Palmate at Texas A&M University were useful in the development of the methods presented here. Critical review by Lee Harrison at the University of California, San Diego and recommendations by three anonymous reviewers and the Academic Editor at Water significantly improved the quality of this manuscript. The lead author acknowledges the Environmental and Water Resources Institutes Remote Sensing Task Committee, Total Maximum Daily Load Analysis and Modeling Task Committee and the Watershed Management Technical Committee whose deliberations motivated this study.

Conflicts of Interest: The authors declare no conflict of interest.

References

1. Read, E.K.; Carr, L.; De Cicco, L.; Dugan, H.A.; Hanson, P.C.; Hart, J.A.; Kreft, J.; Read, J.S.; Winslow, L.A. Water quality data for national-scale aquatic research: The Water Quality Portal. *Water Resour. Res.* **2017**, *53*, 1735–1745. [CrossRef]
2. Ross, M.R.V.; Topp, S.N.; Appling, A.P.; Yang, X.; Kuhn, C.; Butman, D.; Simard, M.; Pavelsky, T.M. AquaSat: A data set to enable remote sensing of water quality for inland waters. *Water Resour. Res.* **2019**, *55*, 10012–10025. [CrossRef]
3. United States Geological Survey. National Hydrography Dataset. 2019. Available online: https://www.usgs.gov/national-hydrography/access-national-hydrography-products (accessed on 23 October 2019).
4. USEPA. ATTAINS. 2020. Available online: https://www.epa.gov/waterdata/attains (accessed on 30 April 2020).

5. Sridharan, V.K.; Quinn, N.W.T.; Kumar, S.; McCutcheon, S.C.; Ahmadisharaf, E.; Fang, X.; Zhang, H.X.; Parker, A. Selecting Reliable Models for Total Maximum Daily Load Development: Holistic Protocol. *J. Hydrol. Eng.* **2021**, *26*, 04021031. [CrossRef]
6. Kuenzer, C.; Ottinger, M.; Wegmann, M.; Guo, H.; Wang, C.; Zhang, J.; Dech, S.; Wikelski, M. Earth observation satellite sensors for biodiversity monitoring: Potentials and bottlenecks. *Int. J. Remote Sens.* **2014**, *35*, 6599–6647. [CrossRef]
7. Yang, X.; Qin, Q.; Yésou, H.; Ledauphin, T.; Koehl, M.; Grussenmeyer, P.; Zhu, Z. Monthly estimation of the surface water extent in France at a 10-m resolution using Sentinel-2 data. *Remote Sens. Environ.* **2020**, *244*, 111803. [CrossRef]
8. Yang, F.; Qi, C.; Su, D.; Ding, S.; He, Y.; Ma, Y. An airborne LiDAR bathymetric waveform decomposition method in very shallow water: A case study around Yuanzhi Island in the South China Sea. *Int. J. Appl. Earth Observ. Geoinf.* **2022**, *109*, 102788. [CrossRef]
9. Legleiter, C.J.; Harrison, L.R. Remote sensing of river bathymetry: Evaluating a range of sensors, platforms, and algorithms on the upper Sacramento River, California, USA. *Water Resour. Res.* **2019**, *55*, 2142–2169. [CrossRef]
10. Chawla, I.; Karthikeyan, L.; Mishra, A.K. A review of remote sensing applications for water security: Quantity, quality, and extremes. *J. Hydrol.* **2020**, *585*, 124826. [CrossRef]
11. Jafarbiglu, H.; Pourreza, A. A comprehensive review of remote sensing platforms, sensors, and applications in nut crops. *Comput. Electron. Agric.* **2022**, *197*, 106844. [CrossRef]
12. Wu, D.; Johansen, K.; Phinn, S.; Robson, A.; Tu, Y.H. Inter-comparison of remote sensing platforms for height estimation of mango and avocado tree crowns. *Int. J. Appl. Earth Obs. Geoinf.* **2020**, *89*, 102091. [CrossRef]
13. Lechner, A.M.; Foody, G.M.; Boyd, D.S. Applications in remote sensing to forest ecology and management. *One Earth* **2020**, *2*, 405–412. [CrossRef]
14. Fuentes, S.; Tongson, E.; Gonzalez, V.C. Urban green infrastructure monitoring using remote sensing from integrated visible and thermal infrared cameras mounted on a moving vehicle. *Sensors* **2021**, *21*, 295. [CrossRef] [PubMed]
15. Gholizadeh, M.H.; Melesse, A.M.; Reddi, L. A comprehensive review on water quality parameters estimation using remote sensing techniques. *Sensors* **2016**, *16*, 1298. [CrossRef] [PubMed]
16. Topp, S.N.; Pavelsky, T.M.; Jensen, D.; Simard, M.; Ross, M.R. Research trends in the use of remote sensing for inland water quality science: Moving towards multidisciplinary applications. *Water* **2020**, *12*, 169. [CrossRef]
17. Steinhausen, M.J.; Wagner, P.D.; Narasimhan, B.; Waske, B. Combining Sentinel-1 and Sentinel-2 data for improved land use and land cover mapping of monsoon regions. *Int. J. Appl. Earth Obs. Geoinf.* **2018**, *73*, 595–604. [CrossRef]
18. Sagan, V.; Peterson, K.T.; Maimaitijiang, M.; Sidike, P.; Sloan, J.; Greeling, B.A.; Maalouf, S.; Adams, C. Monitoring inland water quality using remote sensing: Potential and limitations of spectral indices, bio-optical simulations, machine learning, and cloud computing. *Earth-Sci. Rev.* **2020**, *205*, 103187. [CrossRef]
19. Ammenberg, P.; Flink, P.; Lindell, T.; Pierson, D.; Strombeck, N. Bio-optical modelling combined with remote sensing to assess water quality. *Int. J. Remote Sens.* **2002**, *23*, 1621–1638. [CrossRef]
20. Karki, S.; Sultan, M.; Elkadiri, R.; Elbayoumi, T. Mapping and forecasting onsets of harmful algal blooms using MODIS data over coastal waters surrounding Charlotte County, Florida. *Remote Sens.* **2018**, *10*, 1656. [CrossRef]
21. Apostel, A.; Kalcic, M.; Dagnew, A.; Evenson, G.; Kast, J.; King, K.; Martin, J.; Muenich, R.L.; Scavia, D. Simulating internal watershed processes using multiple SWAT models. *Sci. Total Environ.* **2021**, *10*, 143920. [CrossRef]
22. Mueller-Warrant, G.W.; Whittaker, G.W.; Banowetz, G.M.; Griffith, S.M.; Barnhart, B.L. Methods for improving accuracy and extending results beyond periods covered by traditional ground-truth in remote sensing classification of a complex landscape. *Int. J. Appl. Earth Obs. Geoinf.* **2015**, *38*, 115–128. [CrossRef]
23. Scudiero, E.; Corwin, D.L.; Anderson, R.G.; Skaggs, T.H. Moving forward on remote sensing of soil salinity at regional scale. *Front. Environ. Sci.* **2016**, *4*, 65. [CrossRef]
24. Nagai, S.; Nasahara, K.N.; Akitsu, T.K.; Saitoh, T.M.; Muraoka, H. Importance of the collection of abundant ground-truth data for accurate detection of spatial and temporal variability of vegetation by satellite remote sensing. In *Biogeochemical Cycles: Ecological Drivers and Environmental Impact*; American Geophysical Union: Hoboken, NJ, USA, 2020; pp. 223–244. [CrossRef]
25. Schaider, L.A.; Swetschinski, L.; Campbell, C.; Rudel, R.A. Environmental justice and drinking water quality: Are there socioeconomic disparities in nitrate levels in US drinking water? *Environ. Health* **2019**, *18*, 1–15. [CrossRef] [PubMed]
26. QGIS Development Team. QGIS Geographic Information System. *Open Source Geospatial Foundation Project*. 2022. Available online: http://qgis.osgeo.org (accessed on 20 April 2022).
27. The MathWorks Inc. *MATLAB R2020b*; The MathWorks Inc.: Natick, MA, USA, 2020.
28. Wilson, A.M.; Jetz, W. Remotely sensed high-resolution global cloud dynamics for predicting ecosystem and biodiversity distributions. *PLoS Biol.* **2016**, *14*, e1002415. [CrossRef] [PubMed]
29. Weiss, D.J.; Nelson, A.; Gibson, H.S.; Temperley, W.; Peedell, S.; Lieber, A.; Hancher, M.; Poyart, E.; Belchior, S.; Fullman, N.; et al. A global map of travel time to cities to assess inequalities in accessibility in 2015. *Nature* **2018**, *553*, 333–336. [CrossRef] [PubMed]
30. Kalkan, K.; Maktav, M.D. A cloud removal algorithm to generate cloud and cloud shadow free images using information cloning. *J. Indian Soc. Remote Sens.* **2018**, *46*, 1255–1264. [CrossRef]
31. Alvarez-Mendoza, C.I.; Teodoro, A.; Ramirez-Cando, L. Improving NDVI by removing cirrus clouds with optical remote sensing data from Landsat-8–a case study in Quito, Ecuador. *Remote Sens. Appl. Soc. Environ.* **2019**, *13*, 257–274. [CrossRef]
32. Pickard, B.R.; Daniel, J.; Mehaffey, M.; Jackson, L.E.; Neale, A. EnviroAtlas: A new geospatial tool to foster ecosystem services science and resource management. *Ecosyst. Serv.* **2015**, *14*, 45–55. [CrossRef]

33. Panlasigui, S.; Davis, A.J.S.; Mangiante, M.J.; Darling, J.A. Assessing threats of non-native species to native freshwater biodiversity: Conservation priorities for the United States. *Biol. Conserv.* **2018**, *224*, 199–208. [CrossRef]
34. Walsh, P.; Wheeler, W. *Water Quality Index Aggregation and Cost Benefit Analysis*; NCEE Working Paper No. 12-05; USEPA: Washington, DC, USA; National Center for Environmental Economics: Washington, DC, USA, 2012.
35. Sridharan, V.K.; Kumar, S.N.; Madhur Kumar, S. GitHub Repo remoteSensingPotentialMapping. 2022. Available online: https://github.com/vamsiks2003/remoteSensingPotentialMapping (accessed on 20 April 2022).
36. Sridharan, V.K.; Kumar, S.N.; Google Earth Engine Remote Sensing Potential Application. 2022. Available online: https://vamsiks2003.users.earthengine.app/view/remotesensingpotential (accessed on 20 April 2022).
37. Gorelick, N.; Hancher, M.; Dixon, M.; Ilyushchenko, S.; Thau, D.; Moore, R. Google Earth Engine: Planetary-scale geospatial analysis for everyone. *Remote Sens. Environ.* **2017**, *202*, 18–27. [CrossRef]
38. Dressing, S.A.; Meals, D.W. *Monitoring and Evaluating Nonpoint Source Watershed Projects*; United States Environmental Protection Agency: Washington, DC, USA, 2016; p. 40. Available online: https://www.epa.gov/sites/default/files/2016-06/documents/chapter_9_may_2016_508.pdf (accessed on 14 April 2022).
39. Cooley, S.W.; Smith, L.C.; Stepan, L.; Mascaro, J. Tracking dynamic northern surface water changes with high-frequency planet CubeSat imagery. *Remote Sens.* **2017**, *9*, 1306. [CrossRef]
40. Tucker, G.E.; Hutton, E.W.; Piper, M.D.; Campforts, B.; Gan, T.; Barnhart, K.R.; Kettner, A.J.; Overeem, I.; Peckham, S.D.; McCready, L.; et al. CSDMS: A community platform for numerical modeling of Earth surface processes. *Geosci. Model Dev.* **2022**, *15*, 1413–1439. [CrossRef]
41. United States Geological Survey. USGS EROS Archive—Aerial Photography—National Agriculture Imagery Program (NAIP). 2018. Available online: https://www.usgs.gov/centers/eros/science/usgs-eros-archive-aerial-photography-national-agriculture-imagery-program-naip (accessed on 14 April 2022).

water

MDPI

Article

Development and Demonstration of an Endocrine-Disrupting Compound Footprint Calculator

Rachel Taylor [1], Kathryn Hayden [1], Marc Gluberman [1], Laura Garcia [1], Serap Gorucu [1,2], Bryan Swistock [3] and Heather Preisendanz [1,4,*]

[1] Department of Agricultural and Biological Engineering, The Pennsylvania State University, University Park, State College, PA 16802, USA; rrt5065@psu.edu (R.T.); kxh5242@psu.edu (K.H.); mqg5358@gmail.com (M.G.); laura.garcia5@upr.edu (L.G.); serapgorucu@ufl.edu (S.G.)

[2] Department of Agricultural and Biological Engineering, The University of Florida, Gainesville, FL 32611, USA

[3] Department of Ecosystem Science and Management, The Pennsylvania State University, University Park, State College, PA 16802, USA; brs@psu.edu

[4] Institute for Sustainable Agricultural, Food and Environmental Science, The Pennsylvania State University, University Park, State College, PA 16802, USA

* Correspondence: hpreisen@psu.edu; Tel.: +1-814-863-1817

Abstract: Chemicals in personal care products used in everyday lives become part of the wastewater stream. Wastewater treatment plants were not designed to remove these chemicals; therefore, these products and their metabolites persist in the effluent. Many of these chemicals are known, or suspected to be, endocrine-disrupting compounds (EDCs) and can cause adverse impacts to aquatic organisms at trace concentrations. Here, we developed a publicly available EDC footprint calculator to estimate a household's EDC footprint. The calculator prompts users to input the number of products they own in each of three categories: health and beauty, laundry, and cleaning. The calculator, which is programmed with average values of EDCs in each product, outputs an estimate of the user's EDC footprint (mass) and ranks the contribution of each product to the footprint. When used by a group of 39 citizen scientists across the Susquehanna River Basin in the northeastern United States, the average household EDC footprint was ~150 g. Results of this tool aid in decision making by providing users with the information necessary to reduce the household's footprint through product selection that avoids specific ingredients or by replacing the top-ranking products with greener alternatives.

Keywords: chemical footprint; emerging contaminants; endocrine-disrupting compounds; footprint calculator; personal care products; water quality

Citation: Taylor, R.; Hayden, K.; Gluberman, M.; Garcia, L.; Gorucu, S.; Swistock, B.; Preisendanz, H. Development and Demonstration of an Endocrine-Disrupting Compound Footprint Calculator. *Water* **2022**, *14*, 1587. https://doi.org/10.3390/w14101587

Academic Editors: Nigel W. T. Quinn, Ariel Dinar, Iddo Kan and Vamsi Krishna Sridharan

Received: 26 February 2022
Accepted: 12 May 2022
Published: 16 May 2022

1. Introduction

The personal care product (PCP) industry is a significant contributor to the global economy, accounting for 3.9 million direct and indirect jobs in 2018, representing 1.9% of total US employment and USD 267.3 billion in the United States' gross domestic product [1]. Many of the compounds used in PCPs are referred to as "endocrine-disrupting compounds" (EDCs) because they can mimic or alter hormones, leading to complications in growth, development, and reproduction [2]. Given that many PCPs are applied topically, compounds classified as EDCs have been found in human tissues [3]. Triclosan, an antibacterial agent used in PCPs, has been monitored and detected in blood, breast milk, urine, adipose tissue, and liver [4]. Other EDCs used in cosmetics, including bisphenol A (BPA), phthalates, and parabens, have been studied in human urine samples to better understand potential toxicological effects [2,4]. As more EDCs are identified in consumer products, there is concern over the ability of some EDCs to lower spermatozoid production and potentially increase the risk of breast cancer and other anomalies in human bodies [5].

Globally, EDCs are increasingly considered major contributors to a wide array of ecotoxicological impacts on non-target aquatic organisms. In the United States, EDCs have been found in surface water, particularly downstream of wastewater treatment plants [6], and have the potential to impact the endocrine systems of aquatic species at environmentally relevant concentrations [2]. In Pennsylvania, the smallmouth bass (*Micropterus dolomieu*) decline in the Susquehanna and Juniata Rivers triggered a significant amount of research and heightened awareness around the issues of EDCs in these river networks [7]. For example, the United States Geological Survey (USGS) reported that 60–100% of large- and smallmouth bass populations sampled near wildlife refuges in the Northeast exhibit intersex characteristics [8]. A causal analysis conducted by the Pennsylvania Department of Environmental Protection (PA DEP) indicated the presence of EDCs, such as pesticides, pharmaceuticals, and ingredients in PCPs as likely contributors to the decline [7]. Although a virus was ultimately determined to be the greatest contributor to the decline, the presence of intersex characteristics in the fish populations could not be explained by the virus, and therefore, a better understanding of the presence and impacts of EDCs is still needed [9]. In the Susquehanna River Basin, an increased presence of steroidal hormones was observed with the increased feminization of the local smallmouth bass population, where male fish were developing female sexual characteristics [7].

Beyond the well-documented impacts on fish populations from EDCs in surface water bodies, EDCs in wetlands and vernal pools pose potential threats to amphibians, which are also known to be declining globally at alarming rates [10–12]. The amphibian decline, which has been referred to as the sixth mass extinction [13], is attributed to a wide range of factors, including diseases, invasive species, climate change, and habitat loss. Water quality contaminants, including EDCs, are also suspected to be contributing to the amphibian decline [14,15]. In studies monitoring human wastewater contaminants from septic tanks and wastewater irrigation activities in critical amphibian habitats (vernal pools), EDCs were present at levels high enough to elicit intersex characteristics in native frog species [16–18]. Aside from causing intersex characteristics, EDCs such as triclosan have been shown to affect tadpole hindlimb development at concentrations as low as 0.15 μg/L [19]. Understanding the effects of EDCs on amphibians and other sensitive aquatic organisms is critically important for prioritizing conservation efforts.

The mechanisms through which domestic wastewater introduces EDCs into the environment are generally well-understood, with human sources of EDCs most commonly associated with the usage of PCPs and pharmaceuticals. Despite nearly two decades of research since the seminal Kolpin et al. [6] study promoted an exponential growth of research on the impacts of EDCs on drinking water quality and aquatic ecosystem health, little clear evidence remains of the reduced presence or quantity of these contaminants in the environment. These chemicals are introduced into the environment during various stages of the life cycle of PCPs, including manufacturing, use, and disposal. In each of these stages, EDCs may be present in the influent water to wastewater treatment plants (WWTPs) [20–24]. WWTPs were not designed to remove these chemicals, and therefore, the chemicals and their metabolites, which can retain potency, often persist in the wastewater effluent. This wastewater effluent is typically discharged to surface water bodies but may also be land-applied or used to recharge groundwater aquifers. Although wastewater must be treated to meet permit requirements, EDCs are currently not regulated, and therefore, the extent to which treatment plants remove EDCs prior to discharging their effluent varies widely across treatment technologies and types of EDCs [23].

Given that these chemicals do not currently have water quality regulations, one way to reduce their presence in the environment is by reducing their sources. However, it is difficult for consumers to make informed decisions about the PCPs they purchase because labels are often insufficient for determining whether or not a product may contain

EDCs, or which type of EDCs may be present. In the United States, the Food and Drug Administration (FDA) regulates some PCPs such as toothpaste, deodorant, sunscreen, and antibacterial hand soap. However, the FDA only requires that active ingredients be listed on the product's label, and not all of the EDCs found in these products are considered to be "active ingredients". The FDA requires that cosmetic labels list all ingredients from highest to lowest concentration in the product [25,26], but given that some ingredients may be more potent or exhibit higher endocrine-disrupting potential than others, this method of labeling may not provide information in an easily accessible manner for making informed decisions about product choices.

A large study was conducted by the Silent Spring Institute to quantify the presence of EDCs in commonly used PCPs [27]. The authors selected 66 target compounds that included EDCs and compounds suspected to trigger asthma that were expected to be present in PCPs, and they analyzed 85 samples for these compounds. The samples were composites of up to seven products in each product category and represented more than 200 products. They classified the results of their EDC analysis into four main categories: not detected, 1–100 μg/g, 100–1000 μg/g, and >1000 μg/g. The highest levels of EDCs detected varied by product type, with the highest UV filters in sunscreen; highest cyclosiloxanes in sunscreen and car interior cleaners; highest glycol ethers in floor and carpet cleaners, polish/wax, and sunscreen; highest fragrances in surface cleaners, car fresheners, dryer sheets, air fresheners, and perfume/cologne; highest alkylphenols in shower curtains and car interior cleaners; highest ethanolamines in glass cleaners and laundry detergent; highest antimicrobials in hand and bar soaps; highest BPA in detergent, soap, shampoo, conditioner, detergent, shaving cream, face lotion, toilet bowl cleaners, body wash, and nail polish; highest phthalates in foundation, car fresheners, and perfume/cologne; and highest parabens in face lotion, mascara, hair spray, and sunscreen. In addition to conventional products, composites of alternative products that were advertised as "greener" were also analyzed in the study. The results of the analysis demonstrated the widespread presence of EDCs in commonly used household products and the co-occurrence of multiple compounds in the same products, raising concerns regarding their biological activity and potential toxicological and ecotoxicological implications of the use of these products. Further, the study revealed multiple compounds in the products that were not listed on the product labels, highlighting concerns regarding the ability of consumers to make informed choices should they wish to select products without specific EDCs.

The goal of this research was to develop a tool that the general public could use to estimate their "footprint" (i.e., the mass) of EDCs in products that they currently or typically own and use in their personal hygiene, household cleaning, and laundry routines. Although the footprint does not provide temporal context, it serves as a "snapshot" of the EDC footprint for the products used by members of a household at the time the calculator is taken. The footprint tool was inspired by online water and carbon footprint calculators, which prompt users to answer questions about their daily activities. These types of tools are useful in increasing awareness of complex environmental issues, such as water pollution and climate change. Here, we present an example output of the EDC calculator to demonstrate the utility of its graphical outputs in helping individual users make decisions about ways they could lower their household EDC footprint through their consumer choices. Further, we analyzed EDC footprint results for 39 citizen scientists in the northeastern United States to better understand the dominant product categories and individual products each contributing to total EDC footprints. Overall, we hope that users of the EDC footprint calculator become more aware of the issue of emerging contaminants in the water cycle and feel empowered to reduce their contribution to this global environmental concern.

2. Materials and Methods

2.1. EDC Footprint Tool Development

The EDC footprint calculator was developed by conducting a review of existing databases and papers to identify the masses of EDCs in various commonly used PCPs. The three product categories that were selected are health and beauty products, household cleaning products, and laundry products. The individual products included in the health and beauty products category are hand soap, hand sanitizer, bar soap, body wash, shampoo, conditioner, shaving cream, body lotion, face lotion, facial cleanser, toothpaste, deodorant, hair products, lipstick, mascara, foundation, nail polish, sunscreen, and perfume/cologne. The products included in the household cleaning products category are surface cleaner, floor cleaner, glass cleaner, bathtub and tile cleaner, toilet bowl cleaner, air freshener, carpet cleaner, floor polish/wax, dishwasher detergent, and dish liquid. The individual products in the laundry category are laundry bleach, laundry detergent, and dryer sheets. The EDCs that can be found in these products and are included in the calculator are outlined in Table 1. Although pharmaceuticals are often grouped with personal care products as sources of EDCs, we decided not to include them and instead focused only on products used for personal hygiene and other household cleaning activities, as these are likely to be the dominant contributors to a user's total EDC footprint and are less intrusive than prompting a user to input potentially sensitive or confidential medical information.

Table 1. Categories of endocrine-disrupting compounds (EDCs) and specific compounds within each category that were included in the EDC footprint calculator.

EDC Category	Compounds
UV Filters	Octinoxate, Benzophenone, Benzophenone-1, Benzophenone-3
Cyclosiloxanes	Dodecamethylcyclohexylsiloxane, Decamethylcyclopentasiloxane, Octamethylcyclotetrasiloxane
Parabens	2-Butyl paraben, Methyl paraben, Ethyl paraben,
Glycol Ethers	2,2-Butoxyethoxyethanol, 2,2-Methoxyethoxyethanol, 2-Phenoxyethanol, 2-Butoxyethanol
Antimicrobials	Triclosan, Triclocarban
Ethanolamines	Monoethanolamine, Diethanolamine
Phthalates	Diethyl phthalate, Di-n-propyl phthalate, Di-n-octyl phthalate, Di-n-hexyl phthalate, Di-n-butyl phthalate, Di-isononyl phthalate, Di-isobutyl phthalate, Di- cyclohexyl phthalate, Benzylbutyl phthalate, Bis(2-ethylhexyl) phthalate, Bis(2-ethylhexyl)adipate
Fragrances	Phenethyl alcohol, Musk xylene, Musk ketone, Methyl ionone, Isobornyl acetate, Methyl salicylate, Hexyle cinnemal, 1,3,4,6,7,8-Hexahydro-4,6,6,7,8,8-hexamethylcyclopenta [g]-2-benzopyran (HHCB), 6,7-Dihydro-1,1,2,3,3-pentamethyl-4(5H)-indanone (DPMI), Bucinal, 6-acetyl-1,1,2,4,4,7 hexamethyltetraline (AHTN), Terpineol, Pinene, Eugenol, Methyl salicylate, Methyl eugenol, Benzylacetate, Diphenyl ether, Limonene, Linalool
Alkylphenols	Nonylphenol diethoxylate, Nonylphenol monoethoxylate, 4-t-Nonylphenol, Octylphenol diethoxylate, Octylphenol monoethoxylate, 4-t-Octylphenol

Numbers used in the calculator were taken as an average of the results in the Dodson et al. [27] study when the concentrations were <1000 μg/g. For the products containing EDCs above 1000 μg/g, concentrations were obtained from the Consumer Product Information Database (CPID) [28]. The concentrations of EDCs in perfume and cologne were further improved using concentrations reported by Peters [29]. This report analyzed phthalates and fragrances in 36 perfume products.

The EDC footprint calculator was initially developed in Microsoft® Excel™ as a downloadable spreadsheet-based tool (the blank version is shown in Figure 1). The calculator was divided into sections for each of the three major product categories: cleaners, laundry, and health and beauty. The user interface column accepts input in milliliters (mL) for liquid products and grams (g) for solid products (shown in green on the calculator; Figure 1). If a user has multiple containers of a product (e.g., two tubes of toothpaste), the user can add the quantities together and enter a total amount in the appropriate column. The calculator uses the information (i.e., masses and volumes) inserted by the user and multiplies the concentration of EDCs (and the density of the product if the input is in mL) to estimate the mass (in mg) of EDCs in each product:

$$x\frac{\mu g_{EDC}}{g_{product}} \times y\frac{g_{product}}{cm^3_{product}} \times z\ mL_{product} \times 1\frac{cm^3_{product}}{mL_{product}} \times \frac{1\ mg_{EDC}}{1000\ \mu g_{EDC}} = mg_{EDC} \quad (1)$$

where x is the concentration of EDC in a product from the existing literature [27–29], y is the density of that product from CPID [28], and z is the volume of product the user entered. The calculator then sums the mass of EDCs calculated for each product and determines the total mass (in g) of EDCs estimated to be in all of the products the user entered. This total mas is the EDC footprint and is shown in red in the bottom right corner of the calculator (Figure 1). The footprint tool was made publicly available through Penn State Extension [30].

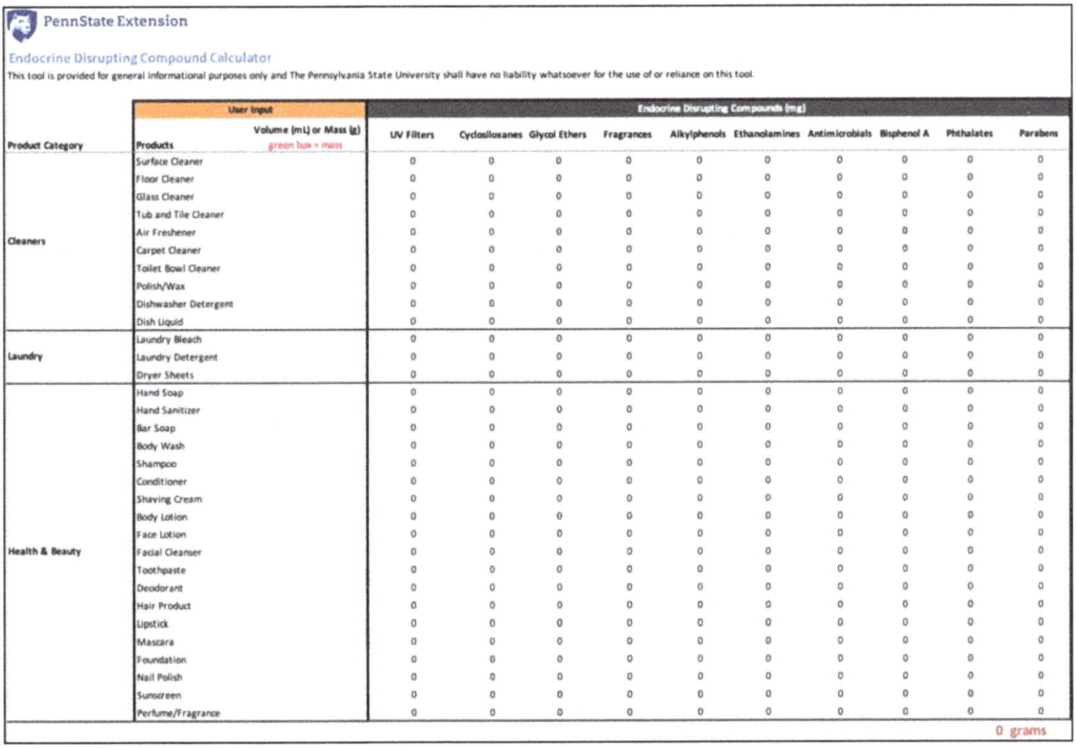
PennState Extension

Endocrine Disrupting Compound Calculator

This tool is provided for general informational purposes only and The Pennsylvania State University shall have no liability whatsoever for the use of or reliance on this tool.

	User Input		Endocrine Disrupting Compounds (mg)									
Product Category	Products	Volume (mL) or Mass (g) green box = mass	UV Filters	Cyclosiloxanes	Glycol Ethers	Fragrances	Alkylphenols	Ethanolamines	Antimicrobials	Bisphenol A	Phthalates	Parabens
Cleaners	Surface Cleaner		0	0	0	0	0	0	0	0	0	0
	Floor Cleaner		0	0	0	0	0	0	0	0	0	0
	Glass Cleaner		0	0	0	0	0	0	0	0	0	0
	Tub and Tile Cleaner		0	0	0	0	0	0	0	0	0	0
	Air Freshener		0	0	0	0	0	0	0	0	0	0
	Carpet Cleaner		0	0	0	0	0	0	0	0	0	0
	Toilet Bowl Cleaner		0	0	0	0	0	0	0	0	0	0
	Polish/Wax		0	0	0	0	0	0	0	0	0	0
	Dishwasher Detergent		0	0	0	0	0	0	0	0	0	0
	Dish Liquid		0	0	0	0	0	0	0	0	0	0
Laundry	Laundry Bleach		0	0	0	0	0	0	0	0	0	0
	Laundry Detergent		0	0	0	0	0	0	0	0	0	0
	Dryer Sheets		0	0	0	0	0	0	0	0	0	0
Health & Beauty	Hand Soap		0	0	0	0	0	0	0	0	0	0
	Hand Sanitizer		0	0	0	0	0	0	0	0	0	0
	Bar Soap		0	0	0	0	0	0	0	0	0	0
	Body Wash		0	0	0	0	0	0	0	0	0	0
	Shampoo		0	0	0	0	0	0	0	0	0	0
	Conditioner		0	0	0	0	0	0	0	0	0	0
	Shaving Cream		0	0	0	0	0	0	0	0	0	0
	Body Lotion		0	0	0	0	0	0	0	0	0	0
	Face Lotion		0	0	0	0	0	0	0	0	0	0
	Facial Cleanser		0	0	0	0	0	0	0	0	0	0
	Toothpaste		0	0	0	0	0	0	0	0	0	0
	Deodorant		0	0	0	0	0	0	0	0	0	0
	Hair Product		0	0	0	0	0	0	0	0	0	0
	Lipstick		0	0	0	0	0	0	0	0	0	0
	Mascara		0	0	0	0	0	0	0	0	0	0
	Foundation		0	0	0	0	0	0	0	0	0	0
	Nail Polish		0	0	0	0	0	0	0	0	0	0
	Sunscreen		0	0	0	0	0	0	0	0	0	0
	Perfume/Fragrance		0	0	0	0	0	0	0	0	0	0
												0 grams

Figure 1. Endocrine-disrupting compound (EDC) footprint calculator user interface. User input is accepted in either volume (mL) or mass (g). The calculator then calculates and displays the estimated mass of EDCs in each product. The total EDC footprint for all products entered by the user is displayed in red on the bottom right side of the tool's interface.

To facilitate a more user-friendly version of the calculator that did not require a Microsoft® Excel™ license, we developed a web-based version of the tool using Qualtrics™ (Qualtrics, Provo, UT, USA). This version made the calculator more widely accessible to a broader audience since it does not require users to have the software or the technical skills to use it. Further, the online Qualtrics version allows data to be collected anonymously, enabling research to be conducted to better understand what typical ranges of EDC footprints are for households at various scales from local to regional to national and potentially even global. Between 2018 and 2022, the EDC footprint tool has been used more than 1400 times, with users in the United States, Mexico, Canada, India, Taiwan, and Europe.

The tool provides several graphical outputs to assist the user in interpreting the results. The first visual output is a pie chart that provides the percent contribution of each of the three product categories to the total EDC footprint, enabling the user to easily assess how much of the total footprint is from products in the health and beauty, cleaners, and laundry categories. The second visual output is a pie chart that provides the percent contribution of each EDC category to the overall EDC footprint. This allows the user to easily assess the percent contribution of the following categories to the overall EDC footprint: UV filters, cyclosiloxanes, glycol ethers, fragrances, alkylphenols, ethanolamines, antimicrobials, BPA, phthalates, and parabens (see Table 1 for the specific chemicals included in each of these categories). Finally, the calculator ranks the percent contribution of each product to the overall EDC footprint and visualizes the contribution (mass) of the top 10 products in a bar chart. This chart is intended to help the user determine which specific products are contributing the most to the total EDC footprint, thereby providing the user with easily accessible information regarding which products have the highest potential to reduce the user's EDC footprint if they were to be exchanged for a "greener" product.

A "green" version of the EDC footprint calculator was developed in the same manner described for the original version of the calculator; however, it has not yet been made publicly available. This "green" version is based on EDC concentrations for products marketed as "eco-friendly" or "green" in the study by Dodson et al. [27]. Specifically, we replaced the original concentrations of each EDC in each product reported by Dodson et al. [27] with the concentrations of the products that Dodson et al. [27] reported for the "green" products.

It should be noted that the EDC footprints generated from this tool are based on the products in a household at the time the footprint tool was used and is not meant to provide an estimate of usage of these products over a specific amount of time. For example, some products, such as hand soap, are likely used more frequently than other products. We did not attempt to capture the time frame over which the products contributing to the footprint would be consumed. Rather, at the scale of an individual household, the results are meant to serve the following purposes: (1) informing the user of the total EDC footprint of the products in the household at the time the calculator was used; (2) visualization of the results to enable the user to understand the contribution of products used in each of the three product categories to the total footprint; (3) determination of the individual products contributing the most to the total footprint; and (4) understanding the individual ingredients categorized as EDCs that are present in the products used in the household. The user is then empowered with the information necessary to make changes in product selection to reduce the household's footprint through product selection that avoids specific ingredients or by replacing the top-ranking products with greener alternatives. We recognize that a footprint could be reduced by having less of a product in the household at the time the footprint calculator is used, such that someone who buys products in bulk may have a higher "snapshot" footprint than someone who buys products as needed; however, the user knows what the shopping habits of the household are and can interpret the results in the context of that knowledge (e.g., a higher snapshot footprint does not necessarily mean a higher annual footprint).

2.2. Citizen Science Demonstration of EDC Footprint Calculator

In June 2021, we recruited 58 citizen scientists to take part in a project that sought to establish links between EDCs in household products they use and the presence of these EDCs in surface water samples from various locations across the Pennsylvania portion of the Susquehanna River Basin, which is the largest tributary to the Chesapeake Bay watershed in the northeastern United States. The citizen scientists were volunteers and filled out the EDC footprint calculator as part of their participation in the project. Results of the footprint tool were distributed to participants in a final report that included their household's footprint, as well as a summary of the results for the other 57 participants to provide context regarding their own household's footprint. After final reports were distributed, 39 citizen scientists responded to a post-study survey, which included some general demographic questions. In this research, we used the results of the demographic questions (Table 2) to interpret the results of the EDC footprint tool based on household size.

Table 2. Demographic information for each of the 39 citizen scientists located across the Susquehanna River basin in the northeastern United States who volunteered to calculate and share their endocrine-disrupting compound (EDC) footprint.

Participant Demographics	Number of Participants	Percentage of Participants
1 Person household	10	26%
2 Person household	12	31%
3 Person household	9	23%
4 Person household	8	21%

3. Results

3.1. Example Results

To demonstrate the results that an individual user receives by using the EDC footprint tool, example product inputs are provided in Figure 2, and the output is shown in Figures 3 and 4. Based on the user's input (Figure 2), the first pie chart shows that EDCs in the household cleaners category contributed to nearly half (48%) of the total EDC footprint, while products in the laundry and health and beauty categories each contributed to 26% of the total EDC footprint (Figure 3a). Glycol ethers, fragrances, and ethanolamines together contributed to more than 85% of the total EDC footprint (Figure 3b). The top individual product that contributed the most to the user's EDC footprint was glass cleaner (Figure 4). The other products contributing the most to the example user's footprint were three household cleaners (floor cleaner, surface cleaner, and dish liquid); five health and beauty products (hand soap, bar soap, perfume/fragrance, body wash, and body lotion); and one laundry product (laundry detergent) (Figure 4).

To facilitate decision making regarding the potential for the example user's EDC footprint to be lowered if individual products were switched to "greener" alternative products (i.e., plant-based laundry detergent, natural air fresheners, etc.), product inputs were recalculated using the "green" version of the calculator. Results show that, by switching to an alternative version of the top five products alone, this example user could reduce the household EDC footprint by more than 75% (Table 3).

Table 3. Potential endocrine-disrupting compound (EDC) footprint reduction that could be achieved by replacing the top five contributing products with a greener alternative for one example user.

Product	EDC Mass (g) Conventional Products	EDC Mass (g) Alternative Products	Percent Reduction (%)
Glass cleaner	31.68	0.44	98.6
Laundry detergent	20.19	0.09	99.6
Hand soap	6.34	3.34	47.3
Bar soap	3.96	0.04	99.0
Floor cleaner	3.52	0.38	89.2
All other products	15.93	NA	0.0
Total Footprint (all products)	**81.62**	**20.21**	**75.2**

NA = not applicable, as only the top five products were switched in this example to alternative products. All other products in this example input (Figure 2) remained the same.

PennState Extension

Endocrine Disrupting Compound Calculator

This tool is provided for general informational purposes only and The Pennsylvania State University shall have no liability whatsoever for the use of or reliance on this tool.

	User Input		Endocrine Disrupting Compounds (mg)											
Product Category	Products	Volume (mL) or Mass (g) green box = mass	UV Filters	Cyclosiloxanes	Glycol Ethers	Fragrances	Alkylphenols	Ethanolamines	Antimicrobials	Bisphenol A	Phthalates	Parabens	EDCs (mg)	% of Grand Total
Cleaners	Surface Cleaner	2000	0	0	0	1706	101	0	0	0	0	0	1,807	2.2%
	Floor Cleaner	1500	0	0	0	3445	77	0	0	0	0	0	3,522	4.4%
	Glass Cleaner	800	0	0	28,800	0	0	2880	0	0	0	0	31,680	39.2%
	Tub and Tile Cleaner	600	0	0	31	152	61	0	0	0	0	0	243	0.3%
	Air Freshener	0	0	0	0	0	0	0	0	0	0	0	0	0.0%
	Carpet Cleaner	0	0	0	0	0	0	0	0	0	0	0	0	0.0%
	Toilet Bowl Cleaner	0	0	0	0	0	0	0	0	0	0	0	0	0.0%
	Polish/Wax	0	0	0	0	0	0	0	0	0	0	0	0	0.0%
	Dishwasher Detergent	0	0	0	0	0	0	0	0	0	0	0	0	0.0%
	Dish Liquid	900	0	0	0	1160	0	0	47	0	47	0	1,254	1.6%
Laundry	Laundry Bleach	1200	0	0	0	421	211	0	0	0	70	0	702	0.9%
	Laundry Detergent	1800	0	0	0	1734	184	18,180	0	92	0	0	20,190	25.0%
	Dryer Sheets	0	0	0	0	0	0	0	0	0	0	0	0	0.0%
Health & Beauty	Hand Soap	1600	0	0	0	3097	0	0	1546	0	850	850	6,342	7.9%
	Hand Sanitizer	500	0	0	0	275	0	0	0	0	25	0	300	0.4%
	Bar Soap	250	0	0	0	775	38	0	3000	13	138	0	3,963	4.9%
	Body Wash	1400	0	0	0	3060	0	0	0	66	66	0	3,191	4.0%
	Shampoo	150	0	0	0	368	8	0	0	8	86	86	556	0.7%
	Conditioner	150	0	0	0	196	0	0	0	7	7	80	290	0.4%
	Shaving Cream	0	0	0	0	0	0	0	0	0	0	0	0	0.0%
	Body Lotion	1200	0	0	0	470	59	0	0	59	59	706	1,354	1.7%
	Face Lotion	200	0	0	0	163	21	10	0	10	10	325	539	0.7%
	Facial Cleanser	150	0	0	8	210	0	0	0	8	83	173	480	0.6%
	Toothpaste	144	0	0	0	86	0	0	79	0	0	0	166	0.2%
	Deodorant	107	0	0	0	279	0	0	0	0	64	0	343	0.4%
	Hair Product	700	0	0	0	212	35	0	0	0	35	0	283	0.4%
	Lipstick	0	0	0	0	0	0	0	0	0	0	0	0	0.0%
	Mascara	9	0	0	0	0	0	0	0	0	0	23	23	0.0%
	Foundation	0	0	0	0	0	0	0	0	0	0	0	0	0.0%
	Nail Polish	45	0	0	0	0	0	0	0	2	5	0	7	0.0%
	Sunscreen	0	0	0	0	0	0	0	0	0	0	0	0	0.0%
	Perfume/Fragrance	180	0	0	0	3072	0	0	0	0	447	0	3,520	4.4%
	Subtotal of each EDC (g)		0.00	0.00	28.84	20.88	0.80	21.07	4.67	0.26	1.99	2.24	81 grams	
	Percent contribution of each EDC to total footprint		0.0%	0.0%	35.7%	25.9%	1.0%	26.1%	5.8%	0.3%	2.5%	2.8%		

Figure 2. Example of one user's input to the endocrine-disrupting compound (EDC) footprint calculator and corresponding footprint calculation results, where the amount of each product input in either volume (mL) or mass (g), depending on whether the product is a solid (green box with red text) or a liquid (white box with black text).

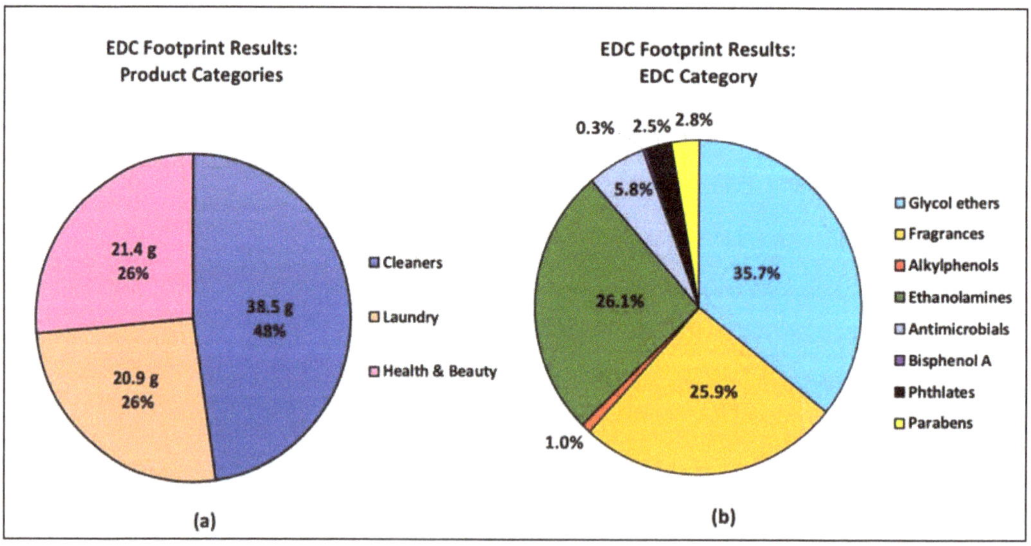

Figure 3. Example of endocrine-disrupting compound (EDC) footprint calculator results for one user with (**a**) the mass and percent contributions of each product category to the total estimated EDC footprint; (**b**) the percent contribution of each EDC to the total estimated EDC footprint.

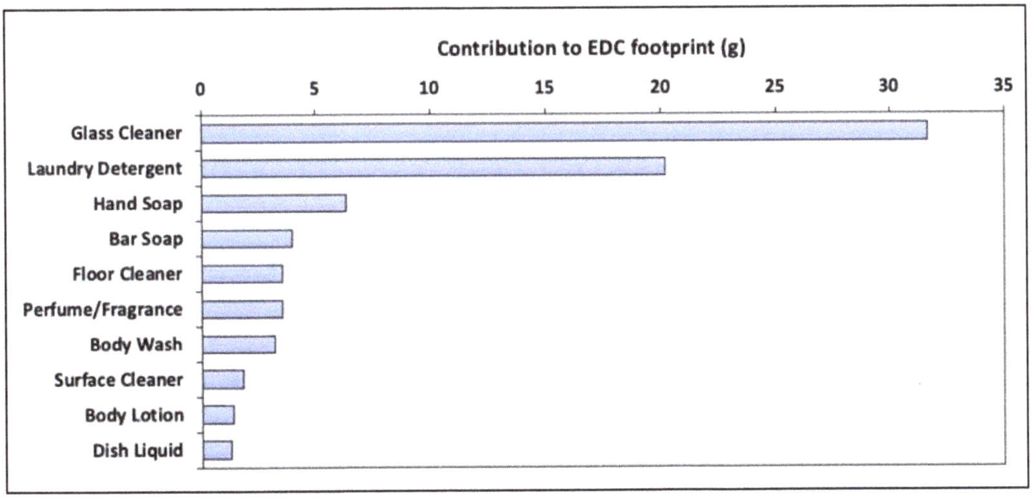

Figure 4. Example of endocrine-disrupting compound (EDC) footprint calculator results for the top 10 products and their contribution to the overall EDC footprint for one example user.

3.2. Citizen Science Footprint Results

The EDC footprints for the 39 citizen scientists who participated in this study ranged from as low as 2.5 g to as high as 720.2 g, with high coefficients of variation (CV > 0.6) for the footprints across all household sizes (Table 4). Household cleaning products contributed nearly half of the total EDCs in the household-based citizen science study. Glass cleaners contain high concentrations of ethanolamines and glycol ether, while air fresheners contain high concentrations of fragrances [27]. Other products that often appeared on the top 10 lists for all 39 footprints were laundry detergent, carpet cleaner, sunscreen, bar soap, and shampoo (Figure 5). Laundry detergents, which generally contain high concentrations of ethanolamines and fragrances [27], are sold and used in larger quantities than are most of the other products.

Table 4. Summary by household size of endocrine-disrupting compound (EDC) footprint calculator results from 39 citizen scientists. Summary results include the average footprint per household size, standard deviation, coefficient of variation, and footprint ranges.

Persons in Household	Number of Footprints	Average Footprint (g)	Standard Deviation (g)	Coefficient of Variation	Footprint Ranges (g)
1	10	112.8	107.2	0.95	15.0–336.7
2	12	151.3	170.3	1.13	2.5–622.1
3	9	183.8	208.8	1.14	44.6–720.2
4	8	144.6	94.6	0.65	31.6–293.4

Evaluation of average masses provides an indication, for this study, of products that are used in most households. In contrast, products with very disparate median and mean values are used relatively in some houses and not at all in others. For example, relatively high values for both mean and median EDC masses from the citizen science portion of this study showed that laundry detergent, glass cleaner, and sunscreen contributed to many of the 39 EDC footprints obtained (Figure 6). In contrast, air freshener and carpet cleaner contributed substantially to the EDC footprints of some households but were not present at all in more than one-half of the households studied, as evidenced by much higher mean than median values for these compounds. Comparison of mean and median values can help future scientists identify potentially different sample populations with targeted outreach techniques. For example, it is likely that nearly all households will have some form of glass cleaner for windows and some laundry detergent. However, efforts to reduce contributions from air fresheners, carpet cleaners, furniture polish, dishwater detergent, and dryer sheets need to target communities where households are likely to have carpets, dishwashers, and dryers.

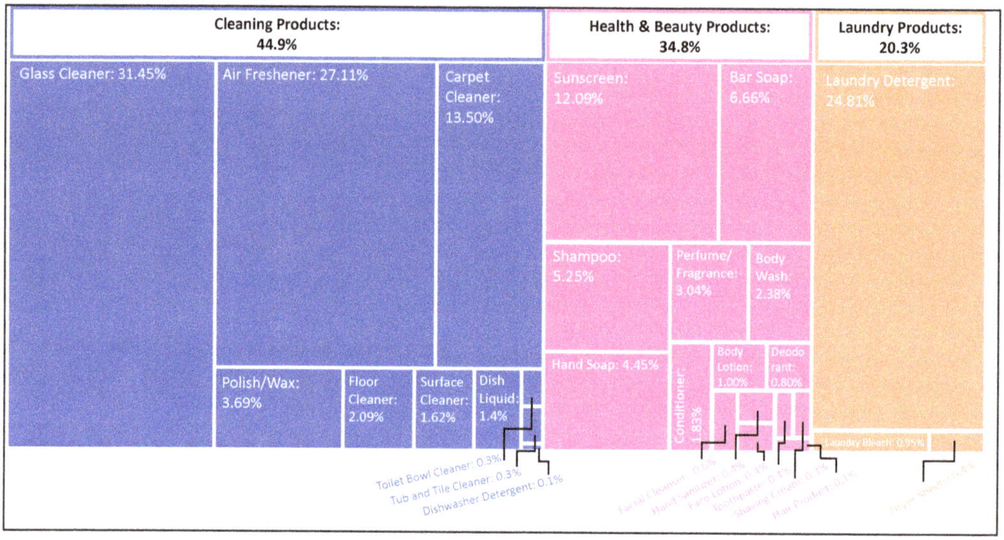

Figure 5. Average percentages of the contribution of each product to the overall footprint of cleaning products, health and beauty products, and laundry products, with products in the cleaning, health and beauty, and laundry categories, contributing an average of 44.9%, 34.8%, and 20.3%, respectively, to the total EDC footprints of the 39 participating citizen scientists. The size of each rectangle provides a visual representation of the extent to which that product contributes to the total footprint.

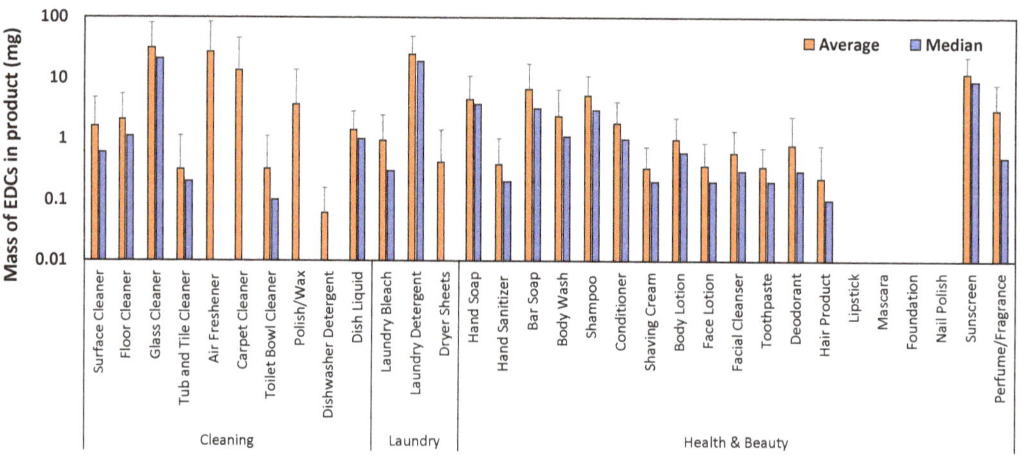

Figure 6. Average and median masses (mg) of EDCs calculated to be in each product for the 39 participating citizen scientists. Error bars represent the standard deviations.

4. Discussion

The average total footprint for each of the households that participated in the citizen science study was 113–184 g (Table 4). This is such a small mass that it can be challenging to interpret the potential for this footprint to inadvertently impact aquatic ecosystems. Therefore, the footprint calculator was programmed to provide the user with perspective for the interpretation of their results by scaling up the per capita footprint (i.e., total household EDC footprint divided by the number of people in the household) and multiplying by the total population of the United States. For example, the average two-person household's footprint of 151 g would estimate that the total EDC footprint for the United States (330 M people) if everyone had the same per capita footprint as the user would be approximately 24,915 metric tonnes, which is the equivalent of approximately 82 commercial airplanes (each plane is approximately 300 tonnes).

In the context of the specific citizen science study conducted here, a more regional context would be the potential impacts of these EDC footprints on the presence of emerging contaminants in the Chesapeake Bay watershed. Given the population in the watershed (18 M people) and an average per capita EDC footprint, the total EDC footprint across the Bay watershed would be 1287 metric tonnes. If this entire footprint reached the Chesapeake Bay, which has a volume of approximately 81.8 km^3, the EDC concentration in the Bay would be 15.7 g/L. Although some of these EDCs can be treated by wastewater treatment plants or septic tanks before ultimately reaching the Bay, mitigation in the treatment facilities may not be effective enough to remove potential ecological risks.

These estimated footprints are significant given the potential impact of the presence of these contaminants in the environment even at trace concentrations (μg–ng/L). For example, triclosan has been shown to affect hindlimb development in amphibians at concentrations as low as 0.15 μg/L [19]. Musks, which are significant contributors to the fragrances commonly found in perfume and cologne [29], have been shown to exhibit estrogenic effects [31,32]. BPA is also known to exhibit estrogenic potential and has been documented to cause gender skewing in flathead minnows at exposure concentrations as low as 0.32 ng/L [33]. Additionally, EDCs have been shown to exhibit synergistic behavior when multiple compounds are present together, such that the total endocrine-disrupting potential of the "cocktail" is greater than simply adding the potential of each individual compound [34]. Although the potential harm to human health is unclear, there is a significant need to understand synergistic interactions and the risks to humans and the environment.

While these results clearly have implications for both human and environmental health, Dodson et al. [27] found that various compounds detected in tested personal care products were not present on the products' labels. This makes decision making more challenging for consumers, as even someone who wants to be an informed consumer may be unable to make satisfying decisions about product selection, especially in real time while shopping. Additionally, the study found that some product labels can be misleading due to labeling standards, with some products advertising to be "fragrance-free" when they indeed contained synthetic fragrances to mask an undesirable chemical odor. Rather than the product being "free" of fragrances, the product has a neutral smell due to a combination of various fragrances ultimately canceling each other out. Therefore, this EDC calculator tool can serve as a mechanism to increase awareness of EDCs and their potential effects on environmental quality, as well as engage the public about the role everyone plays in contributing to the presence of EDCs in the environment.

The rhetoric surrounding the presence of EDCs in the aquatic environment is often filled with fear and uncertainty, particularly because of the near-ubiquitous presence of synthetic chemicals in the environment and because of the lack of water quality standards for EDCs. Here, we sought to provide a tool that could shift the public perception from one of fear to one of empowerment by providing consumers with the knowledge they need to make more informed choices. The results of the tool may help to reduce the consumption of EDC-containing products and ultimately reduce EDC presence in the environment.

Author Contributions: Conceptualization, H.P. and R.T.; methodology, H.P., R.T., K.H., M.G. and L.G.; software, H.P., R.T., M.G., L.G. and S.G.; validation, H.P., R.T. and K.H.; formal analysis, R.T. and K.H.; investigation, H.P., R.T., K.H. and B.S.; resources, H.P. and B.S.; data curation, R.T. and K.H.; writing—original draft preparation, H.P. and R.T.; writing—review and editing, K.H.; visualization, R.T., K.H., M.G. and L.G.; supervision, H.P. and B.S.; project administration, H.P. and B.S.; funding acquisition, H.P. and B.S. All authors have read and agreed to the published version of the manuscript.

Funding: This research was funded by the National Science Foundation (Grant Number 1743981). Rachel Taylor was supported by Penn State's College of Engineering Summer Research Experience for Undergraduates (REU) program in 2014 and is currently supported by the Penn State Department of Agricultural and Biological Engineering. Kathryn Hayden is currently supported by a cooperative agreement from the United States Department of Agriculture's Agricultural Research Service (USDA-ARS). Marc Gluberman was funded by Penn State's College of Agricultural Sciences Summer Undergraduate Research Program in 2015. Laura Garcia was selected to participate in the Summer Research Opportunities Program in 2015 and was supported at Penn State through support from the College of Agricultural Sciences diversity program and from the Department of Agricultural and Biological Engineering. Serap Gorucu was funded, in part, by a grant from the Penn State Office of Physical Plant. Heather Preisendanz is supported, in part, by the USDA National Institute of Food and Agriculture Federal Appropriations under Project PEN04574 and Accession Number 1004448.

Institutional Review Board Statement: The Pennsylvania State University Office of Research Protections determined that this research met the criteria for exempt research according to the policies of the institution and the provisions of applicable federal regulations.

Informed Consent Statement: Informed consent was obtained from all citizen scientists who volunteered to participate in this study.

Data Availability Statement: The calculator tools can be accessed at: https://sites.psu.edu/edccalculator/ (Created on 10 February 2017; accessed on 25 February 2022).

Acknowledgments: The authors wish to thank Jamie Shallenberger from the Susquehanna River Basin Commission for his help in recruiting citizen scientists to participate in this project.

Conflicts of Interest: The authors declare no conflict of interest. The funders had no role in the design of the study; in the collection, analyses, or interpretation of data; in the writing of the manuscript; or in the decision to publish the results.

References

1. Personal Care Products Council. Driving the Economy, Shaping the Future: Economic & Social Contributions Report. 2020. Available online: https://www.personalcarecouncil.org/wp-content/uploads/2020/09/PCPC_EcoReport_2020.pdf (accessed on 23 February 2022).
2. Wee, S.Y.; Aris, A.Z. Endocrine disrupting compounds in drinking water supply system and human health risk implication. *Environ. Int.* **2017**, *106*, 207–233. [CrossRef] [PubMed]
3. Centers for Disease Control and Prevention (CDC). *Fourth National Report on Human Exposure to Environmental Chemicals*; Department of Health and Human Services: Atlanta, GA, USA, 2009. Available online: https://www.cdc.gov/exposurereport/index.html (accessed on 23 February 2022).
4. Karwacka, A.; Zamkowska, D.; Radwan, M.; Jurewicz, J. Exposure to modern, widespread environmental endocrine disrupting chemicals and their effect on the reproductive potential of women: An overview of current epidemiological evidence. *Hum. Fertil.* **2019**, *22*, 2–25. [CrossRef] [PubMed]
5. Gregoraszcuk, E.L.; Rak, A.; Ludewig, G.; Gasinska, A. Effects of estradiol, PCB3 and their hydroxylated metabolites on proliferation, cellcycle, and apoptosis of human breast cancer cells. *Environ. Toxicol. Pharmacol.* **2008**, *25*, 227–233. [CrossRef] [PubMed]

6. Kolpin, D.W.; Furlong, E.T.; Meyer, M.T.; Thurman, E.M.; Zaugg, S.D.; Barber, L.B.; Buxton, H.T. Pharmaceuticals, hormones, and other organic wastewater contaminants in U.S. streams, 1999–2000: A national reconnaissance. *Environ. Sci. Technol.* **2002**, *36*, 1201–1211. [CrossRef]
7. Shull, D.; Pulket, M. *Causal Analysis of the Smallmouth Bass Decline in the Susquehanna and Juniata Rivers*; Pennsylvania Department of Environmental Protection: Philadelphia, PA, USA, 2015. Available online: https://files.dep.state.pa.us/water/Drinking%20Water%20and%20Facility%20Regulation/WaterQualityPortalFiles/SusquehannaRiverStudyUpdates/SMB_CADDIS_Report.pdf (accessed on 23 February 2022).
8. Iwanowicz, L.R.; Blazer, V.S.; Pinkney, A.E.; Guy, C.P.; Major, A.M.; Munney, K.; Mierzykowski, S.; Lingenfelser, S.; Secord, A.; Patnode, K.; et al. Evidence of estrogenic endocrine disruption in smallmouth and largemouth bass inhabiting Northeast U.S. national wildlife refuge waters: A reconnaissance study. *Ecotoxicol. Environ. Saf.* **2016**, *124*, 50–59. [CrossRef]
9. Blazer, V.S.; Gordon, S.; Jones, D.K.; Iwanowicz, L.R.; Walsh, H.L.; Sperry, A.J.; Smalling, K.L. Retrospective analysis of estrogenic endocrine disruption and land-use influences in the Chesapeake Bay watershed. *Chemosphere* **2021**, *266*, 129009. [CrossRef]
10. Baillie, J.E.M.; Hilton-Taylor, C.; Stuart, S.N. *IUCN Red List of Threatened Species: A Global Species Assessment*; International Union for Conservation of Nature: Gland, Switzerland; Cambridge, UK, 2004; Available online: https://www.iucnredlist.org/resources/baillie2004 (accessed on 25 February 2022).
11. Stuart, S.N.; Chanson, J.S.; Cox, N.A.; Young, B.E.; Rodrigues, A.S.L.; Fischman, D.L.; Waller, R.W. Status and trends of amphibian declines and extinctions worldwide. *Science* **2004**, *306*, 1783–1786. [CrossRef]
12. Pyron, R.A. Global amphibian declines have winners and losers. *Proc. Natl. Acad. Sci. USA* **2018**, *115*, 3739–3741. [CrossRef]
13. Wake, D.B.; Vredenburg, V.T. Are we in the midst of the sixth mass extinction? A view from the world of amphibians. *Proc. Natl. Acad. Sci. USA* **2008**, *105*, 11466–11473. [CrossRef]
14. Collins, J.P.; Storfer, A. Global amphibian declines: Sorting the hypotheses. *Divers. Distrib.* **2003**, *9*, 89–98. [CrossRef]
15. Collins, J.P. History, novelty, and emergence of an infectious amphibian disease. *Proc. Natl. Acad. Sci. USA* **2013**, *110*, 9193–9194. [CrossRef] [PubMed]
16. Mina, O.; Gall, H.E.; Carlson, B.; Langkilde, T.L. *A Preliminary Assessment of Endocrine Disrupting Compounds in Vernal Ponds in Central Pennsylvania*; American Society of Agricultural and Biological Engineers: St. Joseph, MI, USA, 2014; Paper No. 1910944. [CrossRef]
17. Mina, O.; Gall, H.E.; Elliott, H.A.; Watson, J.E.; Mashtare, M.L.; Langkilde, T.; Harper, J.P.; Boyer, E.W. Estrogen occurrence and persistence in vernal pools impacted by wastewater irrigation practices. *Agric. Ecosyst. Environ.* **2018**, *257*, 103–112. [CrossRef]
18. Smits, A.P.; Skelly, D.K.; Bolden, S.R. Amphibian intersex in suburban landscapes. *Ecosphere* **2014**, *5*, 1–9. [CrossRef]
19. Veldhoen, N.; Skirrow, R.C.; Osachoff, H.; Wigmore, H.; Clapson, D.J.; Gunderson, M.P.; Van Aggelen, G.; Helbing, C.C. The bactericidal agent triclosan modulates thyroid hormone-associated gene expression and disrupts postembryonic anuran development. *Aquat. Toxicol.* **2006**, *80*, 217–227. [CrossRef] [PubMed]
20. Andrews, W.J.; Masoner, J.R.; Cozzarelli, I.M. Emerging contaminants at a closed and an operating landfill in Oklahoma. *Groundw. Monit. Remediat.* **2012**, *32*, 120–130. [CrossRef]
21. Phillips, P.; Chalmers, A. Wastewater effluent, combined sewer overflows, and other sources of organic compounds to Lake Champlain. *J. Am. Water Resour. Assoc.* **2009**, *45*, 45–57. [CrossRef]
22. Snyder, S.A.; Westerhoff, P.; Yoon, T.; Sedlak, D.L. Pharmaceuticals, personal care products, and endocrine disruptors in water: Implications for the water industry. *Environ. Eng. Sci.* **2004**, *20*, 449–469. [CrossRef]
23. United States Department of Environmental Protection (USEPA). *Treating Contaminants of Emerging Concern: A Literature Review Databas*; Office of Water Engineering and Analysis Division: Washington, DC, USA, 2010; EPA-820-R-10-002. Available online: https://nepis.epa.gov/Exe/ZyPDF.cgi/P1008IK3.PDF?Dockey=P1008IK3.PDF (accessed on 25 February 2022).
24. Verlicchi, P.; Galletti, A.; Petrovic, M.; Barceló, D. Hospital effluents as a source of emerging pollutants: An overview of micropollutants and sustainable treatment options. *J. Hydrol.* **2010**, *389*, 416–428. [CrossRef]
25. Food, Drug, and Cosmetic Act, 21 United States Code §§ 301–392. 1958. Available online: https://www.loc.gov/item/uscode1958-004021009/ (accessed on 23 February 2022).
26. National Archives and Records Administration. Code of Federal Regulations: Regulations Under Section 4 of the Fair Packaging and Labeling Act, 16 CFR. 1995. Available online: https://www.loc.gov/item/cfr1995051-T16CIP500/ (accessed on 23 February 2022).
27. Dodson, R.E.; Nishioka, M.; Standley, L.J.; Perovich, L.J.; Brody, J.G.; Rudel, R.A. Endocrine disruptors and asthma-associated chemicals in consumer products. *Environ. Health Persp.* **2012**, *120*, 935–943. [CrossRef]
28. Consumer Product Information Database (CPID). Available online: https://www.whatsinproducts.com (accessed on 23 February 2022).
29. Peters, R.J.B. *Phthalates and Artificial Musks in Perfumes*; TNO Environment and Geosciences: Apeldoorn, The Netherlands, 2005; R2005/011.
30. Penn State Extension. Endocrine Disrupting Compounds (EDC) Footprint Calculator. Available online: https://extension.psu.edu/endocrine-disrupting-compounds-edc-footprint-calculator (accessed on 23 February 2022).
31. Bitsch, N.; Dudas, C.; Körner, W.; Failing, K.; Biselli, S.; Rimkus, G.; Brunn, H. Estrogenic activity of musk fragrances detected by the E-Screen assay using human MCF-7 cells. *Arch. Environ. Contam. Toxicol.* **2002**, *43*, 0257–0264. [CrossRef]

32. Schreurs, R.H.M.M.; Sonneveld, E.; Jansen, J.H.J.; Seinen, W.; van der Burg, B. Interaction of polycyclic musks and UV filters with the estrogen receptor, androgen receptor, and progesterone receptor in reporter gene bioassays. *Toxicol. Sci.* **2005**, *83*, 264–272. [CrossRef] [PubMed]
33. Parrott, J.L.; Blunt, B.R. Life-cycle exposure of fathead minnows (*Pimephales promelas*) to an ethinylestradiol concentration below 1 ng/L reduces egg fertilization success and demasculinizes males. *Environ. Toxicol.* **2005**, *20*, 131–141. [CrossRef] [PubMed]
34. Wilkinson, J.L.; Hooda, P.S.; Barker, J.; Barton, S.; Swinden, J. Ecotoxic pharmaceuticals, personal care products, and other emerging contaminants: A review of environmental, receptor-mediated, developmental, and epigenetic toxicology with discussion of proposed toxicity to humans. *Crit. Rev. Environ. Sci. Technol.* **2016**, *46*, 336–381. [CrossRef]

water

MDPI

Article

Meeting the Moment: Leveraging Temporal Inequality for Temporal Targeting to Achieve Water-Quality Load-Reduction Goals

Nicole Opalinski [1], Daniel Schultz [1], Tamie L. Veith [2], Matt Royer [3,4] and Heather E. Preisendanz [1,4,*]

1 Department of Agricultural and Biological Engineering, The Pennsylvania State University, State College, PA 16802, USA; nicoleopalinski@gmail.com (N.O.); dp26schultz@gmail.com (D.S.)
2 Pasture Systems and Watershed Management Research Unit, United States Department of Agriculture-Agricultural Research Service, State College, PA 16802, USA; tamie.veith@usda.gov
3 Department of Agricultural Economics, Sociology and Education, The Pennsylvania State University, State College, PA 16802, USA; mzr154@psu.edu
4 Institute for Sustainable Agriculture, Food and Environmental Science, The Pennsylvania State University, State College, PA 16802, USA
* Correspondence: hpreisen@psu.edu; Tel.: +1-814-863-1817

Abstract: Inequality is an emergent property of complex systems. In catchments, variation in hydroclimatic conditions and biogeochemistry cause streamflow and constituent loads to exhibit strong temporal inequality, with most loads exported during "hot moments". Achieving water-quality-restoration goals in a cost-effective manner requires targeted implementation of conservation practices in "hot spots" in the landscape and "hot moments" in time. While spatial targeting is commonly included in development of watershed management plans, the need for temporal targeting is often acknowledged, but no common way to address it has been established. Here, we implement a Lorenz Inequality decision-making framework that uses Lorenz Curves and Gini Coefficients to quantify the degree of temporal inequality exhibited by contaminant loads and demonstrate its utility for eight impaired catchments in the Chesapeake Bay watershed. The framework requires a load-reduction goal be set and then links the degree of temporal inequality in annual nutrient loads to the periods of time during which those loads could be targeted. These results are critical in guiding development of site-specific, cost-effective tools that facilitate load-reduction and water-quality goal attainment for individual catchments. The framework provides valuable insight into site-specific potentials for meeting load-reduction goals.

Keywords: conservation practices; decision making; nutrients; sediment; targeting; water quality

Citation: Opalinski, N.; Schultz, D.; Veith, T.L.; Royer, M.; Preisendanz, H.E. Meeting the Moment: Leveraging Temporal Inequality for Temporal Targeting to Achieve Water-Quality Load-Reduction Goals. *Water* **2022**, *14*, 1003. https://doi.org/10.3390/w14071003

Academic Editors: Nigel W.T. Quinn, Ariel Dinar, Iddo Kan and Vamsi Krishna Sridharan

Received: 15 February 2022
Accepted: 15 March 2022
Published: 22 March 2022

1. Introduction

Inequality is a ubiquitous, emerging property of complex systems. In catchments, the spatial and temporal inequality of hydrologic and biogeochemical responses lead to the emergence of "hot spots" and "hot moments", with the vast majority of these responses occurring during relatively short periods of time and in relatively small locations. While the importance of spatial and temporal inequality is widely recognized, the methods used to identify "hot spots" and "hot moments" are not well established, with the methodology employed to analyze spatial data generally disconnected and inconsistent with the methodology employed to analyze temporal data.

The quantification of "hot spots" has been more consistently reported in the literature than the quantification of "hot moments". By calculating area-normalized loads (or other nutrient-cycle responses, such as gaseous emissions), "hot spots" are identified as the locations over a given spatial extent of interest (i.e., field, catchment, or watershed) that have the highest loads per unit area. If an area of interest needs to be managed for water-quality impairment, for example, then decision makers can direct resources to a relatively small number of places, knowing that implementing conservation practices in those locations will achieve a higher impact on load reduction than placing the same resources and practices in other areas. Crop nutritionists increasingly support this principle through management framework called "4R Nutrient Stewardship": right nutrient source at the right rate at the right time in the right place [1]. Previous research has shown that spatially targeting adoption of agricultural conservation practices at the field scale leads to larger load-reduction goals at the watershed scale [2–6]. It is also important to recognize and manage temporal inequalities, or "hot moments", such that resources can be targeted based on both spatial and temporal inequalities. However, no uniform metric for describing temporal inequality has been widely adopted despite the prevalence of temporal inequality documentation across small and large watersheds [7–10].

The need to quantify the degree of inequality in a system is not new. Perhaps nowhere has the degree of inequality been more routinely quantified than in economics. For more than a century, Lorenz Inequality and the corresponding Gini Coefficient (*G*) have been used to determine wealth distribution by quantifying the degree of income inequality in a population. Lorenz Inequality analysis was first applied to quantify the degree of inequality for streamflow hydrology and water quality in 22 locations in the Lake Okeechobee watershed [11] and has since been utilized globally at the continental scale to better understand how climate change is likely to affect flow regimes [12]. Additionally, the analysis has been applied to time series data for geogenic constituents, nutrients, sediment, and pesticides in more than 100 watersheds ranging from 2.5 km^2 to 70,000 km^2 and at time scales ranging from daily to annually [13–15].

Water-quality degradation of coastal water bodies due to the presence of excess nutrients is a leading global environmental concern [16], with agricultural activities identified as common contributors to degraded water quality [17]. The Chesapeake Bay is the third largest estuary in the world and has a watershed area spanning 166,000 km^2 across seven jurisdictions. In 2010, a federally mandated Total Maximum Daily Load (TMDL) was established by the United States Environmental Protection Agency, designed to reduce nutrient and sediment loads and restore water quality to be in compliance with the Bay's designated use of fishing and swimming by 2025 [18]. To achieve mandated load-reduction goals, widespread adoption of conservation practices has occurred across the Chesapeake Bay watershed. However, current Chesapeake Assessment Scenario Tool (CAST) estimates of load reductions indicate that the Commonwealth of Pennsylvania (PA) in particular is behind the pace likely needed to meet the 2025 reduction goals [19]. Although a range of factors contribute to the overall water quality of the Chesapeake Bay, we argue that a failure to target load reduction during "hot moments" is a contributing factor. The Commonwealth of PA has established a four-tiered system for prioritizing spatial adoption of conservation practices, with each tier of counties needing to reduce 25% of the state's portion of the overall Chesapeake Bay TMDL in its current Watershed Implementation Plan [20]. Tier 1 consists of the two greatest "hot spot" counties that rank highest in nutrient and sediment loads, Tier 2 consists of five counties, whereas Tiers 3 and 4 consist of 16 and 20 counties, respectively. However, no efforts have been documented towards effectively target "hot moments".

The goal of this study is to demonstrate the impact that temporal variability from year to year can have on achieving load-reduction goals in an impaired watershed through the development of a decision-making framework for temporal targeting of "hot moments" during which the targeted load is exported. The framework consists of a novel application of Lorenz Inequality to link the temporal inequality of contaminant loads to the specific "windows of opportunity" and corresponding flow conditions necessary to target to achieve the desired load-reduction goals. The framework is demonstrated here using daily load and discharge data from eight impaired catchments in the Chesapeake Bay watershed. By comparing these loads on an annual basis with established load-reduction goals for each catchment, we determine the catchment-specific variability in percent reduction needed from year to year and discuss how this framework enables watershed planners to understand and inform stakeholders of the risk of a watershed conservation plan.

2. Materials and Methods

2.1. Study Site Selection

The Chesapeake Bay Nontidal Network is a monitoring network comprising 123 water-quality-monitoring stations throughout the Chesapeake Bay watershed that provide nutrient and sediment data [21]. The water-quality-monitoring stations are co-located with U.S. Geological Survey (USGS) streamflow gauges, allowing loads to be calculated. While streamflow data were collected at the sub-daily scale, water-quality data were collected monthly and during targeted storm events, providing 20 data points per station per year. Load data are available in the USGS database at monthly and annual time scales back to 1985; however, daily-scale data were estimated by USGS using the weighted regression on time, discharge, and season (WRTDS) load-estimation technique [21,22].

Here, we analyze eight stations from the Chesapeake Bay Nontidal Network (Table 1; Figure 1). These stations were selected because they are located in Tier 1 or Tier 2 counties in PA, which are the counties with the greatest nutrient loads to the Chesapeake Bay watershed [20]. Further, the drainage areas for each of these monitoring stations are within one county's boundaries, enabling more accurate calculation of the specific load-reduction goals that need to be met at each point (Table 1). The load reduction for TN and TP that each county needs to meet is specified in the PA Watershed Implementation Plan [20], and the load reduction needed for each selected study site was calculated based on the size of the drainage area relative to the county. For example, a drainage area that spans half of a county would need to meet half of the county's mandated load reduction.

Table 1. Information regarding drainage area, percentage of land use (agricultural, forested, and urban), and annual load-reduction goals for total nitrogen (TN) and total phosphorus (TP) for each selected study site in the Pennsylvania portion of the Chesapeake Bay watershed. Land use is based on USGS 2016 National Land Cover Data (https://www.mrlc.gov/data/nlcd-2016-land-cover-conus; accessed on 30 June 2020).

Station ID	Stream Name	County	Drainage Area (km^2)	Land Use (%) Forested/ Developed/ Agriculture	TN Load Reduction (kg-N/y)	TP Load Reduction (kg-P/y)
1570000	Conodoguinet Creek	Cumberland	1217.29	39/20/24	847,735	11,439
1573160	Quittapahilla Creek	Lebanon	192.18	15/33/51	204,356	6916
1573695	Conewago Creek	Lebanon	53.09	42/15/41	56,454	1910
1574000	West Conewago Creek	York	1320.89	57/16/26	1,016,787	-
1575585	Codorus Creek	York	691.53	37/34/28	532,322	-
1576754	Conestoga River	Lancaster	1217.29	17/42/39	2,483,862	101,460
1576787	Pequea Creek	Lancaster	383.32	48/13/39	782,158	31,949
1614500	Conococheague	Franklin	1279.45	19/16/64	839,974	28,985

-, Mandated load-reduction goal has already been met.

2.2. Lorenz Inequality Analysis

The extent of temporal inequality in the TN and TP loads observed at each selected monitoring station were determined using Lorenz Curves and corresponding Gini coefficients (G). The Lorenz Curves were created by sorting the daily loads in ascending order and graphing the fractions of the cumulative loads as a function of the fractions of cumulative time (Figure 2). Lorenz Curves were generated for each station from 2010 through 2018 given that the TMDL began in 2010 and data are available from the Chesapeake Bay Nontidal Network through 2018. In a few cases, where data were not available from 2010–2012, data were analyzed from 2013 through 2018.

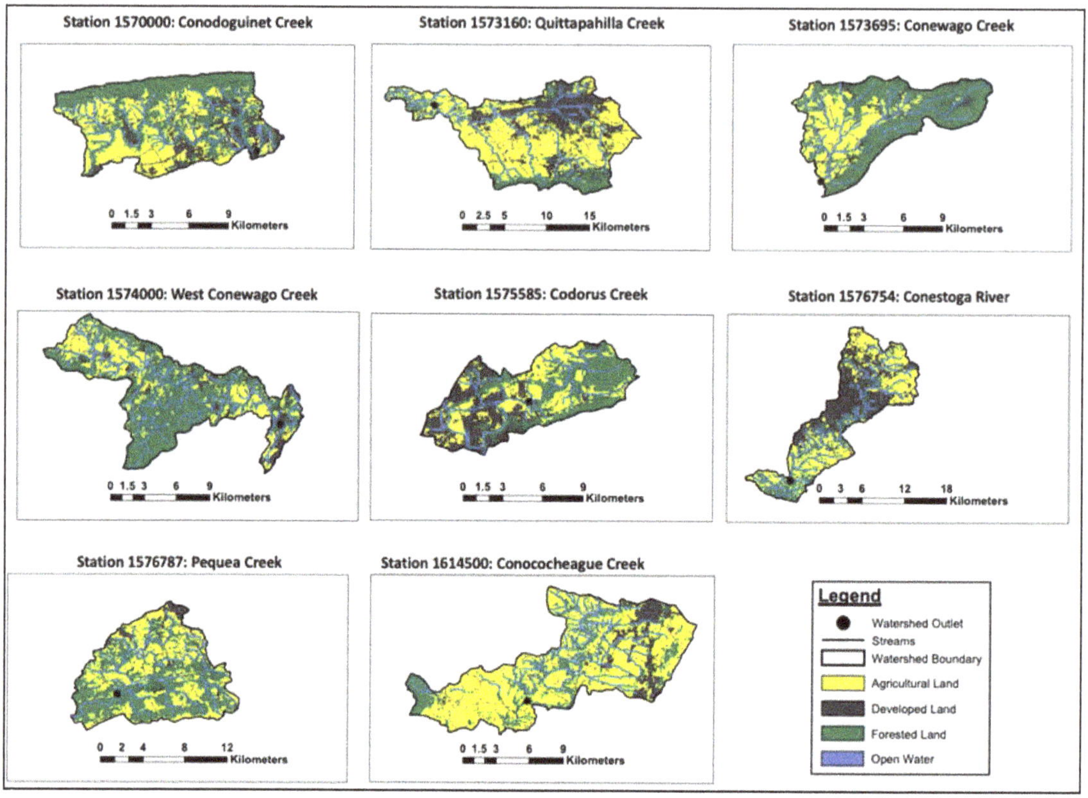

Figure 1. Land-use composition of the eight selected study site in the Pennsylvania portion of the Chesapeake Bay watershed. Base layer for the maps is the USGS 2016 National Land Cover Data (Available at: https://www.mrlc.gov/national-land-cover-database-nlcd-2016; accessed on 30 June 2020).

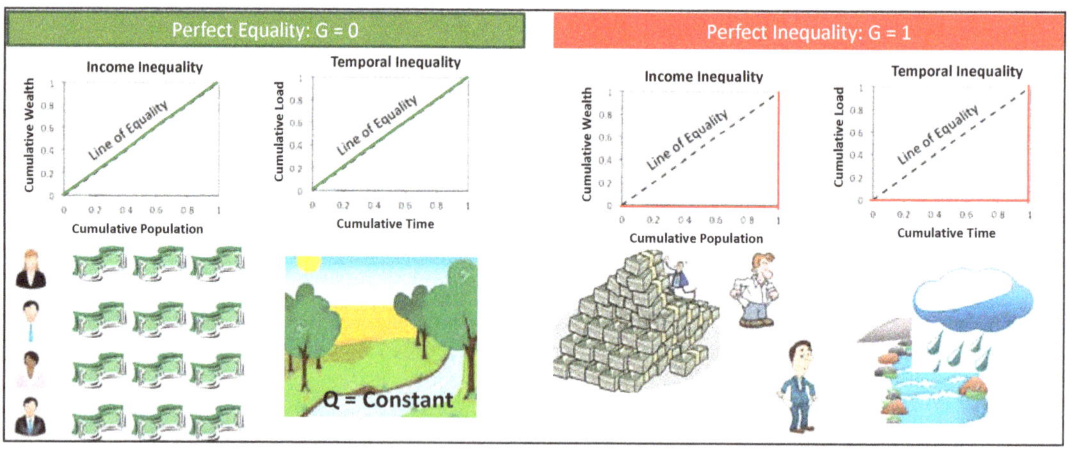

Figure 2. Visual representation of Lorenz Curves for perfect equality ($G = 0$) and perfect inequality ($G = 1$) scenarios in economics and their hydrologic analogues, where G is the Gini Coefficient.

Lorenz Curves plot on or below a line of equality (Figure 2), with the value of G quantifying the extent to which the curve plots below the line of equality. G was calculated as the ratio of the area between the line of equality and the Lorenz Curve to the entire area under the line of equality (Figure 2). These metrics are commonly used in economics and have more recently been applied to hydrology and water-quality data [11–15]. In the economics scenario, G would equal 0 (i.e., perfect equality) if everyone in the population had the same amount of wealth, while G would equal 1 (i.e., perfect inequality) if one person in the population had the entirety of the wealth, while everyone else in the population had none. In the water-quality scenarios, $G = 0$ if every moment contributes equally to observed biogeochemical and/or hydrologic responses, whereas $G = 1$ if one moment contributes to the entirety of the observed responses. The appeal of this analysis is its applicability to data over any duration of time or spatial extent in exactly the same mathematical manner.

2.3. Temporal Targeting Decision-Making Framework

The results of the temporal inequality analyses were used to develop a decision-making framework to identify the time and flow conditions under which the targeted load was exported. The Lorenz Curves were used to identify the fraction of time during which the cumulative TN and TP loads were exported, respectively, during low-flow and high-flow conditions. The framework then links the targeted loads to a flow-duration curve (FDC) for each site, which enables the flow conditions during the periods of time that the targeted loads were exported to be specified. The decision-making framework produces the specific flowrates for high- and low-flow conditions that export the loads that need to be mitigated to meet the desired load-reduction goals. If the annual load is greater than the targeted load, there are generally two "windows of opportunity" for achieving the load-reduction goals, with low-flow targeting resulting in a longer period of time over which opportunities arise to effectively mitigate the load, while high-flow targeting provides a shorter, more targeted period of time to achieve the same load reduction.

3. Results

3.1. Load-Reduction Goals

The percentage of the annual TN and TP loads that must be reduced to meet the load-reduction goals mandated for each county are reported in Table 2. They range from less than 10% of the annual TP load to more than 100% of the annual loads for both TN and TP. For the years in which the load reduction was more than 100% of the annual load, the load was smaller than the mandated load reduction (see Tables 1 and 2), meaning that even if the entire annual load had been effectively mitigated, the annual load reduction would not have been met. Further, the percentage of the annual TN and TP loads that need to be reduced varied across years for each site, with the load reduction needed for some catchments ranging from less than 30% in some years to more than 70% in others (Table 2). The range was particularly large for the West Conewago Creek site, which only needed to reduce its load by approximately 40% in 2011 but, in other years, could have reduced 100% of its load and still not met its annual load-reduction goal (Table 2).

3.2. Temporal Inequality Results

The degree of temporal inequality exhibited by each of the selected study sites was generally lower for TN than TP (Figure 3; Table 3), with an average value of G for TN (G_{TN}) across all years at all sites of 0.44, while the average G for TP (G_{TP}) across all years of all sites was 0.67. Across all sites, the range of G values exhibited by TN was 0.19 to 0.73, while the range for TP was 0.32 to 0.90. The study site with the lowest degree of temporal inequality for TN and TP was Quittapahilla Creek, which had average G_{TN} and G_{TP} values of 0.24 and 0.41, respectively. The Conewago and West Conewago Creek study sites had the highest average G_{TN} and G_{TP} values of >0.60 and >0.75, respectively.

Table 2. Annual loads for total nitrogen (TN) and total phosphorus (TP) for each selected study site, along with the calculated percent reduction of TN and TP needed to meet the annual load-reduction goals provided in Table 1.

Station ID	Stream Name	Year	Annual Load (kg/y)		Reduction Needed in Annual Load (%)	
			TN	TP	TN	TP
1570000	Conodoguinet Creek	2010	2,519,351	46,921	33.65	24.38
		2011	3,020,707	84,057	28.06	13.61
		2012	2,626,795	43,936	32.27	26.04
		2013	1,714,809	21,728	49.44	52.65
		2014	2,257,449	39,676	37.55	28.83
		2015	1,595,774	22,924	53.22	49.90
		2016	1,480,014	20,259	57.28	56.46
		2017	1,329,605	16,788	63.76	68.14
		2018	3,208,919	73,436	26.42	15.58
1573160	Quittapahilla Creek	2013	615,651	16,176	33.19	42.75
		2014	727,136	20,475	28.10	33.77
		2015	505,997	8,768	40.39	78.87
		2016	555,417	10,311	36.79	67.07
		2017	512,714	7,548	39.86	91.62
		2018	888,834	24,005	22.99	28.81
1573695	Conewago Creek	2013	51,752	5,825	109.08	32.80
		2014	81,874	9,607	68.95	19.89
		2015	47,854	3,601	117.97	53.06
		2016	58,677	5,335	96.21	35.81
		2017	47,514	3,185	118.82	59.98
		2018	137,565	24,338	41.04	7.85
1574000	West Conewago Creek	2010	1,696,470	133,314	59.94	-
		2011	2,683,395	357,276	37.89	-
		2012	1,766,520	157,856	57.56	-
		2013	1,451,354	148,154	70.06	-
		2014	2,017,352	225,083	50.40	-
		2015	1,156,909	97,761	87.89	-
		2016	1,455,303	135,015	69.87	-
		2017	781,830	55,144	130.05	-
		2018	2,310,778	270,731	44.00	-
1575585	Codorus Creek	2013	1,320,054	77,114	40.33	-
		2014	1,835,336	108,968	29.00	-
		2015	950,050	41,905	56.03	-
		2016	1,245,531	59,141	42.74	-
		2017	740,603	31,663	71.88	-
		2018	1,598,243	123,154	33.31	-
1576754	Conestoga River	2010	4,296,086	144,816	57.82	70.06
		2011	5,284,077	431,397	47.01	23.52
		2012	3,909,384	152,379	63.54	66.58
		2013	4,023,552	221,673	61.73	45.77
		2014	5,027,739	298,263	49.40	34.02
		2015	3,311,859	138,998	75.00	72.99
		2016	3,281,919	138,687	75.68	73.16
		2017	2,473,277	87,942	100.43	115.37
		2018	4,636,651	325,283	53.57	31.19
1576787	Pequea Creek	2010	1,569,087	87,343	49.85	36.58
		2011	1,482,631	120,837	52.75	26.44
		2012	1,246,702	63,093	62.74	50.64
		2013	1,238,306	137,569	63.16	23.22
		2014	1,605,497	146,393	48.72	21.82
		2015	1,006,888	69,547	77.68	45.94
		2016	958,201	52,977	81.63	60.31
		2017	598,215	16,240	130.75	196.73
		2018	1,218,037	136,156	64.21	23.47

Table 2. *Cont.*

Station ID	Stream Name	Year	Annual Load (kg/y)		Reduction Needed in Annual Load (%)	
			TN	TP	TN	TP
1614500	Conococheague Creek	2010	2,721,257	78,048	30.87	37.14
		2011	3,251,720	131,625	25.83	22.02
		2012	2,839,020	64,177	29.59	45.16
		2013	2,115,025	47,024	39.71	61.64
		2014	2,603,312	82,177	32.27	35.27
		2015	1,511,072	32,248	55.59	89.88
		2016	1,700,538	37,061	49.39	78.21
		2017	1,704,557	43,349	49.28	66.86
		2018	3,647,005	137,505	23.03	21.08

-, Mandated load-reduction goal has already been met.

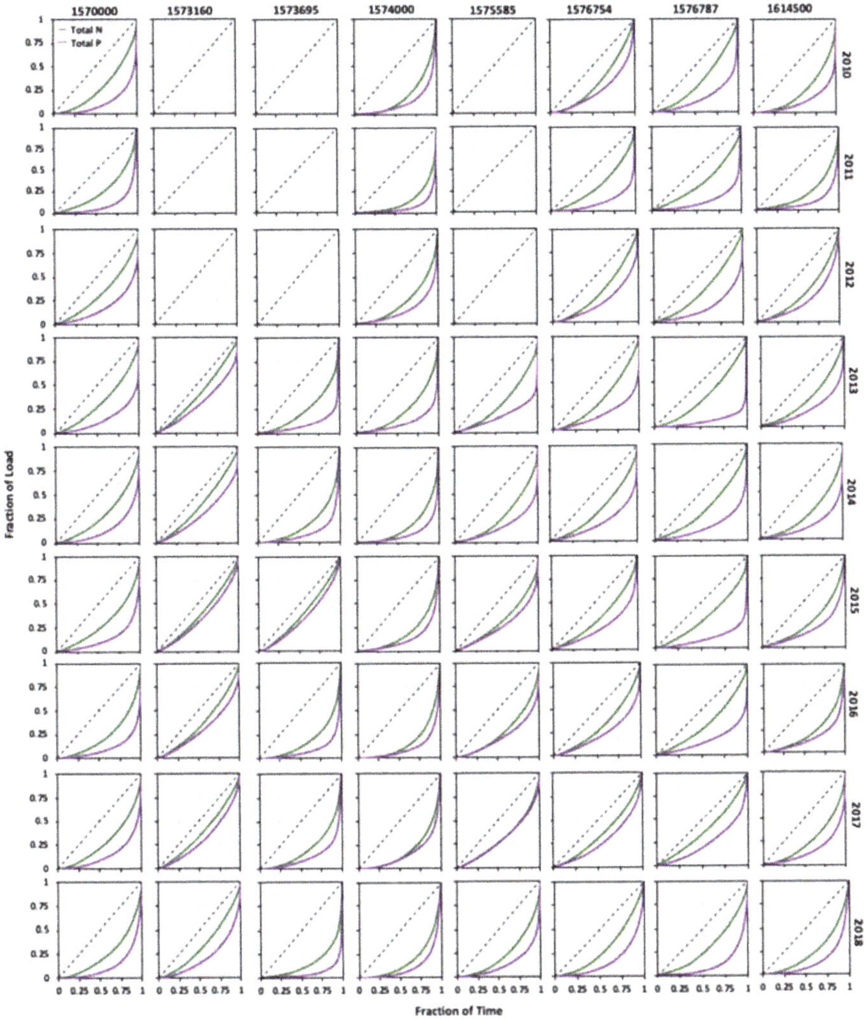

Figure 3. Lorenz Curves for total nitrogen (TN) and total phosphorus (TP) loads exported for each year for each selected study site. The dotted line represents the "Line of Equality", with temporal inequality signified by the degree to which the Lorenz Curve plots below the line. Blank plots are shown when no load data are available for those years for specific gauging stations.

Table 3. Gini Coefficients for total nitrogen (G_{TN}) and total phosphorus (G_{TP}) loads exported for each year for each selected study site, along with percentages of time that each load is exported during low- and high-flow conditions.

Station ID	Stream Name	Year	G_{TN}	G_{TP}	% Time TN		% Time TP	
					Low Flow	High Flow	Low Flow	High Flow
1570000	Conodoguinet Creek	2010	0.42	0.74	63.8	11.0	82.2	0.5
		2011	0.49	0.81	63.6	5.2	78.4	0.3
		2012	0.34	0.64	55.7	12.6	74.6	0.8
		2013	0.42	0.70	77.5	21.6	98.1	3.6
		2014	0.42	0.71	67.7	12.9	83.3	1.4
		2015	0.42	0.72	82.5	22.7	95.3	4.7
		2016	0.52	0.77	89.6	19.4	98.1	3.8
		2017	0.43	0.69	88.8	32.9	98.1	15.1
		2018	0.48	0.72	60.5	5.8	71.2	0.8
1573160	Quittapahilla Creek	2013	0.19	0.38	46.3	22.5	69.9	15.6
		2014	0.22	0.42	43.0	16.7	61.9	6.8
		2015	0.22	0.32	55.6	26.3	94.0	61.1
		2016	0.26	0.40	55.7	20.8	92.1	40.2
		2017	0.22	0.36	55.3	25.5	98.9	79.7
		2018	0.31	0.60	42.2	9.9	69.0	3.8
1573695	Conewago Creek	2013	0.54	0.82	100.0	100.0	97.3	0.5
		2014	0.60	0.82	96.4	23.6	85.5	0.5
		2015	0.61	0.78	100.0	99.7	74.8	31.0
		2016	0.62	0.81	99.7	73.5	95.6	1.1
		2017	0.58	0.75	100.0	100.0	97.8	5.5
		2018	0.7	0.90	91.5	2.7	80.5	<0.1
1574000	West Conewago Creek	2010	0.64	0.76	92.9	16.2	-	-
		2011	0.73	0.86	89.3	3.3	-	-
		2012	0.55	0.72	90.2	19.1	-	-
		2013	0.62	0.79	97.2	25.5	-	-
		2014	0.68	0.82	92.1	8.2	-	-
		2015	0.62	0.74	99.2	47.9	-	-
		2016	0.68	0.79	96.7	19.1	-	-
		2017	0.63	0.69	100.0	100.0	-	-
		2018	0.64	0.76	87.1	7.9	-	-
1575585	Codorus Creek	2013	0.4	0.67	68.5	15.9	-	-
		2014	0.46	0.67	61.1	7.1	-	-
		2015	0.38	0.49	81.9	28.8	-	-
		2016	0.45	0.55	75.1	13.7	-	-
		2017	0.36	0.37	91.5	47.7	-	-
		2018	0.49	0.69	68.8	7.7	-	-
1576753	Conestoga River	2010	0.37	0.53	79.2	32.6	96.7	31.0
		2011	0.40	0.78	74.8	20.3	86.6	0.3
		2012	0.34	0.56	82.5	39.3	96.7	25.4
		2013	0.32	0.66	82.2	38.9	95.3	2.2
		2014	0.36	0.63	73.2	25.8	81.9	1.6
		2015	0.34	0.55	91.8	53.4	98.4	33.4
		2016	0.36	0.54	91.5	51.9	98.6	36.1
		2017	0.30	0.46	100.0	100.0	100.0	99.7
		2018	0.39	0.67	78.4	26.8	81.4	2.2
1576787	Pequea Creek	2010	0.34	0.69	72.1	27.7	90.1	1.4
		2011	0.35	0.78	77.0	27.4	89.6	0.5
		2012	0.32	0.68	80.6	40.2	95.4	5.2
		2013	0.32	0.85	84.4	41.6	96.7	<0.1
		2014	0.35	0.72	72.1	25.8	76.2	0.3
		2015	0.33	0.77	94.2	58.1	98.4	1.1
		2016	0.33	0.69	94.3	62.8	99.5	6.6
		2017	0.24	0.51	100.0	100.0	100.0	99.7
		2018	0.39	0.76	86.6	37.0	83.0	1.1
1614500	Conococheague Creek	2010	0.54	0.75	69.9	5.5	93.7	1.1
		2011	0.60	0.81	71.0	3.6	87.1	0.5
		2012	0.40	0.56	58.2	10.1	84.7	10.1
		2013	0.46	0.63	71.8	12.3	98.1	13.2
		2014	0.48	0.72	66.3	7.7	90.4	0.8
		2015	0.46	0.62	85.5	22.5	100.0	60.0
		2016	0.55	0.68	87.2	12.0	99.2	26.0
		2017	0.48	0.68	81.6	17.5	98.1	14.5
		2018	0.50	0.71	57.8	4.7	76.7	1.6

-, Mandated load-reduction goal has already been met; <0.1%, targeted load exported in a single day.

In general, sites with a higher percentage of agricultural and developed area had lower values of G_{TN} and G_{TP}. The Conodoguinet Creek drainage area comprises approximately 15% forested land, with nearly 85% of the drainage area either developed or agricultural land use (Table 1). Conodoguinet Creek's average values of G_{TN} and G_{TP} were 0.24 and

0.41, respectively, which were the lowest of all the selected sites (Table 3). West Conewago Creek, which was the site with the highest percentage of forested land use (Table 1), had the highest average value of G_{TN} and the second highest average value of G_{TP} (0.64 and 0.77, respectively; Table 3).

3.3. Decision-Making Framework Results

The decision-making framework links annual load-reduction goals to the specific percentage of days within the year needed to fully mitigate the targeted load under either high- or low-flow conditions (Figure 4; Table 3). Thus, through the framework, a watershed planner can identify two "windows of opportunity" in which to mitigate exported loads sufficiently to meet the annual reduction goals for a particular pollutant. To illustrate the framework's utility, data from 2014 at the Conococheague Creek site were selected and analyzed to determine the specific periods of time and corresponding flow conditions during which the targeted loads were exported. The framework shows that to achieve a 32% TN load-reduction goal, either 66.3% of low-flow conditions or 7.7% of high-flow conditions must be targeted. Based on the flow-duration curve for the site, the flowrates when those loads were exported were less than 20 m^3/s if lower flowrates were targeted for treatment. However, if high flowrates were targeted, loads exported were greater than 42 m^3/s (Figure 4). The extent of temporal inequality exhibited by the TP loads meant that either nearly all flow conditions (i.e., flowrates observed less than 90.4% of the time) needed to be targeted to achieve the 35% annual load-reduction goal, or the highest 0.8% of flow conditions (i.e., flowrates higher than 150 m^3/s; Figure 4) could be targeted and achieve the same load-reduction goals.

The temporal targeting analyses across all sites and all years reveal TN and TP loads equivalent to the mandated load-reduction goals can be exported within as little 2.7% of the year for TN and $< 0.5\%$ for TP. These results suggest that mandated load-reduction goals could sometimes be achieved by effectively targeting loads exported over less than ten days of the year. In several cases, the effects of a single storm event were so high that the targeted load under high-flow conditions was exported in a single day (Table 3). The effects of these extreme events on the shape of the Lorenz Curve can be seen for Conewago Creek in 2018 and Pequea Creek in 2013 (Figure 3). However, mitigating the loads of such extreme events requires similarly extreme conservation practices that are designed far beyond those intended for everyday mitigation.

Conversely, during lower flow conditions, temporal targeting results showed that loads equivalent to the annual load-reduction goals were never exported less than 42% of the time for TN or less than 62% of the time for TP across any of the sites (Table 3). To mitigate TN and TP loads during low-flow conditions, conservation practices or best management practices need to be effective in treating TN and TP loads over longer stretches of consecutive days during and between small storm events. However, conservation practice effectiveness depends not only on storm patterns but also on crop-rotation cycles and land-use management. Thus, evaluating seasonal patterns of time-series graphs across several historical years in context of a specific watershed's typical cropping and land-cover patterns may help estimate expected effectiveness and lifecycles of long-term agricultural conservation practices.

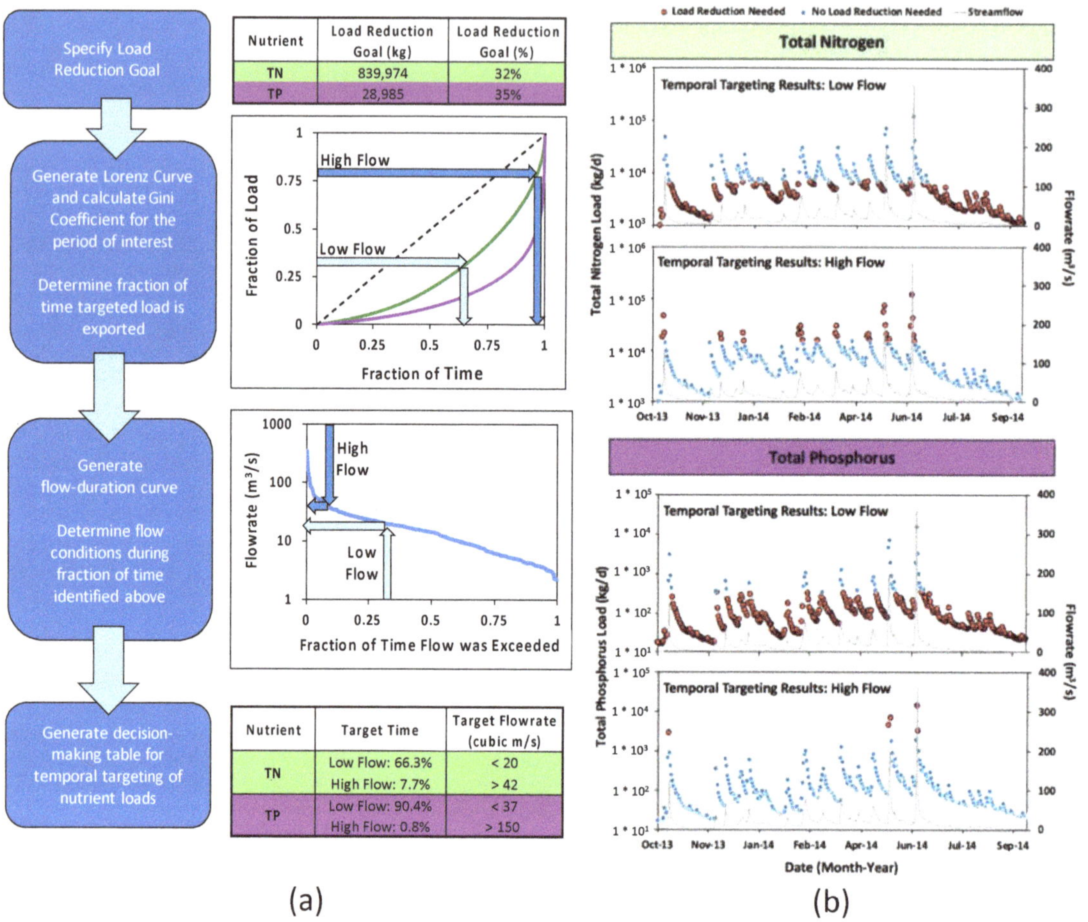

Figure 4. (**a**) Decision-making flow chart demonstrated for Conococheague Creek (2014) to determine site-specific fractions of time during which the targeted loads are exported and identify the corresponding flow conditions; arrows on the Lorenz Curve and flow-duration curve are shown for targeting total nitrogen (TN) load. (**b**) Time-series graphs highlight specific events during which targeted loads for TN and total phosphorus (TP) are exported during low- and high-flow conditions.

4. Discussion

This novel application of Lorenz Inequality and corresponding *G* demonstrates the utility of leveraging an analysis commonly used in economics for quantifying income inequality for quantifying the temporal inequality of contaminant loads and establishing a framework for temporal targeting of "hot moments" to achieve load-reduction goals. Earlier efforts to apply Lorenz Inequality to hydrology and water-quality data have helped to explain the effects of scale (both temporal and spatial) on the degree of temporal inequality exhibited by discharge and loads, with higher inequality in smaller headwater catchments and lower inequality in larger watersheds as well as higher inequality when finer temporal resolution data (e.g., daily scale) are used to generate the Lorenz Curves compared to coarser temporal scales (e.g., monthly) [11]. Further, across the Chesapeake Bay, a wide range of temporal inequality was documented for 108 stations in the Chesapeake Bay Nontidal Network, with *G* ranging from 0.24–0.60 for flow, 0.18–0.69 for TN, 0.36–0.92 for

TP, and 0.39–0.90 for total suspended sediment [15]. However, the results of the previous analysis across the Chesapeake Bay watershed were for the entire 2010–2018 period since the TMDL was enacted, limiting the utility of those results for decision making at the annual scale.

Here, the results of this temporal inequality analysis demonstrate the potential effectiveness of targeting "hot moments" to achieve load-reduction goals in impaired surface water bodies. For catchments with a high degree of temporal inequality (i.e., G approaching 1), this temporal targeting is especially important, as failing to adopt conservation practices that do not adequately reduce loads during high-flow conditions may prevent load-reduction goals from being met. Conversely, in catchments with low degrees of temporal inequality, the period of time over which the targeted load is exported is longer, and spatial targeting may be more effective than temporal targeting for meeting load-reduction goals, with more opportunities available to effectively reduce the load over the course of a year. The implications of this analysis may be helpful in understanding difficulties in meeting water-quality-restoration goals in long-impaired watersheds, such as the Chesapeake Bay.

The results demonstrate that while in some years, load-reduction goals are only a small portion of the overall load exported and may easily be met by targeting a few storm events, in other years, loads are actually less than the targeted load, and even if 100% of the annual load were effectively mitigated, the annual load-reduction goal could not be met. Viewing these expectations through the lens of temporal inequality can be helpful in understanding how easy or difficult achieving load-reduction goals will be in a given watershed since the higher the value of G, the more difficult it will be to reduce loads without capturing and treating high-flow conditions. When the G is relatively low, as is often the case for TN [15], it is because export of the constituent of interest occurs largely during baseflow conditions, and therefore, achieving load reduction without effectively treating high-flow conditions may be possible. In these cases, conservation practices that help reduce groundwater concentrations, such as cover crops and other nonstructural best management practices (BMPs), may be most effective in meeting load-reduction goals. However, when G is high, as is often the case for TP [15], it is because export of the constituent of interest largely occurs during high-flow conditions, and therefore, if these events are not effectively treated, meeting the load-reduction goal may not be possible. In these cases, conservation practices, such as riparian buffers, vegetated filter strips, and detention basins, may be most effective in meeting the load-reduction goals. Water-quality BMPs are often vegetative and therefore are mainly effective in managing low-flow events over only the portion of the year when the plants are actively growing and are not dormant. The portions of the year during which water-quality BMPs are effective should align with months during which large portions of annual loads are exported. This may require landowners to consider designing BMPs to manage high-flow events or a combination of low-flow BMPs that are effective over a longer time interval.

5. Conclusions

Overall, the results of this research demonstrated a decision-making framework that can be applied at any temporal or spatial scale to quantify the importance of targeting "hot moments" to achieve specific load-reduction goals. The results of our analysis demonstrate the site-specific nature of the results such that even across a watershed with a single TMDL, the implementation of conservation practices that will achieve the load-reduction goals is likely to be heterogeneous, with the success of field-scale implementation of appropriate conservation practices relying on local knowledge of hydrology and contaminant transport.

Author Contributions: Conceptualization, H.E.P., T.L.V., and M.R.; methodology, H.E.P. and T.L.V.; software, N.O., D.S., and T.L.V.; formal analysis, N.O. and D.S.; investigation, N.O., D.S., H.E.P., and T.L.V.; resources, H.E.P.; writing—original draft preparation, N.O., D.S., and H.E.P.; writing—review and editing, T.L.V. and M.R.; visualization, N.O., D.S., H.E.P., and T.L.V.; supervision, H.E.P. and M.R.; project administration, H.E.P.; funding acquisition, H.E.P. All authors have read and agreed to the published version of the manuscript.

Funding: Nicole Opalinski worked on this project as part of an independent studies course. Daniel Schultz was funded through support from the Penn State College of Engineering Summer Research Experience for Undergraduates Program and from the Penn State Erickson Discovery Undergraduate Research Program. Heather E. Preisendanz is funded, in part, by the Penn State Institutes of Energy and the Environment and by the USDA National Institute of Food and Agriculture Federal Appropriations under Project PEN04574 and Accession number 1004448. Matt Royer was supported, in part, by the Penn State Center for Nutrient Solutions, which is funded by an EPA STAR Grant (RD835568). All entities involved are equal opportunity providers and employers.

Institutional Review Board Statement: Not applicable.

Informed Consent Statement: Not applicable.

Data Availability Statement: All data are publicly available through the United States Geological Survey and can be accessed at: https://cbrim.er.usgs.gov/introduction.html or https://doi.org/10.5066/P931M7FT.

Conflicts of Interest: Authors Preisendanz and Veith reside in the same household. The funders had no role in the design of the study; in the collection, analyses, or interpretation of data; in the writing of the manuscript, or in the decision to publish the results.

References

1. Fixen, P.E. A brief account of the genesis of 4R nutrient stewardship. *Agron. J.* **2020**, *112*, 4511–4518. [CrossRef]
2. Veith, T.L.; Wolfe, M.L.; Heatwole, C.D. Optimization procedure for cost effective BMP placement at a watershed scale. *J. Am. Water Resour. Assoc.* **2003**, *36*, 1331–1343. [CrossRef]
3. Gitau, M.W.; Veith, T.L.; Gburek, W.J.; Jarrett, A.J. Watershed level best management practice selection and placement in the Town Brook Watershed, New York. *J. Am. Water Resour. Assoc.* **2006**, *42*, 1565–1581. [CrossRef]
4. Ghebremichael, L.T.; Veith, T.L.; Hamlett, J.M. Integrated watershed- and farm-scale modeling framework for targeting critical source areas while maintaining farm economic viability. *J. Environ. Manag.* **2013**, *114*, 381–394. [CrossRef] [PubMed]
5. Amin, M.G.; Veith, T.L.; Shortle, J.S.; Karsten, H.D.; Kleinman, P.J.A. Addressing the spatial disconnect between national-scale total maximum daily loads and localized land management decisions. *J. Environ. Qual.* **2020**, *49*, 613–627. [CrossRef] [PubMed]
6. Karki, R.; Srivastava, P.; Veith, T.L. Application of the Soil and Water Assessment Tool (SWAT) at the field-scale: Categorizing methods and review of applications. *Trans. Am. Soc. Agric. Biol. Eng.* **2020**, *63*, 513–522. [CrossRef]
7. Donner, S.D.; Scavia, D. How climate controls the flux of nitrogen by the Mississippi River and the development of hypoxia in the Gulf of Mexico. *Limnol. Oceanogr.* **2007**, *52*, 856–861. [CrossRef]
8. Richards, P.R.; Baker, D.B.; Kramer, K.W.; Ewing, D.E.; Merryfield, B.J.; Miller, N.L. Storm discharge, loads, and average concentrations in Northwest Ohio rivers, 1975-1995. *J. Am. Water Resour. Assoc.* **2001**, *37*, 423–438. [CrossRef]
9. Royer, T.V.; David, M.B.; Lowell, E.G. Timing of riverine export of nitrate and phosphorus from agricultural watersheds in Illinois: Implications for reducing nutrient loading to the Mississippi River. *Environ. Sci. Technol.* **2006**, *40*, 4126–4131. [CrossRef] [PubMed]
10. Hirsch, R.M. Flux of Nitrogen, Phosphorus, and Suspended Sediment from the Susquehanna River Basin to the Chesapeake Bay during Tropical Storm Lee, September 2011, as an Indicator of the Effects of Reservoir Sedimentation on Water Quality. U.S. Geological Survey Scientific Investigations Report. 2012; 17p. Available online: https://pubs.usgs.gov/sir/2012/5185/ (accessed on 15 February 2022).
11. Jawitz, J.W.; Mitchell, J. Temporal inequality in catchment discharge and solute export. *Water Resour. Res.* **2011**, *47*, W00J14. [CrossRef]
12. Masaki, Y.; Hanasaki, N.; Takahashi, L.; Hijioka, Y. Global-scale analysis on future changes in flow regimes using Gini and Lorenz asymmetry coefficients. *Water Resour. Res.* **2014**, *50*, 4054–4078. [CrossRef]
13. Gall, E.H.; Park, J.; Harman, C.J.; Jawitz, J.W.; Rao, P.S.C. Landscape filtering of hydrologic and biogeochemical responses in managed catchments. *Landsc. Ecol.* **2013**, *28*, 651–664. [CrossRef]
14. Veith, L.T.; Preisendanz, H.E.; Elkin, K.R. Characterizing transport of natural and anthropogenic constituents in a long-term agricultural watershed in the northeastern United States. *J. Soil Water Conserv.* **2020**, *75*, 319–329. [CrossRef]
15. Preisendanz, E.H.; Veith, T.L.; Zhang, Q.; Shortle, J.S. Temporal inequality of nutrient and sediment transport: A decision-making framework for temporal targeting of load reduction goals. *Environ. Res. Lett.* **2021**, *16*, 014005. [CrossRef]
16. Selman, M.; Sugg, Z.; Greenhalgh, S. *Eutrophication and Hypoxia in Coastal Areas: A global Assessment of the State of Knowledge*; World Resources Institute: Washington, DC, USA, 2008; ISBN 978-1-56973-681-4. Available online: https://www.wri.org/research/eutrophication-and-hypoxia-coastal-areas (accessed on 15 February 2022).
17. Mateo-Sagasta, J.; Marjani, S.; Turral, H. *Water Pollution from Agriculture: A Global Review—Executive Summary*; Food and Agriculture Organization of the United Nations, Rome and the International Water Management Institute: Colombo, Sri Lanka, 2017.

18. United States Environmental Protection Agency (USEPA). Chesapeake Bay Total Maximum Daily Load for Nitrogen, Phosphorus and Sediment. 2010. Available online: https://www.epa.gov/chesapeake-bay-tmdl/chesapeake-bay-tmdl-document (accessed on 15 February 2022).
19. Chesapeake Bay Program. 2025 Watershed Implementation Plans (WIPs). 2022. Available online: https://www.chesapeakeprogress.com/clean-water/watershed-implementation-plans (accessed on 15 February 2022).
20. Pennsylvania Department of Environmental Protection (PADEP). Pennsylvania Phase 3 Chesapeake Bay Watershed Implementation Plan. 2021. Available online: https://www.dep.pa.gov/Business/Water/Pennsylvania\T1\textquoterights%20Chesapeake%20Bay%20Program%20Office/WIP3/Pages/PAs-Plan.aspx (accessed on 15 February 2022).
21. Moyer, L.D.; Langland, M.J. Nitrogen, Phosphorus, and Suspended-Sediment Loads and Trends Measured at the Chesapeake Bay Nontidal Network Stations: Water Years 1985–2018, Version 2.0. U.S. Geological Survey Data Release, 2020. Available online: https://www.usgs.gov/data/nitrogen-phosphorus-and-suspended-sediment-loads-and-trends-measured-chesapeake-bay-nontidal (accessed on 15 February 2022).
22. Hirsch, M.R.; Moyer, D.L.; Archfield, A.S. Weighted regression on time, discharge, and season (WRTDS), with an application to Chesapeake Bay River inputs. *J. Am. Water Resour. Assoc.* **2010**, *46*, 857–880. [CrossRef] [PubMed]

water

MDPI

Article

Blending Irrigation Water Sources with Different Salinities and the Economic Damage of Salinity: The Case of Israel

Yehuda Slater [1], Ami Reznik [1], Israel Finkelshtain [1,2] and Iddo Kan [1,2,*]

1 Department of Environmental Economics and Management, The Robert H. Smith Faculty of Agriculture, Food and Environment, The Hebrew University of Jerusalem, P.O. Box 12, Rehovot 761001, Israel; yehuda.slater@mail.huji.ac.il (Y.S.); ami.reznik@mail.huji.ac.il (A.R.); israel.finkelshtain@mail.huji.ac.il (I.F.)
2 The Center for Agricultural Economics Research, P.O. Box 12, Rehovot 761001, Israel
* Correspondence: iddo.kan@mail.huji.ac.il

Abstract: Israel's water and vegetative agriculture sectors are interdependent, as the latter constitutes the solution for wastewater disposal. We employ a dynamic mathematical programming model that captures this interdependence for evaluating the economic damage of irrigation water salinity under two strategies of blending water sources with different salinities: field blending, which enables farmers to assign water with a specific salinity to each crop, and regional blending, under which all crops experience similar water salinity. Relative to field blending, the buildup rate of desalination under regional blending is slightly expedited; nevertheless, reallocations of water sources across sectors and crops increase the average irrigation water salinity, and the overall welfare decreases by USD 0.08 per cubic meter of irrigation water—about 20% of the water's average value of marginal product. Salinity-sensitive crops will face the largest per hectare production reduction if regional blending replaces field blending; however, the combined variations in the prices of irrigation water and agricultural outputs may motivate farmers to move irrigation water to these crops. Under equilibrium conditions in the two sectors, a 1% increase in the average salinity of the irrigation water supplied to a region reduces the value of the marginal product of that water by 2.4% and 1.6% under field and regional blending, respectively.

Keywords: irrigation; salinity; agriculture; policy; water; economics; model

Citation: Slater, Y.; Reznik, A.; Finkelshtain, I.; Kan, I. Blending Irrigation Water Sources with Different Salinities and the Economic Damage of Salinity: The Case of Israel. *Water* **2022**, *14*, 917. https://doi.org/10.3390/w14060917

Academic Editor: Guido D'Urso

Received: 10 February 2022
Accepted: 14 March 2022
Published: 15 March 2022

1. Introduction

For thousands of years, man has been coping with salinization processes in irrigated agriculture [1], which is the main consumer of water worldwide, accounting for nearly 70% of the total global water withdrawal [2]. This problem continues to worsen, and today, 25–30% of the world's irrigated lands in more than 100 countries are affected by salt [3,4]. Population growth, which increases the demand for both food and freshwater for domestic use, further contributes to this growing challenge, as it incentivizes the expansion of irrigated lands and the use of non-freshwater sources such as brackish water and treated wastewater (TWW) for irrigation [5]. These processes are further augmented by climate change, which increases irrigation needs due to higher vapor–pressure deficits [6] and reduces the natural enrichment of freshwater sources [7]. Moreover, the common agronomic solution to salinization is to apply excess amounts of irrigation water to leach the salts below the root zone [8]. However, that method gradually increases the salinity of groundwater bodies [9] and consequently counteracts its original purpose. The use of desalination, which is a remedy for both the growing water shortage and salinization, is steadily increasing [10], but it consumes a great deal of energy and entails high brine disposal costs [11].

The processes described above reflect a strong linkage between agricultural irrigation and the supply of water to different users—a link that should be accounted for in the design of sustainable and economically viable solutions to the problems of water shortages and salinization. This study focuses on irrigation practices in agricultural regions with access to several water sources of different salinities and evaluates the impact of these practices on water management in a multiregional water distribution network. Specifically, we evaluate the economic damage caused by salinity under different strategies for blending irrigation water sources with different salinities. We focus on the case of Israel—a country equipped with a complex water supply system and a large agricultural reliance on non-freshwater sources such as TWW and brackish water, which together, constitute about 60% of the country's total irrigation water. Our economic analysis accounts for the impact of blending options on agricultural cropping patterns, optimal long-term management policies, and the development of the Israeli water supply system.

Despite the vast agronomic literature on the production impacts of irrigation water mixtures (e.g., recent agronomic studies explore the impact of conjunctive use of water resources [12,13], employ GIS for assessing salinity impacts under different irrigation practices [14], and measure the impacts of water irrigation strategies on soil and plant properties [15]), economic analyses of that issue are scarce. Parkinson et al. [16] were probably the first to economically evaluate water blending options. Knapp and Dinar [17], Dinar et al. [18], Kan et al. [19], and Kan [20] employed field-level models to study the profitability of mixing water sources with different salinities for the irrigation of specific crops. Feinerman and Yaron [21] and Kan and Rapaport-Rom [22] incorporated blending options in regional-scale models, in which the land allocation across crops was endogenous. However, all of these studies assumed exogenous water supplies and therefore overlooked the implications of water management strategies within agricultural regions on the water economy as a whole. The contribution of this paper is the introduction of the nexus between the agricultural and water sectors into the economic analysis of water blending strategies.

The linkage between the intraregional management of irrigation water and the design of economy-wide water-supply systems is of particular importance in water economies that supply water to different users from multiple sources and/or where the recycling of domestic TWW in irrigated agriculture creates a strong interdependence between the two sectors. In such water economies, the optimal allocation of water across users depends on their demands for the various water sources, where the demand of any farming region for different water types depends on the irrigation practices in that region. This is the case in Israel, where the water distribution network connects almost all users and sources in the country. That connectivity implies that water usage at a particular place and time may have opportunity costs, as it cannot be used for other purposes at alternative locations and times [23].

Hydroeconomic models provide a powerful tool to analyze water management problems on different scales and under various spatiotemporal conditions (see [24–34]). However, to the best of our knowledge, the only hydroeconomic model that incorporates salinity considerations in the allocation of water to urban and agricultural users is the MYWAS-VALUE (Multi Year Water Allocation System-Vegetative Agricultural Land Use Economics) model, which was developed by Slater et al. [35] for the case of Israel. Slater et al. [35] employed MYWAS-VALUE to evaluate the societal deadweight loss entailed by overlooking the impact of salinity on agricultural production in the design of water infrastructures. However, the model presumes regional irrigation water blending; that is, all of the inflows of water sources into any agricultural region are mixed before they are applied to the irrigated crops. This assumption has two drawbacks: first, compared to field-level blending, regional water blending may increase the detrimental impact of salinity on agricultural production because it affects all of the irrigated crops in any given region, including both salinity-tolerant and salinity-sensitive ones. Consequently, the exaggerated salinity damage may motivate the faster-than-optimal expansion of desalination capacities. Second, it turns out that farmers in Israel rarely blend irrigation water from different sources (personal communication; Anat Levingert, Senior Manager of the Consulting and Professional Service of the Israeli Ministry of Agriculture (Shaham)). Thus, designing the long-term development of water infrastructures under the assumption of the regional blending of irrigation water sources may yield results that both inflate the agricultural damage incurred due to salinity and that are inconsistent with reality.

In this paper, we analyze two irrigation water mixing scenarios: field blending (FB) and regional blending (RB). The difference between the scenarios with respect to the intraregional water supply system is illustrated in Figure 1 for a hypothetical region, in which farmers grow five crops and have access to three water sources with different salinities: freshwater, TWW, and brackish water. Under FB, farmers can select a specific combination of the three sources for each crop, whereas the RB scenario implies one combination for all crops. Note that, while both scenarios do not preclude the non-blending option, avoiding blending in the RB case implies that only one water source is used in the entire region, whereas the FB scenario enables farmers to use all of the water sources that are available to them by assigning a single water source to each crop.

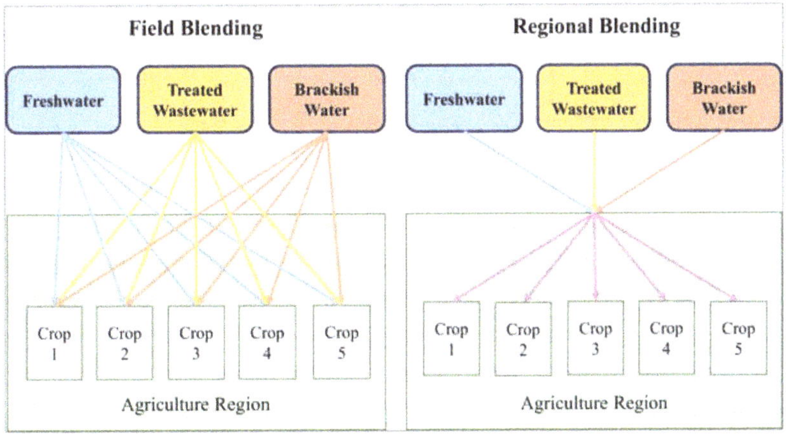

Figure 1. Schematic illustration of the field and regional blending scenarios in an agricultural region with five crops that can be irrigated by three water sources with different salinity levels: freshwater, treated wastewater, and brackish water.

Our analysis is based on the MYWAS-VALUE framework. We first calibrate the model under the FB assumption to reproduce the observed situation in a baseline year (2019). Then, we run the model under the FB and RB scenarios for a period of 30 years. We found that switching from FB to RB slightly expedites the development of desalination plants, but the average irrigation water salinity increased due to the reallocation of water sources across sectors and crops. Although salinity-sensitive crops face the largest reduction in per hectare production, the combined impact of changes in the (endogenously determined) prices of irrigation water and agricultural outputs motivates farmers to shift more water and land to the production of these crops.

We consider three measures of the economic damage caused by salinity under the two blending scenarios. The first measures the reductions in the agricultural production value caused by the design of water infrastructures that ignore the impact of irrigation-water salinity on agricultural production. This reduction amounts to USD 1195 and USD 1326 per hectare under the FB and RB scenarios, respectively (all monetary values are in terms of the 15th year of the 30-year planning horizon). The second measure is based on the negative relationship between the irrigation water value of the marginal product (VMP) in an agricultural region and the average salinity of the region's irrigation water; on average, the VMP decreased with the salinity by USD 0.39 and USD 0.30 per dS m^{-1} per cubic meter of irrigation water for the case of FB and RB, respectively, or by −2.4 and −1.6 in terms of elasticity (note that both VMP and salinity are endogenous in MYWAS-VALUE). The last measure computes the marginal damage caused by salinity based on the shadow values of the salt-balance constraints along the water delivery system; a salinity increase of 1 dS m^{-1} costs USD 525 and USD 534 million a year for the whole country under FB and RB, respectively. Per cubic meter of irrigation water, we achieved USD 0.42 per dS m^{-1}, with minor differences between the blending scenarios being observed.

The following section briefly describes the MYWAS-VALUE model; Section 3 compares the results under the two blending scenarios and discusses the three measures of the economic damage of irrigation water salinity; Section 4 concludes the paper, and Section 5 discusses the limitations of the analysis and avenues for future research.

2. Methods

The Israeli water supply system was designed to cope with challenges associated with temporal and spatial water distribution. Natural freshwater sources are enriched during the winter, whereas most of the consumption occurs in the summer; this pattern requires water storage. The water delivery system was originally designed to transfer water from the rainier northern parts of the country to the populated center and for irrigating the large agricultural lands in the south. Since 2005, with the installation of desalination plants on the Mediterranean coast, the supply has been gradually shifted to a west–east direction. As a public property, water is centrally managed by the government, which designs the supply and controls consumption through a set of prices, quotas, and pumping licenses [36]. These physical and legislational structures imply that the government is facing a water management optimization problem that integrates dynamic and spatial dimensions. The MYWAS-VALUE model was developed to solve such problems.

MYWAS is a dynamic model of the Israeli water system, and VALUE represents the activities in the vegetative agricultural regions as incorporated into MYWAS. MYWAS encompasses 21 urban regions that consume freshwater for domestic and industrial uses and 18 agricultural regions that can consume freshwater, TWW, and brackish water. The water sources are represented in the model by 19 naturally enriched freshwater stocks, 5 seawater desalination plants, 4 non-enriched brackish water aquifers, 19 wastewater treatment plants, 163 freshwater pipelines, and 74 pipelines for sewage, TWW, and brackish water. MYWAS determines the socially optimal allocation of water types to the demand regions during each period throughout a predetermined planning horizon while also accounting for the welfare of the water users in those regions, the variable supply costs, the constraints associated with water availability in the sources, and the infrastructural capacity constraints. In addition, the model determines the extent to which each infrastructural water element is extended during each period while weighing the investment costs versus the net benefits associated with the extended capacities in future periods.

VALUE is a positive mathematical programming (PMP) model of MYWAS's 18 agricultural regions. Each region incorporates 55 crops whose output prices constitute equilibrium in the statewide markets for industrial, export, and local fresh vegetative products that are assumed to be competitive. The crop production functions account for the salinity of the water supplied to each crop. The land allocations to the crops in the regions maximize social welfare subject to regional input constraints, where social welfare incorporates the surpluses of the consumers of agricultural products minus the production costs. The constrained inputs in each region include land, foreign workers (who are allocated to farmers based on cropping patterns), and the amounts of water delivered to the region from accessible sources; the latter is determined by MYWAS.

Population growth shifts the demand for water in urban zones and the country-wide demand for vegetative agricultural products to the right, thereby driving the dynamic expansion of water supply infrastructures. The model tracks the salinity concentrations along the water supply system and can control the salinity of the irrigation water in each agricultural region by increasing desalination capacities and/or changing the shares of allocated water from the different sources accessible to the region. The model reports the efficiency water prices at any node of the water distribution network, which are equal to the shadow price of the water at this node. In addition, it reports the irrigation water's

VMP, which depends on its salinity. Of course, for an optimal allocation, the efficiency water price and the VMP are equal. Based on the efficiency prices, the model reports the allocation of welfare among the urban and agricultural water users, the water suppliers, and the consumers of agricultural outputs (this presumes that prices are the exclusive water allocation instrument in the water economy; in practice, this is the case in Israel, although prices are higher than optimal because of cost recovery regulations [29]). Our scenarios span a 30-year period, which we divide into 10 equal sub-periods to reduce the computational burden; each period is referred to by its last year.

The version of MYWAS-VALUE that is employed in this study was calibrated based on 2019 data (the model is available as a Supplementary Material to this paper). A detailed description of the model is provided by Slater et al. [35]; the rest of this section describes the recalibration of the VALUE model under the FB assumption, which replaces the RB specification based on which the version of the model in Slater et al. [35] is calibrated.

Our challenge is to calibrate VALUE in the absence of field-level information regarding the actual allocation of water types to the crops in each region—an allocation that is assumed to be optimal in the prevailing situation. To that end, we employ a multistage calibration procedure that involves the optimal assignment of the water sources accessed by a region to the crops grown therein while accounting for the crops' relative salinity tolerance and profitability. Specifically, we introduce a preliminary stage to the commonly used two-stage procedure applied for the calibration of classical PMP models [37]. In this preliminary stage, water types are optimally allocated to each crop subject to their respective regional water availability constraints, where the land allocated to each crop is kept constant at its baseline level. Note that our production function for each crop is a nonlinear function that relates the per hectare yield to the per hectare annual water application and the salinity level of the applied water (as in Slater et al. [35], the per hectare annual amount of water applied to each crop is constant, and therefore only changes in the salinity of the water assigned to each crop affect its per hectare yield); thus, changes in the type of water applied to a crop vary its per hectare outputs. Therefore, the preliminary calibration stage optimally assigns the water sources to the crops and sets the production function parameters so as to reproduce the per hectare yield reported in crop budget reports (see Appendix A for a formal description of the preliminary calibration stage). Then, we apply the first stage of the PMP calibration procedure, which elicits the dual values of the perturbed crop-specific land constraints. Note that to obtain the correct dual values, one should also incorporate the water allocations to the crops as decision variables, which renders the optimization problem of that stage nonlinear (in contrast to the first-stage linear problem of the original PMP procedure). The rest of the calibration process follows the second stage of the PMP procedure as well as the calibration of the demand functions for agricultural products and urban water usage (see [35]).

The outcome of the preliminary calibration stage with respect to the optimal allocation of irrigation water types to crops involves minimal blending; that is, each crop is irrigated by only one water type, where mixtures are assigned to a few crops to meet the availability constraints associated with the regional water sources. While this qualitative result was already shown by Kan and Rapaport-Rom [22], here, the water allocation to crops is optimal rather than imposed by other criteria (e.g., Kan and Rapaport-Rom [22] employed a hierarchical procedure to assign water types to crops). Figure 2 presents the allocation of the irrigation water types—desalinated freshwater (EC = 0.25 dS m^{-1}), fresh groundwater (EC = 1 dS m^{-1}), TWW (EC = 1–1.77 dS m^{-1}), and brackish water (EC = 2.35–4.0 dS m^{-1})—to four groups of crops classified according to their sensitivity to salinity: sensitive, moderately sensitive, moderately tolerant, and tolerant [38]; as expected, the higher the salinity tolerance of the crops, the higher the salinity of the irrigation water allocated to them.

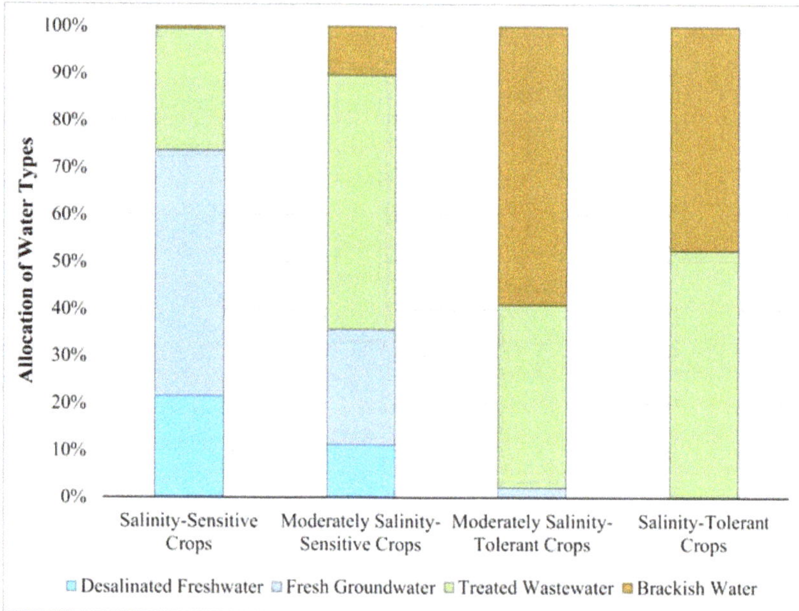

Figure 2. Optimal allocation of the four irrigation water types that differ in terms of their salinity levels to four groups of crops with different salinity tolerances.

3. Results

We used the calibrated MYWAS-VALUE model to evaluate the impact of RB versus FB on the optimal inter-regional water allocation policies, the development of water infrastructures, the agricultural activities of the farmers, and the economic welfare of the country. In the analysis, we refer to the role of adaptation by the farmers through the reallocation of their lands to the crops. We first describe the implications on the water supply patterns associated with the switch from FB to RB. Then, we present the allocation of welfare in the economy under the two scenarios and explain the welfare differences between the two scenarios by analyzing the farmers' adaptations through land and water allocations. Finally, we present three alternative measures to determine irrigation water salinity damage.

3.1. Water Supply

It turns out that the intraregional strategies with respect to irrigation water blending only have a minor impact on the water supply patterns in the country as a whole; the total amount of water supplied to all users varies by less than 1% between the FB and RB scenarios. While minor, these changes correspond to the hypothesis that the salinity impact under the RB assumption is exaggerated: the buildup of seawater desalination capacity under the RB scenario is expedited (Figure 3). Consequently, for a short period of time during the middle of the planning horizon, desalinated water replaced some of the natural freshwater and thereby reduced the salinity of the total supplied freshwater. In addition, the agricultural sector obtains slightly larger amounts of freshwater, whereas the freshwater quantities supplied to the urban sector decrease somewhat (not shown).

3.2. Welfare

The welfare implications of imposing RB instead of FB are summarized in Figure 4, which shows the associated average annual welfare changes (RB minus FB) expected for the various sectors (urban water consumers, consumers of agricultural products, farmers, water suppliers and the overall welfare; the welfare elements represent the annual discounted values, averaged over the planning horizon). In addition, to elicit the impact of agricultural adaptation, we compared the two blending alternatives while assuming that farmers do not adapt to changes over time by reallocating their land to different crops; these scenarios are termed as FBNA and RBNA (NA stands for "no adaptation") in the figure.

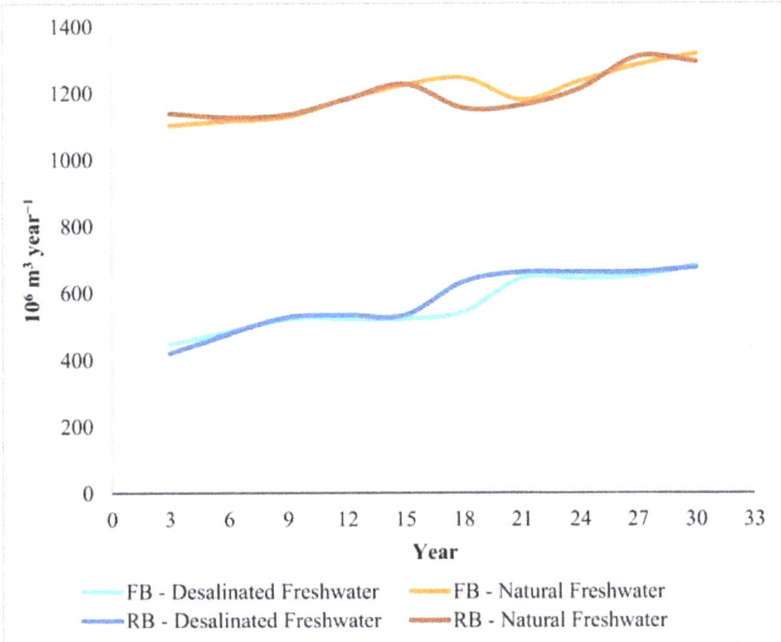

Figure 3. Trajectories of freshwater types supplied from desalination plants and natural sources throughout the simulated 30-year period under the FB and RB scenarios.

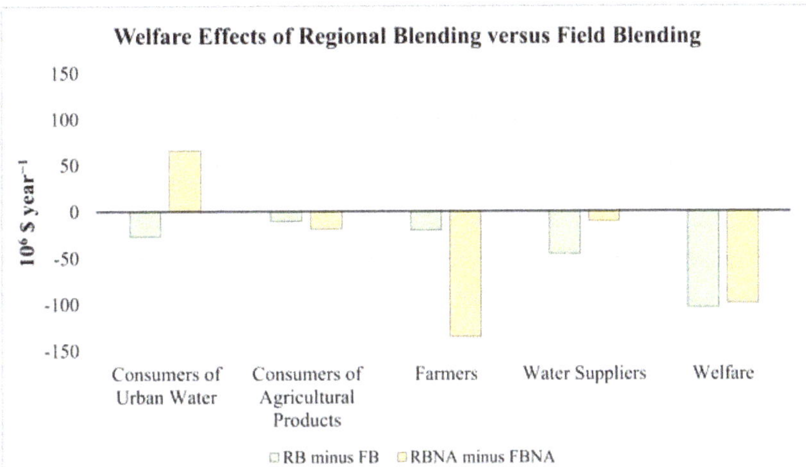

Figure 4. Differences in welfare elements computed under the field blending (FB) and regional blending (RB) scenarios when agricultural land adaptation is allowed (RB minus FB) and not allowed (RBNA minus FBNA).

The overall welfare change is nearly USD 100 million a year—about 5% of the total variable water supply costs. On average, the deadweight loss amounts to USD 0.08 per cubic meter of irrigation water. Most of the burden associated with imposing RB falls on the water suppliers; as it will be shown later, this loss is due to lower efficiency water prices, which stem from the higher salinity of the irrigation water and consequently, its lower VMP. With no adaptation (RBNA minus FBNA), most of the welfare loss caused by RB versus FB is experienced by the farming sector, whereas the consumers of urban water benefit from this situation. Thus, by reallocating agricultural land and irrigation water across crops, farmers manage to reduce their welfare loss by 85% (see Section 3.3).

3.3. Land and Water Management

To understand the welfare changes reported in Figure 4, it is important to study agricultural land and water management decisions with respect to the four salinity tolerance groups presented in Figure 2 as well as the group of rain-fed crops. Figure 5 shows the shares of these five groups in terms of the country's total agricultural land, irrigation water, production value, and profit. While 26% of the land is allocated to salinity-sensitive crops, this group consumes more than 50% of the irrigation water, and accounts for 40% of the total profit. In comparison, the moderately sensitive crops are also responsible for 40% of the profits but consume more land and less water. The other groups of crops produce about 20% of the profits, with relatively little water consumption.

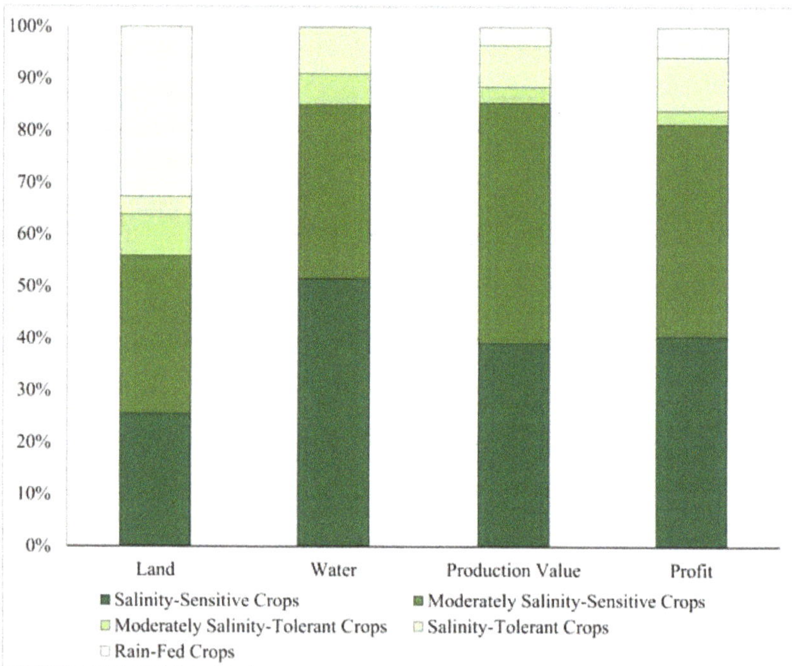

Figure 5. Shares of the groups of salinity-tolerant/sensitive crops and rain-fed crops in the state-wide total agricultural land, irrigation water use, production value, and profit at the calibration stage.

In essence, the shift from FB to RB increases the salinity of the irrigation water for salinity-sensitive crops and reduces that of irrigation water for salinity-tolerant ones. In Figure 6, we report changes (RB minus FB) for a range of measures associated with that shift in relation to the five groups under consideration. Figure 6a shows that the salinity-sensitive crops obtain larger amounts of TWW and less freshwater under RB, whereas all of the other groups face the opposite change. Consequently, the average salinity of the irrigation water applied to the salinity-sensitive crops under RB increases compared to FB, and that of the irrigation water for the other groups declines (Figure 6b); this is because the salinity-sensitive crops consume more than half of the irrigation water (Figure 5), and therefore, on average across all groups, the salinity of their irrigation water increases from 1.11 to 1.16 dS m^{-1}. Figure 6b also shows that the changes in the average VMPs of the irrigation water types are opposite to those of the salinity-sensitive crops (except for the moderately salinity-sensitive crops, in which the change in the average VMP is slightly negative). As previously mentioned, the VMP constitutes the efficiency price of the irrigation water in our model such that lower VMPs imply lower prices; because the salinity-sensitive crops consume most of the water, the average water price declines by 10% (from USD 0.42 to USD 0.38 per cubic meter), and the number of payments delivered to the water suppliers by the agricultural sector decrease. Although urban water consumers face a slight price increase—and therefore their total welfare diminishes (Figure 4)—the overall profit of the water suppliers declines (Figure 4).

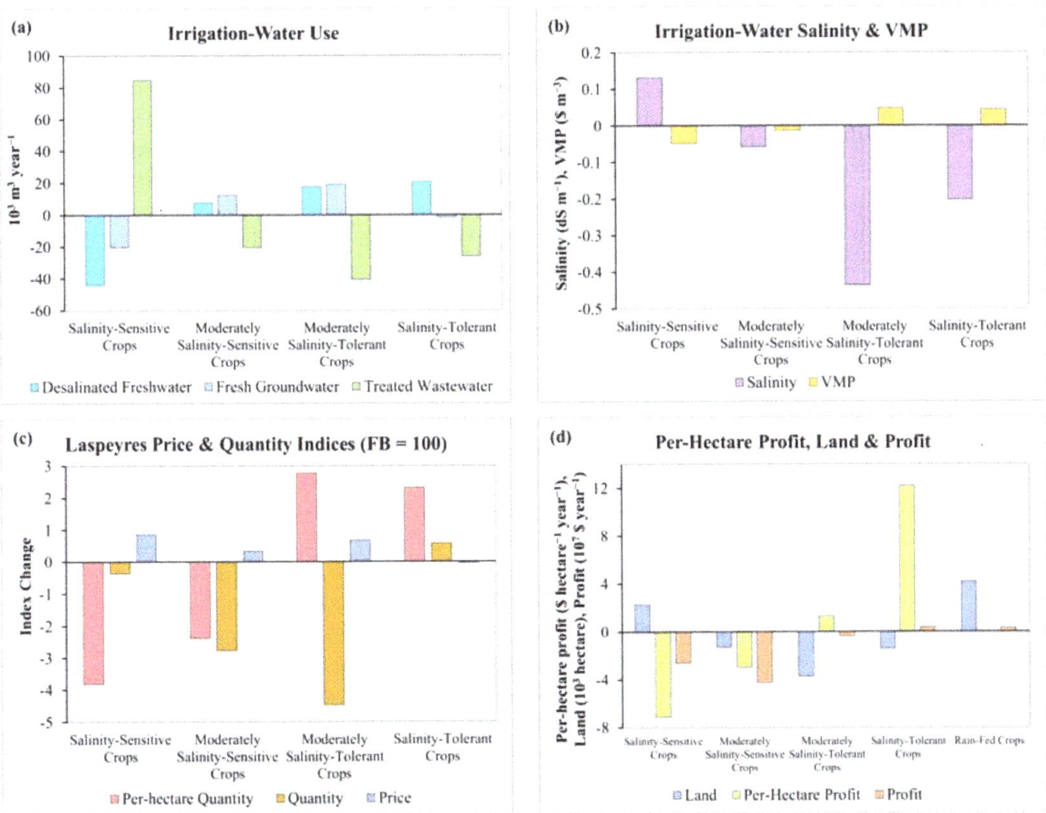

Figure 6. Differences between the RB and FB scenarios (RB minus FB) with respect to (**a**) irrigation water use, (**b**) salinity and VMP of irrigation water, (**c**) changes in Laspeyres quantity and price indices, and (**d**) land allocation, per hectare profit, and total profit—all reported for the groups of crops classified based on their salinity tolerance.

Figure 6c presents changes in the Laspeyres quantity and price indices (FB = 100), and Figure 6d reports the respective changes in land allocation and profits. The per hectare quantity index (computed by holding both the land allocated to the crops and their prices at their values under the FB scenario fixed) of the salinity-sensitive crops exhibits the largest reduction; however, because the land allocated to these crops increases (Figure 6d), the overall quantity of the salinity-sensitive crops' production declines only slightly. In turn, the output prices of these crops increase. Increasing the share of the salinity-sensitive crops in the total agricultural land entails less water for all of the other irrigated crops, and therefore, their share of the land shrinks, and they are replaced by rain-fed crops. Similarly, the combinations of changes in land and the per hectare productivity of the other groups dictate the overall quantity and price changes (Figure 6c), and, in turn, the per hectare profitability and total profit (Figure 6d).

In terms of per hectare profit, even after adaptation through a change in the crop portfolio, the growers of salinity-sensitive crops lose the most from the shift from FB to RB; farmers who grow moderately salinity-sensitive crops show a slight loss, and all other crops benefit. So why is more land allocated to salinity-sensitive crops? We explain this phenomenon using the differences across the crop groups with respect to the relationships between production and output prices, which affect the equilibrium in the markets for agricultural products. On average, the demand elasticity (computed here by dividing the change in the price index by that in the quantity index) of the salinity-sensitive crops is two orders of magnitude larger than that of the other groups; this is because the prices of most of the crops in that group are determined at equilibrium in the agricultural markets for fresh products, which are subjected to import tariffs. Thus, the lower per hectare production of the salinity-sensitive crops will increase the output prices of those crops, thereby increasing their per hectare profitability and motivating farmers to increase their land share; this, in turn, will moderate price changes until equilibrium is reached. As shown in Figure 4, for farmers, the land reallocation benefits amount to USD 115 million a year—about 7% of their profits under the FB scenario.

3.4. Salinity Damage

Here, we discuss three ways to measure irrigation water salinity damage. The first follows Slater et al. [35] and uses the MYWAS-VALUE to evaluate salinity damage in the context of water infrastructure development. In that work, two optimal infrastructural development scenarios were compared: one accounting for changes in irrigation water salinity throughout the planning horizon and the other considering fixed salinity; the difference between the two scenarios reflects the damage associated with salinity when it is ignored when designing water infrastructures. The assessment of that damage in terms of the value of agricultural produce in Slater et al. [35] was USD 1200 per hectare. By repeating the evaluation procedure under the FB and RB scenarios, we obtained per hectare damage of USD 1195 and USD 1326, respectively, i.e., an additional USD 131 per hectare due to RB.

Another way to express the economic damage caused by the salinity of irrigation water is to measure the relationship between salinity and the VMP of the irrigation water. To that end, we use the variability in the water VMPs and salinities across the agricultural regions that were incorporated into MYWAS-VALUE. In panels (a) and (b) of Figure 7, we plot the regional average VMP of the irrigation water versus the respective average salinity levels under the FB and RB scenarios (the data reported in Figure 7 exclude the most southern region, Arava, which is both detached from the country's main water distribution network and is characterized by extremely dry conditions). The regional irrigation water VMPs vary between USD 0.6 per cubic meter to almost zero across regions, with an average of USD 0.34 and USD 0.31 per cubic meter under FB and RB, respectively. Notice the larger variability in the regional average salinities under RB, which stems from the low usage of saline water in some regions with high shares of salinity-sensitive crops. The linear trendlines fitted to the data indicate a clear negative relationship between the water VMP and salinity, with a steeper slope under FB compared to under RB. On average, a salinity increase of 1 dS m^{-1} reduces the water VMP by USD 0.39 and USD 0.30 per cubic meter for the cases of FB and RB, respectively. In terms of elasticity, a 1% increase in the average salinity of the irrigation water supplied to a region reduces the value of marginal product of that water by 2.4% and 1.6% under field and regional blending, respectively (we obtained elasticities by estimating the equation $\ln(VMP_i) = \alpha + \beta \ln(salinity_i) + \varepsilon_i$, in which i is the region index, α is the intercept, and the slope coefficient β represents the elasticity).

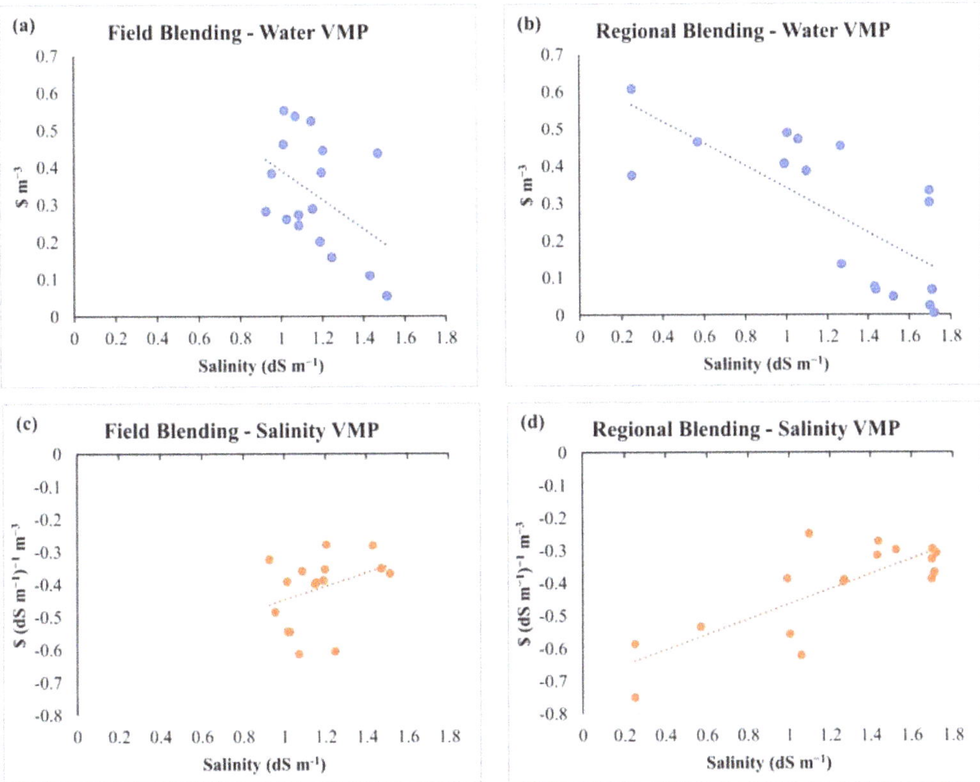

Figure 7. Regional VMPs of irrigation water and its salinity plotted against regional average salinities under the field blending and regional blending scenarios.

Recall that both the regional salinity and VMP of the water are endogenous in the model, and therefore, the curves depicted in Figure 7 represent the socially optimal relations between these measures rather than the marginal impact of salinity on the VMP. Our third measure of salinity damage is the VMP of the salinity itself; that is, the extent to which irrigation water with a higher salinity reduces the value of the agricultural production in a region. The VMP of the salinity is the shadow value of the salt balance constraint, which imposes equality between the amount of salt carried by the irrigation water supplied to a region and the salt content of the irrigation water applied to the crops. We obtained a welfare reduction of USD 525 and USD 534 million a year for a salinity increase of 1 dS m^{-1} under FB and RB, respectively. In panels (c) and (d) of Figure 7, we plotted the regional VMP of salinity divided by the regional amount of irrigation water (the units are USD (dS $m^{-1})^{-1}$ m^{-3}) against the regional average salinity. Per average cubic meter of irrigation water, the VMP of the salinity is similar under both scenarios, amounting to about USD −0.42 per dS m^{-1}. The trendlines fitted to the data indicate that the salinity is characterized by diminishing marginal damage, where a salinity increase of 1 dS m^{-1} reduces the marginal damage by USD 0.21 and USD 0.23 per dS m^{-1} per cubic meter of irrigation water under the FB and RB scenarios, respectively.

4. Conclusions

Based on the context of Israel, this paper shows that the management of irrigation water within the agricultural sector affects the optimal management of water in the water supply sector, and vice versa, and hence, the importance of accounting for the interrelationships between these sectors in the evaluation of economic damage due to irrigation water salinity. We used a mathematical programming model of the Israeli agriculture and water sectors to compare two intraregional irrigation water blending methods: blending at the field level, which enables a specific water salinity to be set for every crop, and regional blending, under which all crops obtain water with the same salinity. We found that enabling field-level blending reduces the land allocated to salinity-sensitive crops and increases welfare by USD 0.08 per cubic meter, which is about 20% of the average VMP for irrigation water. However, blending has been found to be suboptimal; this means that the welfare losses associated with regional blending could be avoided if regions were separated into sub-regions, each assigned to a different water type and a different set of crops. We evaluate the average salinity damage per cubic meter to be in the range of USD 0.30 to USD 0.42 per dS m^{-1} depending on the method employed to evaluate the damage and the irrigation water blending scenario.

5. Discussion

This study focuses on salinity as a single quality measure of irrigation water. However, in water-scarce areas, TWW has become a significant water source that renders salinity but one of many water quality measures. Compared to freshwater irrigation, the reuse of TWW in agricultural applications can harm agricultural production [39], degrade output qualities [40], and threaten the environment [41]. In response, TWW irrigation incentivizes stricter TWW quality standards [42] and attracts the development of new agricultural production technologies [43] and wastewater treatment methods [44]. Moreover, the supply of TWW is more stable than that of natural freshwater [45], and TWW contains nutritional elements that can partially replace fertilizers [46]. These processes have the potential to alter the use of irrigation water sources as well as the damage caused by salinity. To comprehend this, suppose that new regulations impose strict micropollutant standards that can only be met by the desalination of a large fraction of the generated TWW; in this case, the damage caused by salinity would become smaller and less sensitive to salinity changes. This implies that future economic studies of agricultural and water management should account for the interrelations across multiple water-quality measures.

Supplementary Materials: All the data used in this research have been incorporated into the MYWAS-VALUE model, which is available at: https://zenodo.org/record/3702053#.Xx1jpSgzZPZ (accessed on 3 January 2022).

Author Contributions: Conceptualization, Y.S., A.R., I.F. and I.K.; data curation, Y.S. and A.R.; funding acquisition, I.F. and I.K.; investigation, Y.S. and I.K.; methodology, Y.S., A.R., I.F. and I.K.; project administration, I.K.; software, Y.S. and A.R.; writing—original draft, Y.S.; writing—review and editing, A.R., I.F. and I.K. All authors have read and agreed to the published version of the manuscript.

Funding: This study was partially funded by the Israeli Ministry of Agriculture and Rural Development (grant 12-09-0017), the Israeli Water Authority (grant 4501847408), the Israeli Ministry of Energy (grant 222-99-271), and the Center for Agricultural Economics Research at the Hebrew University of Jerusalem.

Data Availability Statement: Not applicable.

Acknowledgments: We thank the Editor and two anonymous reviewers for valuable comments. Thanks go to Yael Kachel from the Israeli Ministry of Agriculture and Rural Development for the data provisions.

Conflicts of Interest: The authors declare no conflict of interest.

Abbreviations

EC	Electrical Conductivity
FB	Field Blending
FBNA	Field Blending No Adaptation
MYWAS	Multi Year Water Allocation System
PMP	Positive Mathematical Programming
RB	Regional Blending
RBNA	Regional Blending No Adaptation
TWW	Treated Wastewater
VALUE	Vegetative Agricultural Land Use Economic
VMP	Value of Marginal Product

Appendix A. Preliminary Calibration Stage

Consider an agricultural region in which farmers grow O crops. The region has access to freshwater, TWW, and brackish water, the regional consumptions of which are limited to the amounts denoted Q^f, Q^h and Q^b, respectively, and their respective prices are p^f, p^h, and p^b, where $p^f > p^h > p^b$. Let q_o^f, q^h, and q_o^b denote the per hectare annual water applications of freshwater, TWW, and brackish water to crop o ($o = 1, \ldots, O$), respectively, where the sets $\mathbf{q}^f = \left(q_1^f, \ldots, q_O^f\right)$, $\mathbf{q}^h = \left(q_1^h, \ldots, q_O^h\right)$, and $\mathbf{q}^b = \left(q_1^b, \ldots, q_O^b\right)$ are defined accordingly. The total per hectare annual application to crop o, w_o, is considered constant. The production function is given by $\theta_o e_o\left(q_o^f, q_o^b, q^h\right)$, in which θ_o is a parameter for calibration, and $e_o(\bullet)$ is the evapotranspiration function of crop o, which is taken from Slater et al. [35]. The salinity of brackish water is higher than that of TWW, the salinity of which is higher than that of freshwater; therefore, $\frac{\partial e_o}{\partial q_o^f} > \frac{\partial e_o}{\partial q_o^h} > \frac{\partial e_o}{\partial q_o^b} > 0$. We denote by x_o, the land allocated to crop o, which is fixed at the preliminary stage. With the above setting, we first solve the nonlinear optimization problem

$$\max_{\mathbf{q}^f, \mathbf{q}^b, \mathbf{q}^h} \pi = \sum_{i=1}^{I} x_o \left[p_o\left(\theta_o e_o\left(q_o^f, q_o^h, q_o^b\right)\right) - p^f q_o^f - p^h q_o^h - p^b q_o^b\right] \tag{A1}$$

s.t.

$$q_o^f + q_o^h + q_o^b = w_o \;\; \forall \;\; o = 1, \ldots, O \tag{A2}$$

$$\sum_{o=1}^{O} x_o q_o^f \leq Q^f \tag{A3}$$

$$\sum_{o=1}^{O} x_o q_o^h \leq Q^h \tag{A4}$$

$$\sum_{o=1}^{O} x_o q_o^b \leq Q^b \tag{A5}$$

$$\mathbf{q}^f, \mathbf{q}^h, \mathbf{q}^b \geq 0 \tag{A6}$$

where the initial values of q_o^f, q_o^h, and q_o^b are set based on the shares of Q^f, Q^h, and Q^b in the total regional water $Q^f + Q^h + Q^b$, and the parameter θ_o is set so as to equate the computed yield to the observed one $\hat{Y}_o$:

$$\theta_o e_o\left(q_o^f, q_o^h, q_o^b\right) = \hat{Y}_o \;\; \forall \;\; o = 1, \ldots, O \tag{A7}$$

The resultant optimal water allocation sets $\mathbf{q}^{f*}$, $\mathbf{q}^{h*}$, and $\mathbf{q}^{b*}$ are then used fto recalibrate θ_o based on Equation (A7).

References

1. Hillel, D. *Salinity Management for Sustainable Irrigation: Integrating Science, Environment, and Economics*; World Bank Publications: Washington, DC, USA, 2000.
2. *Water and Climate Change—The United Nations World Water Development Report*; The United Nations: New York, NY, USA, 2020. Available online: https://www.unwater.org/publications/world-water-development-report-2020/ (accessed on 6 November 2021).
3. Qadir, M.; Quillérou, E.; Nangia, V.; Murtaza, G.; Singh, M.; Thomas, R.J.; Drechsel, P.; Noble, A.D. Economics of salt-induced land degradation and restoration. *Nat. Resour. Forum* **2014**, *38*, 282–295. [CrossRef]
4. Zaman, M.; Shahid, S.A.; Heng, L. Soil salinity: Historical perspectives and a world overview of the problem. In *Guideline for Salinity Assessment, Mitigation and Adaptation Using Nuclear and Related Techniques*; Springer: Cham, Switzerland, 2018; pp. 43–53.
5. Vergine, P.; Salerno, C.; Libutti, A.; Beneduce, L.; Gatta, G.; Berardi, G.; Pollice, A. Closing the water cycle in the agro-industrial sector by reusing treated wastewater for irrigation. *J. Cleaner Prod.* **2017**, *164*, 587–596. [CrossRef]
6. Dai, A.; Zhao, T.; Chen, J. Climate change and drought: A precipitation and evaporation perspective. *Curr. Clim. Change Rep.* **2018**, *4*, 301–312. [CrossRef]
7. Rodell, M.; Famiglietti, J.S.; Wiese, D.N.; Reager, J.T.; Beaudoing, H.K.; Landerer, F.W.; Lo, M.H. Emerging trends in global freshwater availability. *Nature* **2018**, *557*, 651–659. [CrossRef] [PubMed]
8. Russo, D.; Laufer, A.; Bar-Tal, A. Improving water uptake by trees planted on a clayey soil and irrigated with low-quality water by various management means: A numerical study. *Agric. Water Manag.* **2020**, *229*, 105891. [CrossRef]
9. Hopmans, J.W.; Qureshi, A.S.; Kisekka, I.; Munns, R.; Grattan, S.R.; Rengasamy, P.; Ben-Gal, A.; Assouline, S.; Javaux, M.; Minhas, P.S.; et al. Critical knowledge gaps and research priorities in global soil salinity. *Adv. Agron.* **2021**, *169*, 1–191.
10. Darre, N.C.; Toor, G.S. Desalination of water: A review. *Curr. Pollut. Rep.* **2018**, *4*, 104–111. [CrossRef]
11. Jones, E.; Qadir, M.; van Vliet, M.T.; Smakhtin, V.; Kang, S.M. The state of desalination and brine production: A global outlook. *Sci. Tot. Environ.* **2019**, *657*, 1343–1356. [CrossRef]
12. Singh, A. Conjunctive use of water resources for sustainable irrigated agriculture. *J. Hydrol.* **2014**, *519*, 1688–1697. [CrossRef]
13. Gupta, S.K.; Singh, R.B.; Verma, A.K.; Chauhan, C.P.S. Conjunctive use of fresh and saline waters: Comparison of sequential, switching and blending modes. *J. Soil Salin. Water Qual.* **2015**, *7*, 79–89.
14. Peragón, J.M.; Pérez-Latorre, F.J.; Delgado, A.; Tóth, T. Best management irrigation practices assessed by a GIS-based decision tool for reducing salinization risks in olive orchards. *Agric. Water Manag.* **2018**, *202*, 33–41. [CrossRef]
15. Li, J.; Gao, Y.; Zhang, X.; Tian, P.; Li, J.; Tian, Y. Comprehensive comparison of different saline water irrigation strategies for tomato production: Soil properties, plant growth, fruit yield and fruit quality. *Agric. Water Manag.* **2019**, *213*, 521–533. [CrossRef]
16. Parkinson, H.L.; Maletic, J.T.; Wagner, J.P.; Sachs, M.S. Desalting saline ground water for irrigation: A case study, Buckeye Area, Arizona. *Water Resour. Res.* **1970**, *6*, 1496–1500. [CrossRef]
17. Knapp, K.C.; Dinar, A. Reuse of agricultural drainage waters: An economic analysis. *J. Am. Water Resour. Assoc.* **1984**, *20*, 521–525. [CrossRef]
18. Dinar, A.; Letey, J.; Vaux, H.J., Jr. Optimal ratios of saline and nonsaline irrigation waters for crop production. *Soil Sci. Soc. Am. J.* **1986**, *50*, 440–443. [CrossRef]
19. Kan, I.; Schwabe, K.A.; Knapp, K.C. Microeconomics of irrigation with saline water. *J. Agric. Resour. Econ.* **2002**, *27*, 16–39.
20. Kan, I. Yield quality and irrigation with saline water under environmental limitations: The case of processing tomatoes in California. *Agric. Econ.* **2008**, *38*, 57–66. [CrossRef]
21. Feinerman, E.; Yaron, D. Economics of irrigation water mixing within a farm framework. *Water Resour. Res.* **1983**, *19*, 337–345. [CrossRef]
22. Kan, I.; Rapaport-Rom, M. Regional blending of fresh and saline irrigation water: Is it efficient? *Water Resour. Res.* **2012**, *48*, W07517. [CrossRef]
23. Fisher, F.M.; Huber-Lee, A.T. *Liquid Assets: An Economic Approach for Water Management and Conflict Resolution in the Middle East and Beyond*; RFF Press: Washington, DC, USA, 2005.
24. Booker, J.F.; Howitt, R.E.; Michelsen, A.M.; Young, R.A. Economics and the modeling of water resource and policies. *Nat. Resour. Model.* **2012**, *25*, 168–218. [CrossRef]
25. Fisher, F.M.; Huber-Lee, A.T. The value of water: Optimizing models for sustainable management, infrastructure planning, and conflict resolution. *Desalin. Water Treat.* **2011**, *31*, 1–23. [CrossRef]
26. Harou, J.J.; Pulido-Velazquez, M.; Rosenberg, D.E.; Medellín-Azuara, J.; Lund, J.R.; Howitt, R.E. Hydro-economic models: Concepts, design, applications, and future prospects. *J. Hydrol.* **2009**, *375*, 627–643. [CrossRef]
27. Housh, M.; Ostfeld, A.; Shamir, U. Optimal multiyear management of a water supply system under uncertainty: Robust counterpart approach. *Water Resour. Res.* **2011**, *47*, W10515. [CrossRef]
28. Lavee, D.; Ritov, M.; Becker, N. Is desalination the most sustainable alternative for water-shortage mitigation in Israel? *Int. J. Sustain. Econ.* **2011**, *3*, 410–424. [CrossRef]
29. Reznik, A.; Feinerman, E.; Finkelshtain, I.; Kan, I.; Fisher, F.; Huber-Lee, A.; Joyce, B. The cost of covering costs: A nationwide model for water pricing. *Water Econ. Pol.* **2016**, *2*, 1650024. [CrossRef]
30. Reznik, A.; Feinerman, E.; Finkelshtain, I.; Fisher, F.; Huber-Lee, A.; Joyce, B.; Kan, I. Economic implications of agricultural reuse of treated wastewater in Israel: A statewide long-term perspective. *Ecol. Econ.* **2017**, *135*, 222–233. [CrossRef]

31. Saglam, Y. Welfare Implications of Water Shortages: Higher Prices of Desalination. Available at SSRN: https://ssrn.com/abstract=2271865 (accessed on 23 November 2021).
32. Tanaka, S.K.; Zhu, T.J.; Lund, J.R.; Howitt, R.E.; Jenkins, M.W.; Pulido, M.A.; Tauber, M.; Ritzema, R.S.; Ferreira, I.C. Climate warming and water management adaptation for California. *Clim. Change* **2006**, *76*, 361–387. [CrossRef]
33. Howitt, R.E.; MacEwan, D.; Medellín-Azuara, J.; Lund, J.R. *Economic Modeling of Agriculture and Water in California Using the Statewide Agricultural Production Model—Report for the California Department of Water Resources*; University of California: Davis, CA, USA, 2010.
34. Welle, P.D.; Medellín-Azuara, J.; Viers, J.H.; Mauter, M.S. Economic and policy drivers of agricultural water desalination in California's central valley. *Agric. Water Manag.* **2017**, *194*, 192–203. [CrossRef]
35. Slater, Y.; Finkelshtain, I.; Reznik, A.; Kan, I. Large-scale desalination and the external impact on irrigation—Water salinity: Economic analysis for the case of Israel. *Water Resour. Res.* **2020**, *56*, e2019WR025657. [CrossRef]
36. Kislev, Y. *The Water Economy of Israel—Policy Paper No. 2011.15*; Taub Center for Social Policy Studies in Israel: Jerusalem, Israel, 2011. Available online: http://taubcenter.org.il/wp-content/files_mf/thewatereconomyofisrael.pdf (accessed on 23 November 2021).
37. Howitt, R.E. Positive mathematical programming. *Am. J. Agric. Econ.* **1995**, *77*, 329–342. [CrossRef]
38. Maas, E.V.; Hoffman, G.J. Crop salt tolerance—Current assessment. *J. Irrig. Drain. Div. Am. Soc. Civil Eng.* **1977**, *103*, 115–134. [CrossRef]
39. Assouline, S.; Russo, D.; Silber, A.; Or, D. Balancing water scarcity and quality for sustainable irrigated agriculture. *Water Resour. Res.* **2015**, *51*, 3419–3436. [CrossRef]
40. Malchi, T.; Maor, Y.; Tadmor, G.; Shenker, M.; Chefetz, B. Irrigation of root vegetables with treated wastewater: Evaluating uptake of pharmaceuticals and the associated human health risks. *Environ. Sci. Technol.* **2014**, *48*, 9325–9333. [CrossRef] [PubMed]
41. Christou, A.; Agüera, A.; Bayona, J.M.; Cytryn, E.; Fotopoulos, V.; Lambropoulou, D.; Manaia, C.M.; Michael, C.; Revitt, M.; Schröder, P.; et al. The potential implications of reclaimed wastewater reuse for irrigation on the agricultural environment: The knowns and unknowns of the fate of antibiotics and antibiotic resistant bacteria and resistance genes—A review. *Water Res.* **2017**, *123*, 448–467. [CrossRef] [PubMed]
42. Miarov, O.; Tal, A.; Avisar, D. A critical evaluation of comparative regulatory strategies for monitoring pharmaceuticals in recycled wastewater. *J. Environ. Manag.* **2020**, *254*, 109794. [CrossRef] [PubMed]
43. Ogunmokun, F.A.; Wallach, R. Remediating the adverse effects of treated wastewater irrigation by repeated on-surface surfactant application. *Water Resour. Res.* **2021**, *57*, e2020WR029429. [CrossRef]
44. Grandclément, C.; Seyssiecq, I.; Piram, A.; Wong-Wah-Chung, P.; Vanot, G.; Tiliacos, N.; Roche, N.; Doumenq, P. From the conventional biological wastewater treatment to hybrid processes, the evaluation of organic micropollutant removal: A review. *Water Res.* **2017**, *111*, 297–317. [CrossRef]
45. Feinerman, E.; Tsur, Y. Perennial crops under stochastic water supply. *Agric. Econ.* **2014**, *45*, 757–766. [CrossRef]
46. Ofori, S.; Puškáčová, A.; Růžičková, I.; Wanner, J. Treated wastewater reuse for irrigation: Pros and cons. *Sci. Tot. Environ.* **2021**, *760*, 144026. [CrossRef]

water

MDPI

Article

Environmental Decision Support Systems as a Service: Demonstration on CE-QUAL-W2 Model

Yoav Bornstein [1], Ben Dayan [2], Amir Cahn [1], Scott Wells [3] and Mashor Housh [1,*]

1 Department of Natural Resources and Environmental Management, University of Haifa, 2611001 Haifa, Israel; yoavborenst@gmail.com (Y.B.); amir.cahn@gmail.com (A.C.)
2 Independent Researcher, 821735 Hadera, Israel; ben.dayan@gmail.com
3 Department of Civil and Environmental Engineering, Portland State University, Portland, OR 97201, USA; wellss@pdx.edu
* Correspondence: mhoush@univ.haifa.ac.il

Abstract: An environmental decision support system (EDSS) can be used as an important tool for the rehabilitation and preservation of ecosystems. Nonetheless, high assimilation costs (both money and time) are one of the main reasons these tools are not widely adopted in practice. This work presents a low-cost paradigm of *"EDSS as a Service."* This paradigm is demonstrated for developing a water quality EDSS as a service that utilizes the well-known CE-QUAL-W2 model as a kernel for deriving optimized decisions. The paradigm is leveraging new open-source technologies in software development (e.g., Docker, Kubernetes, and Helm) with cloud computing to significantly reduce the assimilation costs of the EDSS for organizations and researchers working on the rehabilitation and preservation of water bodies.

Keywords: CE-QUAL-W2; EDSS; SaaS

Citation: Bornstein, Y.; Dayan, B.; Cahn, A.; Wells, S.; Housh, M. Environmental Decision Support Systems as a Service: Demonstration on CE-QUAL-W2 Model. *Water* **2022**, *14*, 885. https://doi.org/10.3390/w14060885

Academic Editors: Nigel W.T. Quinn, Ariel Dinar, Iddo Kan and Vamsi Krishna Sridharan

Received: 23 January 2022
Accepted: 7 March 2022
Published: 11 March 2022

1. Introduction

The rehabilitation and preservation of ecosystems is an important goal to achieve globally. There are multiple incentives for meeting this goal: preserving biodiversity, mitigating climate change, and assuring future generations can enjoy clean air, land, and water. Water quantity and quality management play an essential part in these conservation efforts. Therefore, environmental agencies usually include in their guidelines the need for stakeholder involvement in the decision-making process [1,2]. A standard tool that can engage stakeholders in decision making is a decision support system (DSS). Specifically, in the case of environmental usage, the tool is often called an environmental decision support system (EDSS). In water-quality-related issues, a water quality model can be used as a kernel in the EDSS that can guide recommendations to stakeholders. The recommendations may include the impacts of changes in flow quantity and water quality, managing the day-to-day operations of hydropower dams (or any water infrastructure) under environmental restrictions, or responding to an unexpected pollution event.

To meet conservation efforts in aquatic environments, agencies often recommend developing EDSSs to guide decision making and engage stakeholders [1]. These EDSSs are part of a holistic system for allocating and managing water to maintain ecosystem functions. Nevertheless, such tools are still not widely adopted in practice. The following factors make an EDSS expensive (in resources and time) to implement: (1) Model assimilation and calibration; (2) Lack of required expertise in the use and interpretation of EDSSs; (3) Software development for the implementation and maintenance of the EDSS; (4) Computer resources needed for the model computation and hosting the EDSS application (the installation of the software on a computer or a server).

This study shows how the last two challenges could be addressed using a low-cost implementation that leverages new open-source technologies in software development (e.g., Docker, Kubernetes, and Helm) and cloud computing. We demonstrate a water quality EDSS that uses the CE-QUAL-W2 model [3] as a kernel to explore the water quality changes resulting from different management decisions.

The choice of using the CE-QUAL-W2 model was based on a simple approach for selecting a water quality model from Mateus [4]. This approach was based on a systematic review of the main available models. The review consisted of the model abilities, dissemination and publications, and the usage experience. In all categories, the CE-QUAL-W2 model was ranked first. CE-QUAL-W2 can simulate the hydrodynamics and water quality of rivers, lakes, reservoirs, and estuaries [3]; thus, when used as a kernel in an EDSS, it allows decision support for multiple types of aquatic environments.

Furthermore, this model is open-source, which means that there is no need to invest money in buying licenses to use the model, and users that are familiar with software programming can add features. Mateus [4] also notes that CE-QUAL-W2 simulations are relatively fast and require low computational power compared to other models. The model input and output are based on text files. The model itself is an executable that can be run without interacting with a graphical user interface (GUI). This model design acts as a simple external application programing interface (API) that allows another software program (e.g., EDSS) to change the inputs, execute the model, and analyze the results. CE-QUAL-W2 was implemented in over 2000 sites in 116 countries [5]. The source code is actively maintained with bug fixes and new features by Portland State University, USA.

As a result, we chose the CE-QUAL-W2 model as the kernel of our EDSS. The main objective of a water quality EDSS is to provide a simple interface for stakeholders to better plan future projects in the wetland or handle the day-to-day operations of the wetland. Using CE-QUAL-W2 for decision making is not new; several attempts have been made over the years to use the model in decision-making contexts.

For example, Eturak [6] used the model as part of an EDSS to understand the impact of the planned "Buyuk Melen" reservoir on its watershed in Turkey. As the reservoir was still in the planning stages, there was no option to calibrate the model. Thus, the model's setup was conducted according to the best knowledge available at the time. Next, a few scenarios with different flow volumes of domestic and industrial wastewater were chosen, and their simulations were executed using the model. Later, the results were compared and graphed for the stakeholders to discuss the implications of different scenarios.

The manual approach of Eturak [6], where a modeler familiar with CE-QUAL-W2 can provide the needed analysis, highlights several drawbacks: (1) Execution of the different scenarios needs continuous involvement of the modeler in the process, including setting up the different inputs, executing model runs, and then comparing the results. Thus, the modeler must be involved in the detailed manual planning of each scenario proposed by the stakeholders. This increases the cost of using the EDSS and makes the discussion/involvement of the stakeholders more difficult; (2) The manual process is time-consuming. For complex models, the stakeholders will need to wait for a report from the modeler for each of their scenario requests. This does not allow for an active discussion in which scenarios are refined rapidly. In practice, it is hard to define the scenarios in advance. Usually, the scenarios are refined during an active discussion between the stakeholders (partly by seeing how the model reacts). As such, an active discussion is critical, and it could be conducted only if the EDSS can be used in real-time without manually performing time-consuming analyses.

Kumar [7] developed a user-friendly web-based EDSS to interact with an existing calibrated CE-QUAL-W2 model in the Occoquan Reservoir in northern Virginia, USA. The purpose was to enhance stakeholders' interaction with the modeling software. The implementation included a multi-part system controlled by the user from a web server. The web server connects to a bridge module that sends the requests for model execution and mines the results. The final part utilizes other computers and servers on the local network to execute the different model runs in parallel. The results are then returned to the user for analysis.

In the experiment of Kumar [7], the downsides of the manual approach in Erturk [6] were addressed. Thus, the modeler was no longer involved in the process, and the parallel execution on multiple computers facilitated fast computation. Still, the work of Kumar [7] had some other downsides: (1) Not all parts of the EDSS were reusable. Furthermore, the code was written specifically for the subject reservoir. Any other user who wants to use this infrastructure will need to change the source code to match their system; (2) On each computer in the network intended to be used for model execution, a piece of supporting software must be installed. This requires information technology (IT) to set up and maintain the software; (3) Part of the implementation uses ArcGIS, which is a licensed program.

Shaw [8] implemented a very different approach for the Cumberland River system. This study described a method for computing hourly power generation schemes for a hydropower reservoir using high-fidelity models, surrogate modeling techniques, and optimization methods. The predictive power of the high-fidelity hydrodynamic and water quality model CE-QUAL-W2 was emulated by an artificial neural network (ANN) then integrated into a genetic algorithm (GA) optimization approach to maximize the hydropower generation subject to constraints on dam operations and water quality. By using the ANN as a surrogate model, Shaw [8] demonstrated a way to address the drawbacks of both Eturak [6] and Kumar [7]. The surrogate model ran within 2 s versus a 6 min runtime in the CE-QUAL-W2 model of the considered system. This allowed running the EDSS on a single six-core computer in a reasonable time frame.

Nevertheless, the surrogate approach still had some drawbacks: (1) There was still a potential and need to implement and train a surrogate model in addition to the CE-QUAL-W2 model. For the training itself, many CE-QUAL-W2 runs were needed. In this case, 729 runs were made; (2) The solution was implemented using MATLAB and its "Neural Network" and "Optimization" toolboxes, which are licensed software.

Given the examples above, there is still a potential to develop a reusable EDSS solely based on open-source tools without investing in expensive computation hardware. This could be achieved by combining the software as a service (SaaS) paradigm with cloud computing technology. According to market analysts, such as the international data corporation (IDC), cloud computing has become more common and accessible over the last decade. They also show a trend of reduced usage costs for the users. They indicate that these factors have made SaaS usage more popular in the last few years. This conclusion is derived from the immense growth in revenues and market share for SaaS in the public cloud [9]. This can be explained by the benefits of this paradigm for the customers and the service providers. Some of these benefits are: (1) Customers do not need to have any computer infrastructure or install software on computers; (2) The company does not need to pay for computers and servers that are not in use and can adopt a "pay as you use" model; (3) The virtually "infinite" parallelization option for on-demand computer power allows the companies to offer efficient solutions for any number of customers, termed scalability; (4) The company eliminates the need for local IT personnel to maintain the computation infrastructure and server rooms; (5) Updates are deployed quickly to all users. Similar to our paradigm, other studies discussed leveraging cloud computing and SaaS for environmental software development. Swain [10] presented an open-source platform for interacting and developing environmental applications. The platform was designed to simplify modern web-based software development over cloud infrastructure for scientists. The study focused on simplifying the development but did not consider computational aspects, such as parallel execution of the models. Ercan [11] developed a cloud-based SaaS to calibrate the Soil and Water Assessment Tool [12]. The proposed solution is based on a single algorithm that utilizes multiple (up to 256) central processing units (CPU) for parallel executions of the model [11]. Recently, Li [13] implemented a Docker-based [14] framework for developing EDSS that can be used on cloud infrastructure [13]. Considering the examples above, they all support the need to simplify environmental software development for computationally intensive applications, making EDSS more accessible to environmental scientists, hydrologists, and stakeholders. This study continues these efforts while focusing on a water quality EDSS that requires intensive computational power. We propose a generic EDSS, which is based on the popular CE-QUAL-W2 model for: (1) Simplifying and reducing the needed amount of software development by requiring only a decision algorithm development instead of a full computational infrastructure; (2) Supporting flexible scaling of computer resources of parallel model runs, which are expected in decision problems that involve water quality simulations. To achieve the above goals, we explore new technologies that simplify and reduce the cost of assimilating a water quality EDSS. Thus, offering a new paradigm of "Water Quality EDSS as a Service", an open-source computationally efficient platform that can support any EDSS algorithm application utilizing the CE-QUAL-W2 model. This paradigm can make these tools more accessible and approachable for use by environmental agencies and organizations, enabling advanced decision making and increasing stakeholder engagement.

The conceptual framework of EDSS as a service is illustrated in Figure 1. The EDSS system disconnects the different levels of complexity between software and algorithm developers and the different end-users. The software developer maintains the computational infrastructure, and, thus, they are responsible for advanced software development (e.g., configuring the computational cluster). In contrast, the algorithm developer interacts only with a higher level of the EDSS using a simple interface; thus, little experience in cloud computing technology is expected from the algorithm developer. Relying on a simple interface, water engineers and/or hydrologists with experience in algorithm development can use the system to distribute heavy computational tasks in the cloud. End-users (e.g., a watershed manager) can define scenarios and initiate runs using a web-based interface without the need to make changes to the EDSS; hence, the arrow shown in Figure 1 does not reach a deep level as in the case of the developers. Lastly, the public stakeholders can view and comment on the published scenarios and results (i.e., they have one-directional arrow in Figure 1).

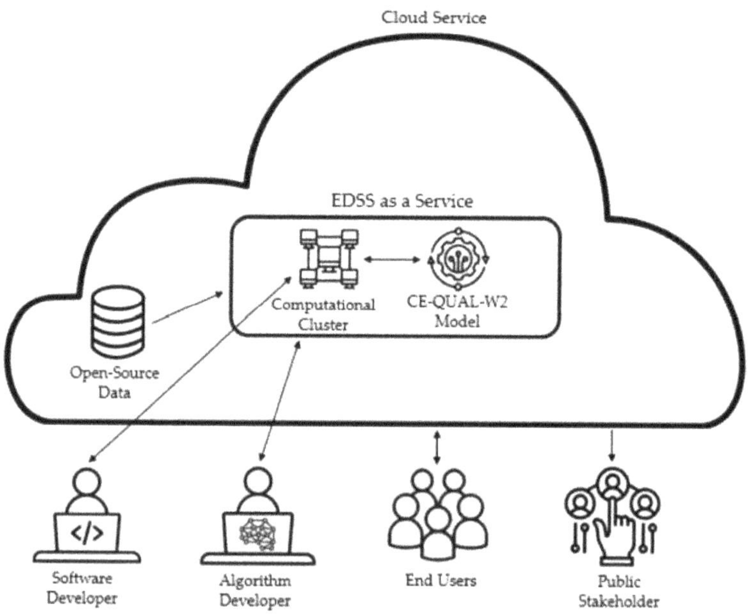

Figure 1. The conceptual framework of EDSS as a service.

2. Methods

The main principles that guide us in designing and constructing the EDSS as a service are: (1) Supplying a low-cost solution and keeping it an open-source service that can be changed and enhanced by future users; (2) Allowing the use of any cloud computing provider, such as AWS, Google, Azure, or Alibaba.

2.1. Solution Architecture

In order to meet the goals above for developing an EDSS as a service, an architectural design based on the latest available technology for SaaS is proposed. Figure 2 shows the encapsulation of the different core layers, followed by a detailed description of each of the core layers.

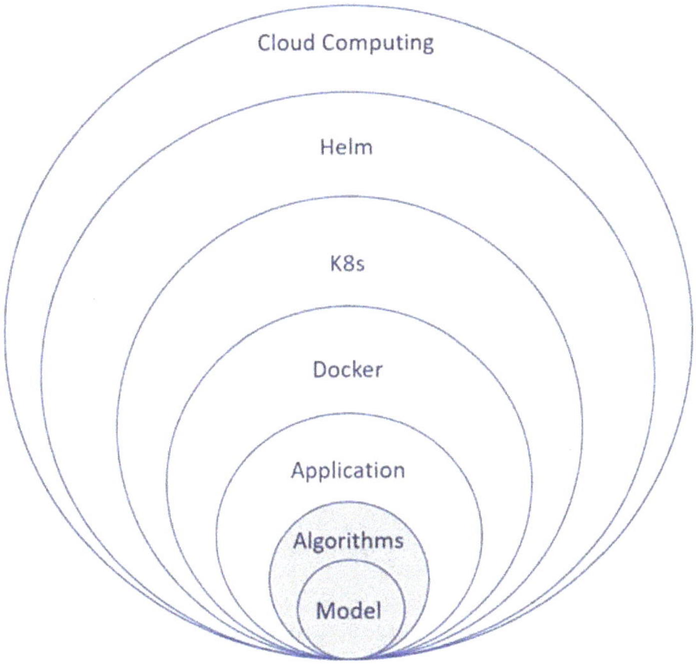

Figure 2. The different layers in the EDSS as a service design. Note: Grey represents layers that will require changes between implementations of the EDSS. White represents the infrastructure layers that do not require changes between different implementations.

Model—The CE-QUAL-W2 model is the heart of the EDSS, as it is used as a kernel for decision making. The model was released only for Windows operating systems in the last few years. Although it is possible to develop this kind of EDSS as a service in a Windows environment [14], we prefer the Linux environment since it is more cloud-environment-compatible and license-free. For that need, we created an open-source GitHub project. This project holds the needed files and instructions to compile the CE-QUAL-W2 source code to be executed in a Linux environment [15]. Besides the model executable, this layer also includes the user-specific input files of a calibrated model ready for simulations. These input files are the template that the application layer changes according to the algorithm requirements.

Algorithm—The layer of the algorithm is responsible for two primary operations: (1) Deciding on the needed permutation for the model simulations and supplying the different parameters needed for the model input files for each of these simulations; (2) Analyzing the results of the simulations according to the developed algorithm. A single EDSS can hold multiple algorithms for the user. For example, an algorithm can conduct a grid search for the best matching simulation output according to the user input targets, while another algorithm can plot a specific model output parameter as recorded in all the different simulations.

Application—This layer has five different responsibilities. The interactions of these responsibilities are described in Figure 3: (1) Provide a simple web-based user interface for the EDSS, allowing users to send requests to the service (Figure 4). An example of a user request is shown later in Section 2.2.1; (2) Pass the user request to the algorithm layer and receive a response from the algorithm to specify the needed simulation permutations list; (3) Initiate parallel model execution requests according to the permutations; (4) Collect all the model simulation results and send them back to the algorithm once the model simulations are completed. (5) Obtain the analyzed results from the algorithm and display them back in the web user interface.

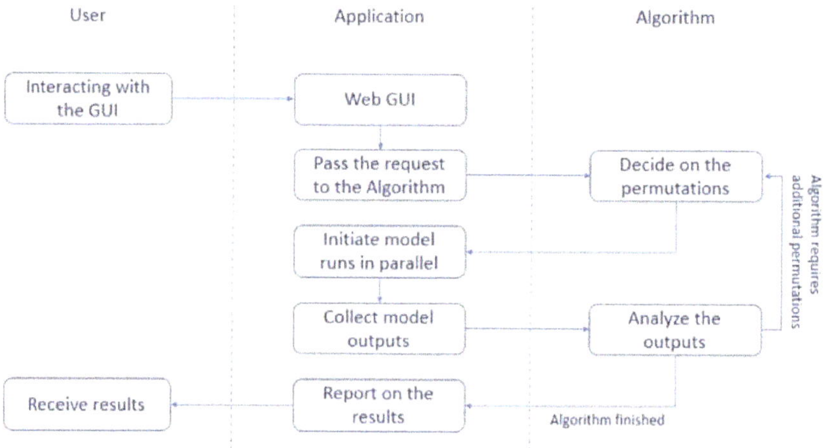

Figure 3. Application layer interactions.

Docker—The Docker container layer was developed to allow the isolation and packaging of the software together with all of its dependencies. A container is an executable that can be run on any computing environment without worrying about the operating system or the hardware infrastructure [14]. The Docker infrastructure is free for use and holds the following benefits for the EDSS implementation: (1) No need to set up the infrastructure or the operating system; (2) It can run on cloud resources as well as on a single computer. The Docker engine can be run on Windows, Linux, or Mac operating systems; (3) Cloud providers supply a cost-effective and straightforward interface for setting up your application using Docker. In this EDSS as a service, a single Docker container was created. This container can be used either for running the application or running individual model simulations in parallel across multiple model executions.

Kubernetes (K8s)—This is a layer developed after organizations started to adopt the Docker solution and were looking for a way to streamline the scaling process and coordinate multiple services encapsulated as Docker containers. Scaling involves initiating more and more Docker containers according to the demand [16]. K8s is also open-source. In this EDSS, K8s is leveraged to manage the runs of multiple Docker containers in parallel, each running a different model simulation. This allows for the automatic scaling of the cloud computing power when needed, allowing the user to benefit from the "pay as you use" cloud computing model while all the simulations are performed in parallel.

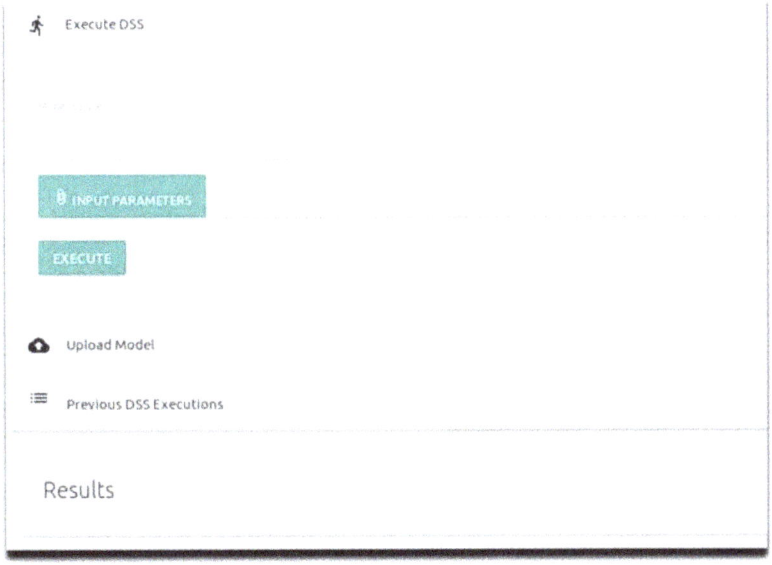

Figure 4. The web-based user interface of the EDSS.

Helm—This is a packaging manager for Kubernetes that was developed to simplify the deployment of K8s applications according to a predefined configuration file [17]. It is free to use and allows the easy and repeatable deployment of this EDSS to the cloud computing environment. Deploying the EDSS to a cloud provider would require many manual configurations without using Helm.

Cloud computing—This final layer allows the EDSS as a service to hold one unit that runs only the user interface during the idle times and scale to multiple computer units when model simulations are needed. To make it generic for any cloud computing provider, the EDSS relies on basic computing building blocks used by all cloud computing providers. However, it could have been easier to design the system for a specific cloud computing provider. Where applicable, current industry standards and best practices were applied to ensure that the APIs and interfaces between the layers will be supported by various third-party tools to simplify the infrastructure layers' deployment, scaling, and management.

2.2. Interfaces

There are two different personas for this EDSS, and dedicated interfaces were created to match their distinct needs. The first interface is for the user who wants to consult the EDSS about a water quality issue. The second persona is the EDSS developer (algorithm developer) who wants to introduce a new algorithm or enhance an existing algorithm.

2.2.1. User Interface

As displayed in Figure 4, the end-user interface is divided into three sections: (1) Initiate the model—The model user sets the model that will be run and the input parameters file for the algorithm. According to this input file, the algorithm determines the needed permutations and analyzes the runs. The input file can be in any format that the algorithm code can read. For the case study, we developed an algorithm that reads JSON format files due to their ease of processing using any computer programming language [18], but CSV or TXT could have been used as well. Once the input file is selected, the user can start the run by pressing the "Execute" button; (2) Upload model—The model user uploads their

CE-QUAL-W2 model input files (not the executable itself) based on their calibration. This allows the user flexibility of usage without needing their CE-QUAL-W2 input files to be part of the source code that is wrapped with the Docker image; (3) Results—These are the results reported by the algorithm once all the model runs are finished and saved to a downloadable zip file. There is also an option to view the results from previous runs, as shown in Figure 4. In the EDSS GitHub repository, we provide a demo video tutorial with instructions on using the EDSS.

2.2.2. Algorithm Developer Interface

An EDSS can use numerous algorithms to support decision making, depending on the issue at hand. This current design of the EDSS as a service allows developing different algorithms and embedding them in the infrastructure for different use cases. The algorithm developer needs basic knowledge in Python programing language, working with Git source code control, and a basic understanding of cloud computer infrastructure to test the code change. The logic for each of the core modules is well isolated, and the linkages between the layers (Application -> Docker -> K8s -> Helm -> Cloud Computer) are clearly defined. The developer does not need to change the infrastructure layers to support a new algorithm.

For developing a new algorithm, the following steps need to be taken:

1. Fork the current GitHub project [19] to a new one;
2. Creating a Travis CI account [20]. The current GitHub project is configured to use Travis CI for running the unit tests and pushing the created Docker images to the Docker repository. The new Travis CI account needs to be configured in the GitHub project;
3. Creating a Docker account [14] to hold the Docker images. All the mentioned accounts (GitHub, Travis CI, and Docker) are free for open-source projects;
4. The "*dss/scripts/build_and_push_to_docker_hub.sh*" script needs to be updated to reference the new Docker repository. The Helm "*dss/chart/wqdss/values.yaml*" file needs to be updated to retrieve the Docker images from the correct repository;
5. As described in Figure A1 (See Appendix A), the entry function of the algorithm is "*execute*" under the "*dss/src/wqdss/processing.py*" Python module in the "*Execution*" class (see EDSS GitHub repository [19]). This is where the code should be changed for a new algorithm. In the case of several algorithms, the input file should include the algorithm name and the applicable function according to the selected algorithm that will be called from the "*execute*" function. In order to request multiple model simulations, the "*execute_run_async*" function needs to be called. Once this function returns, the results can be analyzed. This function can be called several times if needed. In order to publish the results, the class member "*self.result*" should be updated before exiting the "*execute*" function.
6. The changed code then needs to be pushed to the main GitHub branch in order for the Travis CI to trigger the unit tests and publish the Docker image to the Docker repository;
7. Deploying the code using Helm and verifying the EDSS behavior.

2.3. *Cloud Computing Resources Scaling*

When using cloud resources with this EDSS, the scaling is performed in two levels: (1) The horizontal pod autoscaling (HPA), which determines how many different Docker pods (images) are brought up in order to receive a request for model execution. The Docker images are configured to run a single model each. The Helm "values.yaml" file has an "hpa" section with an option to configure the minimum and the maximum number of pods brought up. In addition, the CPU target utilization percentage is also defined in this section. Once the CPU usage crosses this limit, a new pod is created; (2) The cloud provider K8s cluster auto-scaling. Once there are more CPU usage requests than cloud computing provides, the cluster will automatically bring up additional computer resources to run the pods according to the limits defined in the cluster settings.

Multiple factors will affect the scaling efficiency, starting from the chosen type of computer node that brings a different type of CPU generation. The run time of a single model can be cut in half between the newest CPU generation and old CPU generations. As mentioned above, different configurations of the K8s cluster and HPA will also have a significant effect on optimizing the scaling, where optimized scaling is defined as running all the model permutations in parallel. However, having optimized scaling comes with having more cloud computing nodes ready for execution, which results in a higher idle time cost since the user needs to pay for the running computing nodes even if almost no CPU is utilized. Thus, having optimized scaling is not necessarily a primary goal.

Nonetheless, we want to benchmark an optimal scaling performance for reference. The "Spokane river example" was used to demonstrate the scaling. This example is shipped with the CE-QUAL-W2 version 4.1 package. We changed the simulation days from 200-205 to 200-300 to create a longer run and highlight the benefit of the scaling. The Google Cloud provider was chosen to deploy the EDSS. The e2-standard-2 (2 vCPUs and 8GB memory) were used for the computing nodes. To perform ten parallel simulations, both the HPA and K8s cluster configurations were made such that all the needed computing nodes and pods were already up and running at the beginning of the execution. That is ten pods and five computing nodes. This run of ten model permutations took 26 min. Using the same configuration, only changing the HPA to allow a single pod (i.e., ten serial runs without parallelization) took 3 h and 50 min (23 × 10 = 230 min).

3. Case Study

Israel has thirteen different drainage and river authorities, each responsible for streams and rivers in a different part of the country. Among their many duties, these authorities are also responsible for rehabilitating the rivers and streams and adapting them for leisure and recreational purposes [21]. Each authority is independent and relatively small (staffed with less than ten people). One of these authorities is the Yarqon River Authority (YRA), responsible for the Yarqon stream [22]. The Yarqon is a lowland coastal stream, about 28 km long, in central Israel that flows between a mix of agricultural fields and urban areas, ending in an estuary connected to the Mediterranean Sea, as shown in Figure 5. While the historical flow of the stream originated in natural springs, over the years, due to over-exploitation of the aquifer, the natural springs dried out [23]. Today, there are two different water sources for the stream: (1) Pumps supplying water from lower depths in the aquifer for the upstream part instead of the springs that dried out; (2) There are three different Wastewater Treatment Plants (WWTP) which discharge tertiary treated water in two locations in the middle section of the stream [23]. In order to promote preservation and recovery efforts, the YRA works with 18 different stakeholders [22], which poses a significant obstacle to reaching a consensus on rehabilitation efforts. We set out to show how a water quality EDSS as a service can serve as a solution for the YRA in its efforts to rehabilitate the stream.

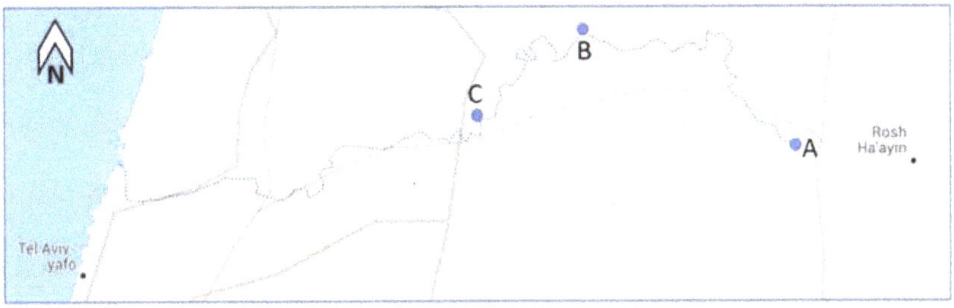

Figure 5. Study area map. (A) Source of the stream, from aquifer water pumps. (B) Discharge points of two WWTP. (C) Discharge point of a third WWTP. Note: background map is adapted from "Michelin maps".

3.1. Applying the CE-QUAL-W2

Water quality models are rarely used in Israel's rivers and streams. Therefore, to show the benefits of the EDSS, a CE-QUAL-W2 model had to be applied for the Yarqon stream. However, the available data on flows, temperature, and water quality in the stream were insufficient to calibrate the model. Therefore, for EDDS demonstration purposes, we applied a non-calibrated model for the stream according to the best data that was available. Although the EDSS cannot supply outputs for decision making under these conditions, we can still demonstrate how the EDSS will assist the YRA if it invests in calibrating a model for the stream.

3.2. The EDSS Algorithm

For demonstration purposes, the example implemented algorithm was a simple recursive grid search, which used two types of inputs as shown in Figure A2 in the Appendix A: (1) *model_run* that specified the parameters needed for the algorithm to set the different run permutations; (2) *model_analysis* that specified the parameters for finding the best run. The JSON format for the input file was chosen over other formats, such as CSV or TXT, due to its flexibility. Under the *model_run* section, there is an option to define any number of model input files that need to be changed in the *input_files* section. For each input file, the user needs to define the following: (1) *name*—Name of the input file; (2) *col_name*—Name of the CSV column that needs to be changed (the parameter that is changed); (3) *min_val*—The minimum value of the parameter that is being changed; (4) *max_val*—The maximum value of the parameter that is being changed; (5) *steps*—The increase interval in the parameter value between the minimum and maximum definitions. This parameter can hold a list of values. If more than one value is defined, a recursive run is conducted in smaller intervals for further rounds of model simulations around the previous result that best matched the target. Although the user can define the smallest interval in the first run and obtain the same results faster, the recursive option was added to allow cloud computing cost reduction by reducing the overall number of permutations without significantly impacting the accuracy of the result. In the example shown in Figure A2, in the first pass of model execution, nine different permutations will be executed as there are (2 − 1) / 0.5 + 1 = (34 − 30) / 2 + 1 = 3 different parameter values to set in each of the 2 input files. In the second pass of model execution, the minimum and maximum range are set from the previous round best-run parameter value +/− the previous step value divided by two. In this case, there are another 0.5 / 0.05 + 1 = 2 / 0.2 + 1 = 11 different parameter values to set in each of the 2 input files.

Under the *model_analysis* section, the model output file that needs to be analyzed is defined in the *output_file* field. Under the section of *parameters*, any number of parameters can be defined. Each parameter has the following definitions: (1) *name*—Column name in the output csv file; (2) *target*—The target value of the parameter we want to reach; (3) *weight*—This is a relative parameter that allows setting a priority between the different defined parameters; (4) *score_step*—This is defined in order to unify the units of different parameters. It is defined as the deviation from the target per unit score. For example, in Figure A2, a deviation of 0.1 g/m^3 in total nitrogen (TN) is considered 1 score, while a deviation of 0.5 g/m^3 in dissolved oxygen (DO) is considered 1 score. As such, dividing the deviations by the corresponding *score_step* will facilitate summing the deviations of different parameters in score units, as shown in Equation (1). Equation (1) defines what score each model simulation will receive according to the distance from the target.

$$Score = \sum_{i=1}^{n} \left(\frac{|\ target_{\ i} - actual\ value\ |\ /\ score\ step}{weight} \right) \quad (1)$$

Since the score represents deviations from targets, we seek the model simulation that minimizes the score. For example, for the run in Figure A2, we seek the model simulation that minimizes the score defined in Equation (2).

$$Score = \left(\frac{|\ 0.6 - Simulated\ TN\ |\ /\ 0.1}{4} \right) + \left(\frac{|\ 11 - Simulated\ DO\ |\ /\ 0.6}{2} \right) \quad (2)$$

3.3. Case Study Results

In order to demonstrate the abilities of the EDSS, we choose a decision-making problem in which the decision maker needs to decide on the pump flow rate that supplies the water from the aquifer in the most upstream section (point A in Figure 5) to achieve the desired goal of water quality. More specifically, the algorithm needs to quantify the impact of changes in the pump flow rate on the downstream section water quality, given the additional inputs from WWTPs in the middle of the stream and the natural processes occurring along the stream's 28 km path (Figure 5). This decision-making problem represents a constant debate among stakeholders on the amount of freshwater allocated to the stream to meet water quality targets.

A scenario of excess ammonia (NH4-N) concentration downstream was simulated. As shown in Figure A3 in Appendix A, the EDSS input file defines a possible range of pump flows (*col_name* = q) between 0.2 m^3/s (*min_val*) and 0.8 m^3/s (*max_val*). The flow changes as defined in the steps were set for two recursive runs, one with 0.1 m^3/sec and the second with 0.01 m^3/s. The "qin_br1.csv" (*name*) is the input file that needs to be changed. The NH4-N concentration target of 0.57 g/m^3 (*target*) was set. The weight is not relevant in this case, as only a single target parameter was used. The simulated NH4-N concentration is extracted from the output file of "tsr_1_seg42.csv" (*name*). The defined *score_step* was set to 0.01 g/m^3. The pump flow that minimizes the score was found by the EDSS as shown in Figure A4 in Appendix A: (1) id—Each EDSS execution has a unique string; (2) status—This field is changed from "RUNNING" to "COMPLETED" once the EDSS publishes the results; (3) result—This shows the results of the best run. In this case, as two recursive runs were defined, we can see two sets of results with different identification strings (*best_run*). The first is for the search between 0.2 m^3/s and 0.8 m^3/s with 0.1 m^3/s interval, in which a flow of 0.6 m^3/s (*params*) had the best relative score of 0.525 out of the seven runs. The second results are for the run between 0.55 m^3/sec and 0.65 m^3/s with a 0.01 m^3/s interval. In this second search, the best score was 0.025, corresponding to a flow of 0.55 m^3/s; (4) A link to download the zip output files from the model executions is available, as shown in Figure A4.

In this example, seven parallel simulations were conducted on cloud computing resources in the first round. In the second round, eleven simulations were conducted in parallel. Altogether, eighteen simulations were conducted in order to reach the result. A more cloud-computing-expensive path could have been taken if a "step" of 0.01 m^3/s was solely defined with no recursive rounds. In that case, 61 simulations would have been conducted, resulting in a shorter run time (in case these runs are performed in parallel), but with more cloud computing charges.

4. Benefits from EDSS as a Service

The developed paradigm of a "Water Quality EDSS as a Service" holds multiple benefits for environmental organizations working on preserving and recovering aquatic ecosystems. This specific study implementation is beneficial to organizations or researchers with a calibrated CE-QUAL-W2 model for the water bodies they manage since this remains the highest cost in the EDSS. We can divide the direct users' benefits into two categories: (1) Benefits to the end-user who consults the EDSS; (2) Benefits to the developer looking to introduce a new algorithm to the user.

4.1. User Benefits

1. Ease of use—The simple web interface allows interaction with the system without the need to understand the format of the CE-QUAL-W2 input files. There is also no need to install any software on the user's computer.
2. Reduce simulation time—As the different model simulations are in parallel, the user can obtain the result within the time frame of a single model simulation.
3. Extensibility—A software developer can embed any algorithm in the service.
4. Low cost—All the software used in this solution is open source. The only needed payments are for the "pay as you use cloud computer" and for a software developer to adjust the needed algorithm to the organization's needs and then bring up the service for use. Next, we review the developer benefits, which also help maintain a low development cost. An additional aspect is that the developed infrastructure is generic and can support any cloud computing provider (such as AWS, Google, Azure, or Alibaba), where some environmental organizations or researchers can obtain grants for usage from one of the providers. Additional cost savings can also be achieved if a prominent environmental organization maintains the service and algorithms for smaller organizations. For example, in the YRA case, if the Israeli water authority takes responsibility for developing the algorithms and maintaining the service, all thirteen drainage authorities will benefit from it without each of the thirteen developing their algorithms.

4.2. Developer Benefits

1. Ease of new algorithm implementation—The architecture of the different software layers described in Figure 2 might look overwhelming. However, it isolates the different core components from the infrastructure layers. Therefore, the algorithm developer would only need to understand the interfaces between the application and the algorithm, be familiar with the Python programming language, and have a basic knowledge of deploying the service to the cloud provider.
2. Flexible user interface—The flexible design for the user input does not require having any web development knowledge. It also allows the creation of an input file that is clear and easily produced by the user.

In addition to the direct user's benefits, the "EDSS as a Service" paradigm also enabled the usage of legacy software in a modern service. Specifically, the CE-QUAL-W2 model has been developed, fine-tuned, and stabilized over the past 40 years. As an open-source project, it will be hard to finance the model redesign into a modern software design. This "EDSS as a Service" allows users to leverage the benefits of a modern web service along with its kernel of a well-established model. The developed infrastructure can also be used for fine-tuning the model calibration. Today, the modeler calibrating the CE-QUAL-W2 model needs to run the model multiple times while changing different tuning parameters. This is conducted to find the best match to the historic observed results in the field. This process is tedious and time-consuming, especially for long-running simulations. An algorithm similar to the grid search algorithm implemented for the YRA can be utilized to perform a parallel search for the best tuning parameters that maximize the goodness of fit between the observed and simulated results.

5. Broader Policy Implications

Water quality management is inherently complex since it involves the interplay of physical, chemical, hydrological, and biological parameters influenced by environmental, socio-economic, and technical factors [24]. In recent decades, this has been compounded by stricter legislation and increased public awareness around the world. As mentioned earlier, EDSS as a service can assist policymakers in understanding these complexities by enabling decision makers to gather, share, and interpret results for different planning and operation scenarios on demand without paying for the necessary license, data storage, or technical support. There are three main policy implications, which are highlighted below.

5.1. Adaptive Management

By combining current water quality data with an EDSS, "what-if" scenario testing can better inform management and policy decisions. However, it is often difficult for decision makers to readily access model information and use models directly to evaluate alternative scenarios [25]. This makes the use of predictive models cumbersome and costly, reinforcing the need for EDSS as a service as a mainstream tool to assist deliberation. Such an adaptive management approach can enable watershed managers to build a flexible, accurate understanding of their system and refine and update management strategies [26].

5.2. Community Engagement

There is a growing recognition that citizens should be actively involved as key stakeholders in the watershed planning process. By engaging communities, policymakers can gain local knowledge, increase fairness and equity in decision making, and avoid failure due to non-acceptance [7]. An effective EDSS can allow citizens to take a participatory approach and highlight the various advantages and disadvantages of different scenarios [27].

5.3. Long-Term Resilience Planning

Growing urban populations, limited budgets, and the impact of climate change are increasingly putting a strain on watershed management. Even a slight increase in temperature can lead to general changes in the microbiological processes in lakes, rivers, and streams [28]. Therefore, resilience planning is essential to cope with and recover from disruption and anticipate trends and variability. For example, when a factory requests permission to dump water that was part of the manufacturing process into a stream, an EDSS can be used to understand the impact on the stream. This can be conducted by plotting the water quality parameters under different flow and quality outputs from the factory as predicted by the model in different points downstream. After such a plot is created, the treatment level of the disposed water can be determined for the factory. In another application using the same example, a government agency could find thresholds related to the stream condition (e.g., flow, temperature, nutrients, etc.), where dumping of water in a certain quality from the factory will not be allowed or the amount will be limited below the regular output. In conclusion, EDSS as a service enables local communities and national regulators alike to take a holistic approach to planning for the future.

Author Contributions: Conceptualization, Y.B. and M.H.; methodology, Y.B.; software, B.D.; validation, Y.B., A.C. and S.W.; investigation, Y.B.; data curation, Y.B.; writing—original draft preparation, Y.B. and A.C.; writing—review and editing, M.H., S.W., A.C. and B.D.; visualization, Y.B.; supervision, M.H.; project administration, Y.B. All authors have read and agreed to the published version of the manuscript.

Funding: This research received no external funding.

Data Availability Statement: EDSS GitHub repository: https://github.com/WQDSS/Evaluna (accessed on 22 May 2021) CE-QUAL-W2 for Linux environment compilation, GitHub repository—https://github.com/WQDSS/CE-QUAL-W2-Linux (accessed on 22 May 2021).

Acknowledgments: This work was supported by the Yarqon River Authority.

Conflicts of Interest: The authors declare no conflict of interest.

Appendix A

```
async def execute(self, params):
  # Start your algorithm here.
  # "params" is the file passed by the user.
  # It can be further processed or changed before passed to "excute_run_async" function.
  # This is an example where "permutations" and "iterations" are set by the algorithm.
  awaitables = [self.excute_run_async(self.model_name, params, p, iterations) for p in permutations
  # The requests for model excution are passed, and now we await for the results.
  await asyncio.gather(*awaitables)
  # "self.runs" is updated with the results and can now be processed.
  # The processing is either for the final results or for another execution round.
  # Write your processing code here.
  self.results.append({"The info you want to return to the user"})
  # Once the function is exited, the info is returned to the user.
```

Figure A1. Code snippet for creating a new algorithm.


```
{
 "model_run":
  {
   "input_files":
   [
    {
    "name": "hangq01.csv",
    "col_name": "Q",
    "min_val": "1",
    "max_val": "2",
    "steps": ["0.5", "0.05"]
    },
    {
    "name": "qin_br8.csv",
    "col_name": "QWD",
    "min_val": "30",
    "max_val": "34",
    "steps": ["2", "0.2"]
    }
   ]
  },
 "model_analysis" :
  {
   "output_file": "tsr_2_seg7.csv",
   "parameters":
   [
    {
     "name": "TN",
     "target": "0.6",
     "weigth": "4",
     "score_step": "0.1"
    },
    {
     "name": "DO",
     "target": "11",
     "weigth": "2",
     "score_step": "0.5"
    }
   ]
  }
}
```


Figure A2. Example input file. Note: Two different model files are changed for the permutations, and a single recursive step is defined. The algorithm analysis considers two output parameters with different "weight" and "score_step" values.

```
{
 "model_run":
  {
   "input_files":
   [
    {
    "name": "qin_br1.csv",
    "col_name": "q",
    "min_val": "0.2",
    "max_val": "0.8",
    "steps": ["0.1", "0.01"]
    }
   ]
  },
  "model_analysis":
  {
   "output_file": "tsr_1_seg42.csv",
   "parameters":
   [
    {
     "name": "NH4",
     "target": "0.57",
     "weigth": "4",
     "score_step": "0.01"
    }
   ]
  }
}
```


Figure A3. Case study input file.


```
{
 "id": "a146e491-91cb-4ee6-b942-7cd998d41a90",
 "status": "COMPLETED",
 "results":
  [
   {
    "base_run": "f308f1a4-2062-4213-b9d2-b5e0d5f4a5c7",
    "params":
     {
     "qin_br1.csv": 0.6000000000000001
     },
     "score": 0.5249999999999977
   },
   {
    "base_run": "3f388682-e699-46ee-b3bb-a98f39e5c73d",
    "params":
     {
     "qin_br1.csv": 0.55
     },
     "score": 0.0250000000000022
   }
  ],
  "link": "best_run/a146e491-91cb-4ee6-b942-7cd998d41a90"
}
```


Figure A4. The output of the EDSS. Note: In the first round of simulations (i.e., steps of 0.1 m^3/s), 0.6 m^3/s flow had the best score. In the second round of simulations (i.e., steps of 0.01 m^3/s around 0.6 m^3/s), 0.55 m^3/s flow had the best score.

References

1. Ramsar Convention Secretariat. *Water Allocation and Management: Guidelines for the Allocation and Management of Water for Maintaining the Ecological Functions of Wetlands*, 4th ed.; Ramsar Convention Secretariat: Gland, Switzerland, 2010.
2. U.S. Environmental Protection Agency. 2001. Stakeholder Involvement & Public Participation at the U.S. EPA: Lessons Learned, Barriers, & Innovative Approaches. Available online: https://www.epa.gov/sites/production/files/2015-09/documents/stakeholder-involvement-public-participation-at-epa.pdf (accessed on 22 May 2021).
3. Cole, T.M.; Wells, S.A. *CE-QUAL-W2: A Two-Dimensional, Laterally Averaged, Hydrodynamic and Water Quality Model, Version 4.1*; Portland State University: Portland, OR, USA, 2003.
4. Mateus, M.; da Silva, R.; Almeida, C.; Silva, M.; Reis, F. ScoRE—A Simple Approach to Select a Water Quality Model. *Water* **2018**, *10*, 1811. [CrossRef]
5. CE_QUAL-W2 Application by Country. Available online: http://cee.pdx.edu/w2/ (accessed on 22 May 2021).
6. Erturk, A.; Ekdal, A.; Gurel, M.; Zorlutuna, Y.; Tavsan, C.; Seker, D.; Tanik, A.; Ozturk, I. Application of Water Quality Modelling as a Decision Support System Tool for Planned Buyuk Melen Reservoir and Its Watershed. Sustainable Use and Development of Watersheds. *Sustain. Use Dev. Watersheds* **2008**, 227–242. [CrossRef]
7. Kumar, S.; Godrej, A.; Grizzard, T. A web-based environmental decision support system for legacy models. *J. Hydroinform.* **2015**, *17*, 874–890. [CrossRef]
8. Shaw, A.; Sawyer, H.; LeBoeuf, E.; McDonald, M.; Hadjerioua, B. Hydropower Optimization Using Artificial Neural Network Surrogate Models of a High-Fidelity Hydrodynamics and Water Quality Model. *Water Resour. Res.* **2017**, *53*, 9444–9461. [CrossRef]
9. Martins, R.; Oliveira, T.; Thomas, M. An empirical analysis to assess the determinants of SaaS diffusion in firms. *Comput. Hum. Behav.* **2016**, *62*, 19–33. [CrossRef]
10. Swain, N.; Christensen, S.; Snow, A.; Dolder, H.; Espinoza-Dávalos, G.; Goharian, E.; Jones, N.; Nelson, J.; Ames, D.; Burian, S. A new open source platform for lowering the barrier for environmental web app development. *Environ. Model. Softw.* **2016**, *85*, 11–26. [CrossRef]
11. Ercan, M.; Goodall, J.; Castronova, A.; Humphrey, M.; Beekwilder, N. Calibration of SWAT models using the cloud. *Environ. Model. Softw.* **2014**, *62*, 188–196. [CrossRef]
12. Gassman, P.W.; Reyes, M.R.; Green, C.H.; Arnold, J.G. The Soil and Water Assessment Tool: Historical development, applications, and future research directions. *Trans. ASABE* **2007**, *50*, 1211–1240. [CrossRef]
13. Li, Y. Towards fast prototyping of cloud-based environmental decision support systems for environmental scientists using R Shiny and Docker. *Environ. Model. Softw.* **2020**, *132*, 104797. [CrossRef]
14. Docker Windows Containers. Available online: https://www.docker.com/products/windows-containers (accessed on 22 May 2021).
15. CE-QUAL-W2-Linux, Port of the CE-QUAL-W2 Model to Linux. Available online: https://github.com/WQDSS/CE-QUAL-W2-Linux (accessed on 22 May 2021).
16. Kubernetes. Available online: https://kubernetes.io/ (accessed on 22 May 2021).
17. Helm. Available online: https://helm.sh/ (accessed on 22 May 2021).
18. JSON. Available online: https://www.json.org/json-en.html (accessed on 22 May 2021).
19. EDSS. Available online: https://github.com/WQDSS/Evaluna (accessed on 22 May 2021).
20. TravisCI. Available online: https://travis-ci.org/ (accessed on 22 May 2021).
21. Israel Drainage Authorities (Hebrew). Retrieved from Kishon Drainage Authority. Available online: https://www.rnkishon.co.il (accessed on 22 May 2021).
22. Yarqon River Authority. Available online: https://www.yarqon.org.il/en/ (accessed on 22 May 2021).
23. Arnon, S.; Avni, N.; Gafny, S. Nutrient uptake and macroinvertebrate community structure in a highly regulated Mediterranean stream receiving treated wastewater. *Aquat. Sci.* **2015**, *77*, 623–637. [CrossRef]
24. Wang, G.; Mang, S.; Cai, H.; Liu, S.; Zhang, Z.; Wang, L.; Innes, J.L. Integrated watershed management: Evolution, development and emerging trends. *J. For. Res.* **2016**, *27*, 967–994. [CrossRef]
25. Uusitalo, L.; Lehikoinen, A.; Helle, I.; Myrberg, K. An overview of methods to evaluate uncertainty of deterministic models in decision support. *Environ. Model. Softw.* **2015**, *63*, 24–31. [CrossRef]
26. Li, Y.; Degener, J.; Gaudreau, M.; Li, Y.; Kappas, M. Adaptive capacity based water quality resilience transformation and policy implications in rapidly urbanizing landscapes. *Sci. Total Environ.* **2016**, *569*, 168–178. [CrossRef] [PubMed]
27. Hewitt, R.J.; Macleod, C.J. What do users really need? Participatory development of decision support tools for environmental management based on outcomes. *Environments* **2017**, *4*, 88. [CrossRef]
28. Bhateria, R.; Jain, D. Water quality assessment of lake water: A review. *Sustain. Water Resour. Manag.* **2016**, *2*, 161–173. [CrossRef]

water

MDPI

Article

A Hybrid Model for Water Quality Prediction Based on an Artificial Neural Network, Wavelet Transform, and Long Short-Term Memory

Junhao Wu [1] and Zhaocai Wang [2,*]

[1] College of Economics and Management, Shanghai Ocean University, Shanghai 201306, China; a2735262574@163.com
[2] College of Information, Shanghai Ocean University, Shanghai 201306, China
* Correspondence: zcwang1028@163.com

Abstract: Clean water is an indispensable essential resource on which humans and other living beings depend. Therefore, the establishment of a water quality prediction model to predict future water quality conditions has a significant social and economic value. In this study, a model based on an artificial neural network (ANN), discrete wavelet transform (DWT), and long short-term memory (LSTM) was constructed to predict the water quality of the Jinjiang River. Firstly, a multi-layer perceptron neural network was used to process the missing values based on the time series in the water quality dataset used in this research. Secondly, the Daubechies 5 (Db5) wavelet was used to divide the water quality data into low-frequency signals and high-frequency signals. Then, the signals were used as the input of LSTM, and LSTM was used for training, testing, and prediction. Finally, the prediction results were compared with the nonlinear auto regression (NAR) neural network model, the ANN-LSTM model, the ARIMA model, multi-layer perceptron neural networks, the LSTM model, and the CNN-LSTM model. The outcome indicated that the ANN-WT-LSTM model proposed in this study performed better than previous models in many evaluation indices. Therefore, the research methods of this study can provide technical support and practical reference for water quality monitoring and the management of the Jinjiang River and other basins.

Keywords: water quality forecast; MLP neural network; Daubechies 5 wavelet; long short-term memory network; hybrid model; decomposition-and-ensemble

Citation: Wu, J.; Wang, Z. A Hybrid Model for Water Quality Prediction Based on an Artificial Neural Network, Wavelet Transform, and Long Short-Term Memory. *Water* **2022**, *14*, 610. https://doi.org/10.3390/w14040610

Academic Editors: Nigel W. T. Quinn, Ariel Dinar, Iddo Kan and Vamsi Krishna Sridharan

Received: 7 December 2021
Accepted: 10 February 2022
Published: 17 February 2022

1. Introduction

Water is one of the most essential natural resources on which all life depends. However, various economic activities have an indispensable impact on the environment through different pathways [1]. Take China as an example: in recent years, along with high-speed economic development and urbanization, China's limited freshwater resources have been drastically reduced and, at the same time, increasing water pollution poses a serious threat to human survival and security and has become a significant obstacle to human health and sustainable socio-economic development. From the perspective of China's actual national conditions, water resources are relatively scarce. In addition, as China is undergoing a period of rapid socio-economic development, the demand for water resources is accelerating. Although China has 2.8 trillion water resources [2], which seems to be very rich, the per capita share of water resources is only 2400 cubic meters due to its large population [3], and account for less than one-quarter of the world's total per capita water resources. In addition, the discharge of industrial wastewater and domestic sewage into water bodies without treatment has led to the severe pollution of various water bodies, including rivers and lakes, thus seriously damaging the ecological environment, biodiversity, and the ecological and service functions of water bodies [4]. According to previous studies, only a small number of rivers worldwide are not affected by water pollution [5]. At present, the pollution and eutrophication of rivers in China are severe. According to the 2019 statistics from China's State Environmental Protection Administration, the seven major water systems in China in descending order of pollution level are listed as follows: the Liaohe River basin, the Haihe River basin, the Huaihe River basin, the Yellow River basin, the Songhua River basin, the Pearl River basin, and the Yangtze River basin, with more than 70% of the Liaohe, Haihe, Huaihe, and Yellow River basins being polluted. Huang et al. [6] conducted an analysis of water quality data from 2424 water quality observation stations in China from 2003–2018 and concluded that the quality of river water in China showed significant spatial differences, with 17.2% of sampling sites in eastern China showing poor water quality during the period of 2016–2018, compared to 4.6% in the western region. Moreover, 24.4% of the sampling sites in coastal areas (buffer zone of 20 km from the coastline) showed poor water quality. Although the Chinese government has invested a great deal of money into the treatment and management of polluted water bodies, the pollution proportion of water resources is still quite impressive, which has brought severe economic and social costs to China's water environment remediation [7]. Water quality prediction is a necessary tool for water environment planning, management, and control; an important element of water pollution research; and a fundamental part of water environmental protection and management. Thus, it is vital to find a reasonable and effective water quality prediction method. At the same time, predicting future water quality is a prerequisite for preventing rapid changes in water quality and proposing countermeasures. Therefore, the accurate prediction of water quality changes can not only effectively ensure the safety of people's drinking water, but can also have a positive impact on guiding fishery production and protecting biodiversity.

Research into water quality prediction dates back to the 1920s. Streeter and Phelps developed a coupled model based on biochemical oxygen demand and dissolved oxygen when they studied pollution sources in the Ohio River. They proposed a one-dimensional steady-state oxygen balance water quality model (the S-P model). Since then, many scholars have supplemented and revised their theories [8–10]. At present, the research methods of water quality prediction are mainly divided into two categories: one is to use theoretical mathematical model and physical model to predict the development trend of water quality mechanism [11], the other is a non-mechanistic prediction method that builds mathematical statistical prediction models based on historical data. The mechanistic prediction method analyses the physical, chemical, and biological changes of each factor in the water resource cycle; establishes a mathematical model reflecting the relationship between the substances; and solves the corresponding mathematical equations to predict the trend of water quality changes. For example, Zhang et al. incorporated the operation rules of dams or sluices into the reservoir regulation module, used an improved SWAT model to simulate the water quantity and quality in the Huaihe River basin, and compared the results with those of the original SWAT model. The results showed that the improved SWAT model was more accurate in simulating the water quantity and quality in the Huaihe River basin [12]. Peng et al. used the Environmental Fluid Dynamics Code (EFDC) model coupled with a geographic information system (GIS) model to simulate the water quality of the lower Charles River, and the results showed that the accuracy of the model was improved compared with the original EFDC model [13]. The mechanistic models of river water quality tend to provide a more comprehensive description of water quality changes, as they consider the effects of physical, chemical, and biological processes on the spatial and temporal transport and transformation patterns of pollutants in river waters; however, at the same time, most of these models are complex and require a great deal of basic information and data (numerical model uses a large amount of water quality data as the basis for calculation), and it is difficult to obtain a continuous distribution of water quality in space and time. This has greatly limited the application of these models [14]. In addition, the mechanics of many water environment systems are not fully understood by scholars; hence, it is difficult to describe them accurately using exclusively mechanistic modelling. In contrast, non-mechanical water quality modelling is a black-box approach to a particular water quality system, which is modelled by mathematical statistics or other mathematical methods to make predictions about water quality. Commonly used non-mechanical water quality simulation prediction methods include regression models, probability statistical models, grey prediction models, time series models, etc.

In recent years, neural networks and other machine learning algorithms have been applied by many researchers in the field of water quality prediction and have achieved good prediction results. The SOTA table of the progress of research based on water quality prediction is shown in Table 1 (distinguishing between mechanistic and non-mechanistic models).

Table 1. Overview of water quality prediction research.

Water Quality Mechanical Prediction Methods			
Research Scholars	**Research Subjects**	**Model Name**	**Model Characteristics**
Lee et al. (2017) [15]	Environmental Fluid Dynamic Code	The Galing River in Kuantan, Pahang, Malaysia	The Environmental Fluid Dynamic Code (EFDC) model considers the effects of temperature, humidity, radiation, cloud cover, evaporation, wind direction, and wind speed, which makes the simulation results closer to the reality. The water quality (TOC, TN, TP) in the upstream section is significantly improved, and the prediction accuracy can be improved by about 37%; if the sewage from the tributary is at the same location, it will increase by about 77%. A total of five water quality management plans for improving the water quality of the Galing River were evaluated using EFDC.
Deus et al. (2013) [16]	Two-dimensional water quality model	The Tucuruí reservoir, Pará, Brazil	The use of CE-QUAL-W2 to model hydrodynamics and water quality can reproduce horizontal and vertical gradients and their temporal changes. The field data of temperature, nitrate, ammonia, phosphorus, total suspended solids (TSS), and dissolved oxygen and chlorophyll a are used to verify the prediction effect of the model, and it has been confirmed that it can be used to simulate the response of water quality to the various management schemes of the fish industry.
Al-Zubaidi and Wells (2018) [17]	Three-dimensional hydrodynamic model	Lake Chaplain, Washington, DC, USA	The 3D hydrodynamic and water quality model is developed by expanding the 2D fully implicit scheme of CE-QUAL-W2 in three dimensions. The governing equations include a continuity equation, free surface equation, momentum equation, and transport equation, and the momentum and transport equations are solved by the time-splitting technique. The hydrodynamic equation and water quality equation are solved at the same time to realize the feedback between water quality and hydrodynamics. The results showed that the solution of the hydrodynamic equation of the model was very consistent with the field data.

Table 1. *Cont.*

Water Quality Mechanical Prediction Methods			
Research Scholars	**Research Subjects**	**Model Name**	**Model Characteristics**
Yang et al. (2017) [18]	Finite volume method	Urban Lake in Tianjin, China	The Navier Stokes equation is used to establish a two-dimensional hydraulic model, the finite volume method is used to calculate the parameters of the two-dimensional uncertain eutrophication model, and the Bayesian method is used to correct the model parameters. The model reflects the interaction between nutrients, phytoplankton, and zooplankton. It can be used to simulate the changes of seasonal and regional water quality indicators (DO, NH_4^+, NO_3^-, and PO_4^{3-}), and can calculate hydrodynamic information and eutrophication dynamics with reasonable accuracy (all relative errors are less than 11%).
Colton et al. (2022) [19]	Mass balance model	the Laurentian Great Lakes	The model calculates the mass balance and dynamic simulation evaluation of some trace metal loads in the Great Lakes basin, summarizes the loads of the tributaries and connecting channels, and estimates the atmospheric input and sedimentation. Among them, the load of conservative elements (Na and Cl) is used to calibrate the black box method. The mass balance of these elements can be accurately reproduced to 90% in a long-term trend.
Wang et al. (2018) [20]	Soft-sensing method based on WASP model	Taihu Lake and Beihai Lake in China	The WASP model is employed as a soft-sensing method and its unknown parameters are estimated by the unscented Kalman filter. The results show that the proposed soft sensing method can describe the changes of relevant water quality indexes (DO, BOD, TN, and Chl_a), and has improved accuracy compared to the nonlinear least square method and traditional trial and error method.
Yang et al. (2021) [21]	MIKE 21 FM model	Dongshan Lake in Guangdong Province of China	By using the MIKE 21 FM model and considering different flow arrangements, several model scenarios were established to predict the impact of diversion on selected water quality parameters. The results showed that the inflow and outflow arrangement was the main factor determining the flow field of the whole lake and the change trend of NH_3-N, and the increase in flow showed an unequal influence in each region. Wind was also shown to be important for the formation of air circulation and the change of pollutants.
Water Quality Non-Mechanical Prediction Method			
Research Scholars	**Research Subjects**	**Model Name**	**Methods andResults**
Najafzadeh et al. (2021) [22]	None	SVM, GEP, MTree, EPR, and MARS models	The d-factor of the SVM model was 0.79 for the *Kx* metric with 95% confidence space. The d-factor value was 0.87 for the *Ky* metric, which is better than the other models in terms of prediction accuracy.
Song et al. (2021) [23]	Haihe River	SWT-ISSA-LSTM	Based on the strong noise immunity of the simultaneous wavelet transform, the simultaneous wavelet transform is used to denoise the dataset, followed by an improved sparrow search algorithm to optimize the hyperparameters of the LSTM. The mean absolute error (MAE) of the model for predicting the water quality of Yongding River was 0.4727, which is much lower than other models.

Table 1. *Cont.*

Water Quality Mechanical Prediction Methods			
Research Scholars	**Research Subjects**	**Model Name**	**Model Characteristics**
Noori et al. (2013) [24]	Sefidrood River Basin	ROANFIS	The Pearson correlation coefficient (R) and root mean square error of the best-fit ROANFIS model were 0.96 and 7.12, respectively. In the test step of the selected ROANFIS model, the uncertainty analysis showed that the 95% confidence interval and the d-factor were predicted as 94% and 0.83, respectively.
Noori et al. (2013) [25]	Sefidrood River Basin	RONNM	The results showed that the best-fit RONNM had a Pearson correlation coefficient (R) and root mean square error of 0.94 and 7.75, respectively. In addition, the accuracy analysis of the model outputs based on the developed difference ratio statistics showed that RONNM was more advantageous.
Noori et al. (2015) [26]	Sefidrood River basin	SVM	The percentage of observed data included by the bandwidth of 95% prediction uncertainty (95ppu) and 95% confidence interval (d-factor) was selected for analysis. The results showed that the support vector machine model was more sensitive to the capacity parameter (C) than kernel parameter (gamma) and fault tolerance (epsilon), and it had acceptable uncertainty in BOD_5 prediction.
Ahmed et al. (2019) [27]	Data from PCRWR	Polynomial regression, random forest, etc.	Multiple linear regression, polynomial regression, random forest, and other machine learning regression models were used to predict WQI separately. The results show that the mean absolute error (MAE) of polynomial regression was 2.7273, which bests the other models in terms of performance.
Liu (2019) [28]	Guazhou automatic water quality monitoring station	LSTM	The mean interpolation and Pearson correlation coefficient were first used to preprocess the dataset, followed by LSTM to predict the PH and CODMn metrics. The mean squared error (MSE) of the model was 0.0017 for the DO dataset, which outperformed the ARIMA and SVR models.
Hu et al. (2019) [29]	None	LSTM	Linear interpolation and Pearson correlation coefficients were first used to preprocess the dataset, followed by LSTM to predict PH, temperature, and other indicators. The results indicated that in the short-term prediction, the prediction accuracy of PH and water temperature could reach 98.56% and 98.97%, and the prediction time lengths were 0.273 s and 0.257 s, respectively. In the long-term prediction, the prediction accuracy of pH and water temperature could reach 95.76% and 96.88%, respectively.
Elias Eze et al. (2021) [30]	South Africa	EEMD-DL-LSTM model	Firstly, EEMD was used to decompose temperature and PH into individual IMF components, and then each IMF component was used as the input of LSTM to train the neural network. The results showed that the average absolute error of this hybrid model was 0.0375, which is much lower than that of the BPNN and DL-LSTM models.

Archana et al. used the depth belief network in unsupervised learning to study the PH, dissolved oxygen, turbidity, and other water quality parameters of the Chaskaman reservoir for prediction and analysis [31]. The results show that this method performs better than the classical method for prediction. Wang et al. introduced the Holt–Winters seasonal model based on the ARIMA model and predicted the total phosphorus and total nitrogen in the reservoir. The results showed that the model had a prediction accuracy of 97.5% and had many advantages, such as fast learning speed [32]. Mohamed et al. analyzed the irrigation water quality index in Egypt by means of an integrated evaluation method and an artificial neural network model. In addition, the ARIMA model was developed to predict IWQI in Bahr El-Baqar drain, Egypt [33]. Shi et al. proposed a combination of the wavelet artificial neural network (WANN) model and the high-frequency alternative measurement of water quality anomaly detection and early warning method [34]. Li et al. proposed an EEMD-SVR water quality prediction model to predict the water quality of Jialing River in China. The model first decomposes water quality indicators, such as DO, into each IMF component by the EEMD algorithm, and then builds the SVR model based on each IMF component. The results showed that the hybrid model outperformed the standard SVR model and BPNN model in a variety of evaluation indicators [35]. Ewaid et al. established a multiple linear regression model according to the specified weight and predicted the water quality of the Euphrates River [36]. Xu combined wavelet transform and BPNN to establish a short-term wavelet neural network water quality prediction model and used the model to predict the water quality of intensive freshwater pearl culture ponds in Duchang County, Jiangxi Province, China. The results showed that the RMSE of the model was 3.822 in DO metrics, which was much lower than that of the BPNN and ELman models, showing desirable performance [37]. Qin et al. developed a PSO-WSVR model and used a particle swarm algorithm to optimize the parameters of the weighted support vector regression machine to predict water quality in Yixing, China. The results showed that the model reduced RMSE, MAE, MAPE, and MSE by 46.74%, 17.86%, 43.62%, and 67.84%, respectively, compared with the standard SVR model [38]. Tizro et al. used the ARIMA model to study nine water quality parameters of Hor Rood River [39]. Faruk established an ARIMA-ANN model with 108 months of water quality data from the Büyük Menderes River in Turkey from 1996–2004. The model consisted of two parts: firstly, the ARIMA model was used to model the linear part of the dataset, and then the artificial neural network was used to model the nonlinear part of the water quality series based on the fact that the ARIMA model could not solve the nonlinear part of the water quality series well. The results showed that the correlation coefficients between the predicted values of the hybrid model and the observed data for boron, dissolved oxygen, and water temperature were 0.902, 0.893, and 0.909, respectively [40]. Zhang et al. developed an ARIMA-RBFNN model to predict the total nitrogen (TN) and total phosphorus (TP) of Chagan Lake. The results showed that the RMSE values of this hybrid model were 0.139 and 0.036 for TN and TP indicators, respectively, which were improved compared to the ARIMA and RBFNN models [41]. Than et al. developed the LSTM-MA model, classified the water quality of Dongnai River from 2012 to 2019, predicted the water quality in the next two years, and proved that the LSTM-MA hybrid model has a quicker training time and more precise prediction than ARIMA, NAR, NAR-MA, and LSTM models [42]. Jian et al. first used an improved grey correlation (IGRA) to extract the features of water quality information and subsequently used LSTM to predict the water quality of Taihu Lake and Victoria Harbor; the results showed that the RMSE values of the model were 0.07 and 0.067, which were lower than those of the BPNN and ARIMA models, showing good performance [43]. Hameed et al. used an RBF neural network (RBFNN) and BPNN model to forecast and compare the water quality in Malaysia, respectively. The results showed that the RMSE of BPNN was 0.867 and the RMSE of RBFNN was 0.0194, and the RBF neural network outperformed the BP neural network model in terms of prediction accuracy [44].

In summary, although scholars have proposed a large number of research methods in the field of water quality prediction, the prediction results of traditional statistical models are not satisfactory for time series with large fluctuations and long-term trends. For example, the regression analysis model is relatively simple, but its requirements for statistical data are high, demanding a large sample and data with a good distribution pattern; the time series model has a relatively sound theoretical basis, but its prediction accuracy is poor;

the grey prediction model is suitable for the case of small and discontinuous historical data, but the model is susceptible to the influence of unstable data, resulting in a large prediction error; the support vector machine is suitable for small samples, but it is more sensitive to the choice of parameters and kernel functions. In addition, traditional single deep learning models, such as back Propagation neural network (BPNN) and RBFNN, lack the memory ability for historical information. Moreover, most of the missing data filling methods cannot effectively handle the time-series information in the dataset, resulting in large errors in the estimation of missing values. Therefore, this study attempts to use an artificial neural network to fill in the missing information of water quality, comprehensively apply wavelet transform and the LSTM model to the field of water quality prediction, and compare the prediction results with ANN-LSTM, ARIMA, NARNN, CNN-LSTM, and DWT-CNN-LSTM models so as to prove the effectiveness of the proposed model.

This study is divided into the following parts: Section 2 introduces the artificial neural network model, wavelet transform, long-short term memory network model, and error evaluation index; Section 3 takes the Jinjiang River Basin as the research object, constructs the ANN-WT-LSTM model for water quality prediction, and compares the prediction results with the NAR neural network model, ANN-LSTM model, and ARIMA model; and the conclusion and research prospects are presented in Section 5.

2. Materials and Methods

2.1. Study Area Description and Dataset Analysis

The Jinjiang River is 182 km long, with a watershed area of 5629 square kilometers, an average slope of 0.19%, and an average annual runoff of 5.13 billion cubic meters. It is the largest river in Quanzhou and the third largest river in Fujian Province. The following Figure 1 shows the geographical location of the Jinjiang River.

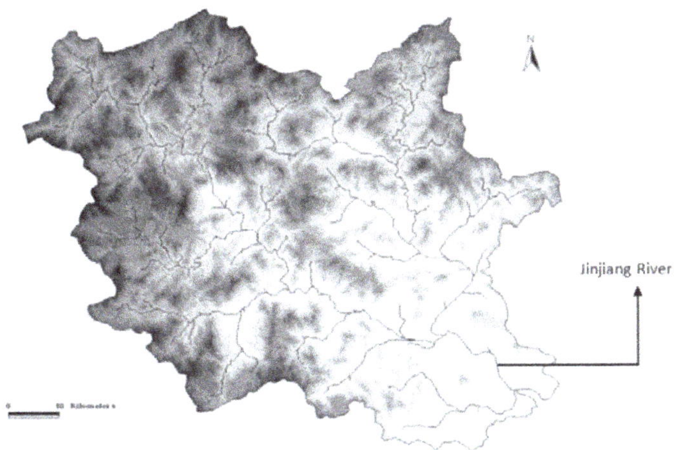

Figure 1. Geographical overview of Jinjiang River basin.

The Jinjiang River is divided into two tributaries, the east stream and the west stream, and the source of the Jinjiang River is the west stream, which is 153 km long with a watershed area of 3101 square kilometers and an average annual runoff of 3.65 billion cubic meters. The east stream of the Jinjiang River originates at the southern foot of Xueshan Mountain in Jindou, Yongchun. The river is 120 km long, with a watershed area of 1917 square kilometers and an average annual runoff of 1.4 billion cubic meters. Quanzhou City, through which the Jinjiang River flows downstream, is one of the most economically developed regions in Fujian Province. Quanzhou, located in the southeastern part of Fujian Province, is one of the three central cities in Fujian Province, and its total economic output has remained the first in Fujian Province for 22 consecutive years. In

2020, the city's population was over 7 million, ranking first in the province in terms of population size. As the Jinjiang River basin covers 53.8% of Quanzhou's land area, water resources are very important for the city's sustainable development. At the same time, there has been a serious pollution problem in the Jinjiang River basin [45,46]. The traditional industrial development model has caused great damage to local sustainable development, the pressure on the water environment is increasing, pollution from some enterprises is rebounding, the construction of environmental protection infrastructure is lagging behind, and the proportion of domestic pollution sources is increasing day by day. Therefore, the accurate prediction of water quality in the Jinjiang River basin will provide crucial decision data support for future pollution control programs.

The dataset used in this study was selected from the weekly report of automatic water quality monitoring at the Shilong section of Jinjiang River basin. Among the many water quality evaluation indexes, we selected dissolved oxygen (DO), permanganate index (CODMn), ammonia nitrogen (NH_3-N), and TP (total phosphorus), which are the four most representative indexes of the research object. The time of data collection was from 7 January 2013 to 21 June 2021. The data update cycle occurred once a week, with a total of 443 groups of data. We used the first 421 groups of data as the training set and the last 22 groups as the test set. The images of the dataset are shown in Figure 2.

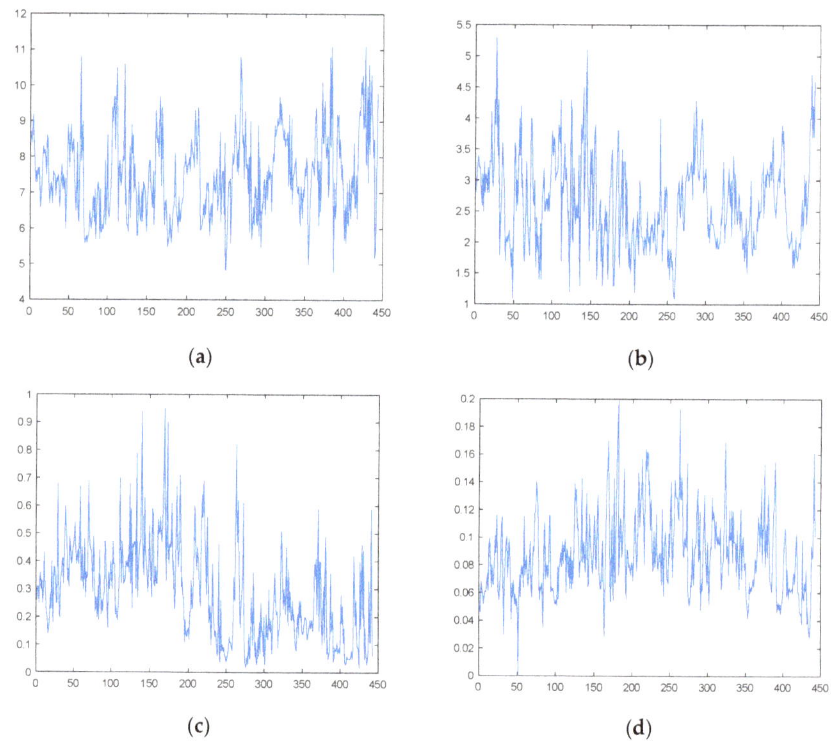

Figure 2. The image of dataset. (**a**) Dissolved oxygen (DO); (**b**) CODMn; (**c**) NH3-N; (**d**) Total phosphorus (TP).

Next, the dataset was analyzed and the missing values were found. The analysis results are shown in Table 2.

Then, we used Pearson's correlation coefficient to analyze the correlation of each dataset. The results are shown in the Table 3. From the above correlation analysis table, it can be seen that the DO dataset was negatively correlated with the CODMn, TP, and NH_3-N datasets; the CODMn dataset showed a weak positive correlation with the TP and a significant positive correlation with the NH_3-N dataset; and the TP dataset showed a significant positive correlation with the NH_3-N dataset.

Table 2. Descriptive statistics of experimental dataset.

Index	Minimum Value	Maximum Value	Mean Value	Standard Deviation	Variance	Number of Missing Values
DO	4.80	11.10	7.49	1.19	1.19	1
CODMn	1.10	5.30	2.64	0.74	0.74	1
TP	0.00	0.19	0.09	0.03	0.03	0
NH_3-N	0.02	0.95	0.29	0.17	0.17	2

Table 3. Correlation coefficients for each dataset.

Correlation Coefficient	DO	CODMn	TP	NH_3-N
DO	1	−0.024	−0.201	−0.136
CODMn	-	1	0.031	0.087
TP	-	-	1	0.448
NH_3-N	-	-	-	1

2.2. The Framework of the Proposed Model

The single neural network model is susceptible to fluctuations in the water quality time series during training, which affects the prediction accuracy. Therefore, this study introduced the signal time and frequency decomposition method for water quality data preprocessing and built a hybrid prediction model based on "decomposition- prediction-reconstruction" to improve the overall prediction accuracy. The hybrid model is made up of five components:

1. Data preprocessing: firstly make a descriptive analysis of the collected water quality data, find the missing value, estimate the missing value by artificial neural network, and then normalize it to eliminate the influence of dimension.
2. Discrete wavelet transform: The db5 wavelet technique is used to decompose the water quality time series datasets.
3. Model training, detection: Split the high-frequency and low-frequency signals of each dataset obtained from the db5 wavelet decomposition into a training set and a test set according to a fixed ratio. In this study, we set the first 421 sets of each dataset as the training set and the last 22 sets as the test set. Subsequently, we used LSTM to train each training set and adjust the relevant parameters of LSTM, such as learning rate and the maximum number of iterations.
4. The predictions obtained from the decomposed test set of each sub-series are superimposed to obtain the final prediction results.
5. Model evaluation: This study used four indicators—MSE, RMSE, MAE and MAPE—to evaluate the model's performance.

The whole algorithm flow chart is shown in Figure 3.

2.3. Data Normalization

Data normalization is a fundamental task for mining data in machine learning. In practical research, different methods and evaluation metrics often have different scales and units, which will produce diverse data analysis results. In order to reduce the relative relationship between quantities and to eliminate the influence of the dimension between indicators, the data must be normalized in order to achieve comparability between data indicators and to achieve the expectation of data optimization. The original data are normalized such that the indicators are in the same order of magnitude, which is convenient for comprehensive comparison and evaluation. Commonly used normalization methods include min-max normalization [47] and Z-score normalization [48]. Minimum-maximum normalization, also known as outlier normalization, is a linear transformation of the original data such that the resulting values map to between 0 and 1. There are also some other data normalization methods, such as the Z-score standardization method. However, the Z-score application also has risks. Firstly, the estimation of the Z-score requires the overall mean and variance, but this value is difficult to obtain in real analysis and mining. In most cases, it is replaced by the sample mean and standard deviation. Secondly, Z-score has certain requirements for data distribution, and normal distribution is the most conducive to Z-score calculation. Therefore, we chose the min-max normalization method. It is more suitable for use on data with relatively concentrated values. The transformation function of the min-max normalization used in this study is as follows:

$$X' = \frac{x - x_{\min}}{x_{\max} - x_{\min}} \tag{1}$$

where $x_{\max}$ is the maximum value of the sample data and $x_{\min}$ is the minimum value of the sample data.

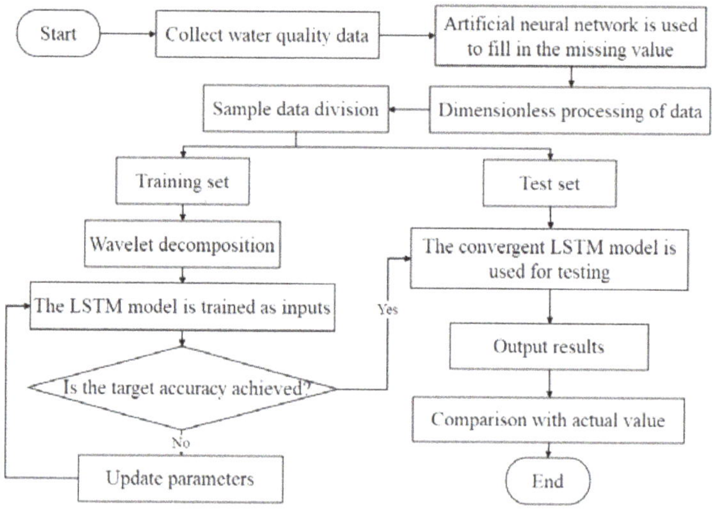

Figure 3. Flow chart of water quality prediction model.

2.4. *Artificial Neural Network (ANN)*

During the collection of time-series data, the loss of single or multiple attributes of some data in the final dataset or the loss of single or multiple records will be caused by acquisition, storage, and human error. These data are called missing data. The lack or incompleteness of data brings many difficulties to data mining, which will lead to the deviation of the analysis results and mislead users' decisions, resulting in adverse consequences. Therefore, filling the missing data completely under certain conditions is of great significance for macro data mining in big data scenarios. Nowadays, there are several ways to deal with missing data, such as the deletion method [49,50], missing value filling method based on a statistical model [51], or the method based on parameter estimation. This method first judges the missing mechanism of the missing value and then establishes a specific model to estimate the missing value. This method is widely used because it is more flexible in application and can be applied to datasets with a large number of missing values [52]. Common methods include the expectation maximization method, multiple filling method [53–57], maximum likelihood estimation method, etc. Austin et al. used multiple interpolations to estimate missing values in clinical medicine [58]. Chang et al.

developed a distributed multiple filling method with communication efficiency to estimate the missing data in distributed health data networks (DHDNs) [59].

In summary, research on interpolation methods for missing values of time series has received increasing attention from scholars in various fields, and although some scholars have considered the correlation characteristics of time series, most of these studies have not quantified the correlation between the observed quantities. Although some scholars consider the correlation characteristics of time series, most of the studies are still based on traditional interpolation or regression analysis methods. Moreover, some traditional models, such as piecewise linear interpolation [60], cannot estimate the missing value well [61,62]. Therefore, with the development of machine learning, researchers can gradually apply various machine learning algorithms to the field of missing value filling, which can to some extent solve the problem of non-linearity that cannot be handled by traditional methods. Machine learning methods for missing value estimation include the KNN method [63], artificial neural network, etc.

Artificial neural network (ANN) is a classical fundamental technique in machine learning. Compared with general multi-factor prediction methods, its prediction method has the advantages of high fault tolerance, high reliability, and fast prediction speed. In addition, ANN is a powerful interpolation tool [64–66]. Artificial neural networks generally have more than three layers of multilayer neural networks, which generally include three-layer structures of input, hidden, and output layers, as shown in Figure 4.

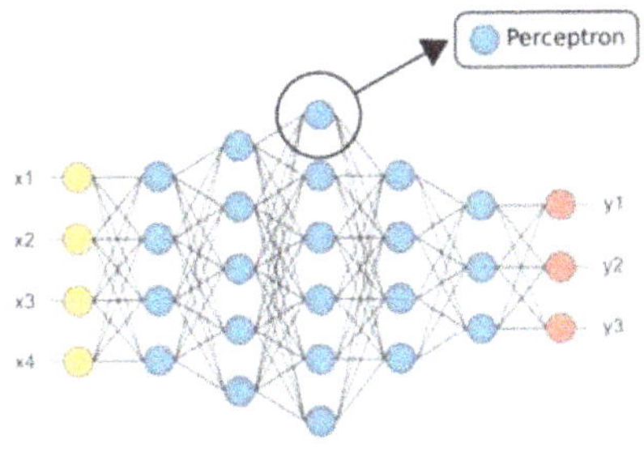

Figure 4. Topology of neural network structure.

The relationship between the input x_i and output y_i of neurons is $y_i = f(net_i)$, where $net_i = XW$ is the net activation, $X = [x_0, x_1, \cdots, x_n]$ is the input vector, $W = [w_{i0}, w_{i1}, \cdots, w_{in}]^T$ is the weight vector, and $f(\cdot)$ is the activation function, which represents the function of mapping the net activation and output. Some commonly used activation functions include $y = kx + c$, $y = \frac{1}{1+e^{-ax}}$, $y = \frac{2}{1+e^{-ax}} - 1$, etc.

A neural network can be divided into two states: learning state and working state. The learning state is used to adjust the weight of the neural network to make the output close to the actual value, while the working state uses the established network for classification and prediction without changing the weight of the neural network. The learning mode of the neural network is tutorial learning. The weight of the network is adjusted by the difference between the actual output and expected output of the network to make the model adapt as accurately as possible.

In this study, the MLP neural network was used to estimate the missing values from the water quality data of the Jinjiang River. The activation function of the output layer is constant. The single-layer perceptron is the simplest neural network, which is composed of input and output layers, and the input and output layers are directly connected. The MLP neural network contains an input layer, output layer, and several hidden layers, which is a kind of multi-layer feed-forward neural network based on BP algorithm training. The input signal is passed forward through the input layer to the hidden layer, and subsequently the neurons in the hidden layer are computationally processed and then passed forward to the output layer, which is a forward transmission process in which the output of the MLP neural network depends only on the current input and not on past or future inputs;

thus, the MLP neural network is also known as a multi-layer feed-forward neural network. Among many neural network architectures, MLP neural networks are simple in structure, easy to implement, and have good fault tolerance, robustness, and excellent nonlinear mapping capability (Figure 5).

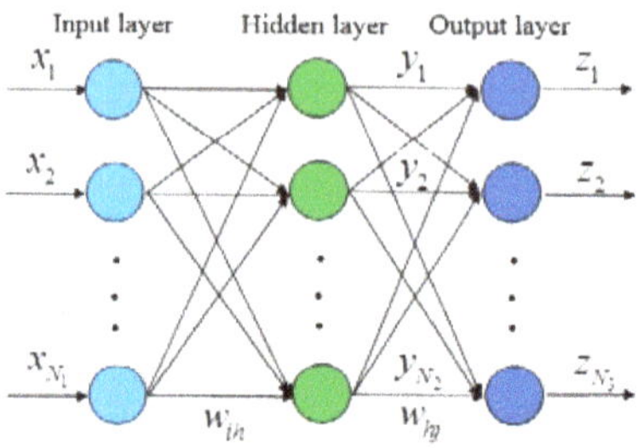

Figure 5. Topology of MLP neural network structure.

2.5. Basic Principle of Wavelet Transform

In the process of time-series data acquisition, there will be some noise in the time series data due to observation error, systematic error, or other reasons, and the noise will seriously affect the data processing results. Therefore, in the data preprocessing stage, different methods should be selected to denoise the data according to the type of noise. Common denoising methods include the Fourier transform [67], the wavelet transform [68], etc.

The Fourier transform is a widely used analysis method in the field of signal processing. It converts a time domain signal into a frequency domain signal. Its basic idea is to decompose the signal into the superposition of a series of continuous sine waves with different frequencies. However, Fourier transform also has many disadvantages. The traditional Fourier transform can only realize the overall transformation between the signal time domain and the frequency domain and cannot distinguish time-domain information. However, Fourier transform is only suitable for stable signals; most signals have variability, which significantly limits the application of Fourier transform.

The basic idea of wavelet transform is to adaptively adjust the time-frequency window according to the signal, decomposing the original signal into a series of sub-band signals with different spatial resolutions, frequency characteristics, and directional characteristics after stretching and translating. These sub-bands have good local characteristics in both the time and frequency domains and can therefore be used to represent the local characteristics of the original signal, thus enabling the localization of the signal in time and frequency. This method can overcome the limitations of Fourier analysis in dealing with non-smooth signals and complex images.

The mathematical definition of wavelet is as follows: let $\psi \in L^2(R) \cap L(R)$, which is almost always 0 on R and satisfies $C_\psi = \int_{-\infty}^{+\infty} \frac{\left|\overset{\wedge}{\psi}(\omega)\right|}{|\omega|} d\omega$, then ψ is the wavelet, where $\overset{\wedge}{\psi}(\omega) = \frac{1}{\sqrt{2\pi}} \int\limits_{-\infty}^{+\infty} \psi(t) e^{-ir} dt$ is the Fourier transform of ψ. Wavelet transform is one order of magnitude faster than fast Fourier transform. When the signal length is M, the computational complexity of Fourier transform is $O_f = M\log_2 M$ and that of wavelet transform is $O_M = M$.

Wavelet transform can be divided into continuous wavelet transform (CWT) and discrete wavelet transform (DWT).

The formula of continuous wavelet transform is:

$$W_f(a,b) = \frac{1}{\sqrt{a}} \int_{-\infty}^{+\infty} f(t) \overline{\psi(\frac{t-b}{a})} dt \tag{2}$$

where $W_f(a,b)$ is the continuous wavelet coefficient, a is the scaling factor, b is the translation factor, $\overline{\psi(\frac{t-b}{a})}$ is the conjugate function of $\psi(\frac{t-b}{a})$, and $f(t)$ represents the original data. The scale of wavelet transform is controlled by adjusting the values of a and b to realize the adaptive time-frequency signal analysis.

The discrete wavelet transform formula is:

$$W_f(j,k) = \int_{-\infty}^{+\infty} f(t) \frac{\overline{\psi(\frac{t}{a_0{}^j} - kb_0)}}{\sqrt{a_0{}^j}} dt \tag{3}$$

where $W_f(j,k)$ is the discrete wavelet coefficient and $f(t)$ is the original data.

The dbn wavelet is the most common wavelet transform and is mainly used in discrete wavelet transform. For wavelets of a finite length, when applied to fast wavelet transform, there will be a sequence composed of two real numbers. One is the coefficient of the high-pass filter, which is called the wavelet filter, and the other is the coefficient of the low-pass filter, which is called the adjustment filter. Firstly, the wavelet transform decomposes the original data into the low-frequency wavelet coefficient cA_n and high-frequency wavelet coefficient $cD_1, cD_2, \ldots, cD_n$ by using the low-pass filter and high-pass filter, respectively. Among them, the low-frequency wavelet coefficient can be further decomposed and iterated several times until the maximum decomposition time is reached. Finally, the decomposed wavelet low-frequency signal and high-frequency signal are added to realize wavelet reconstruction. The formula is:

$$f(t) = cA_n l(\psi_{ik}(t)) + \sum_{i=1}^{n} cD_n h(\psi_{ik}(t)) \tag{4}$$

where $f(t)$ is the restored signal; $l(\psi_{ik}(t))$ and $h(\psi_{ik}(t))$ are the low-pass filter and high pass filter, respectively; cA_n is low-frequency wavelet coefficient; and cD_n is high-frequency wavelet coefficient.

The calculation steps of wavelet transform are as follows:

Step 1. Elect the wavelet function and align it with the starting point of the analysis signal.

Step 2. Calculate the approximation degree between the signal to be analyzed and the wavelet function at this time; that is, the wavelet transform's coefficient C. The larger the coefficient C, If the coefficient C is larger, the more similar the current signal is to the waveform of the selected wavelet function.

Step 3. Move the wavelet function to the right one-unit time along the time axis, and then repeat Steps 1 and 2 to calculate the transformation coefficient C until it covers the whole signal length.

Step 4. Scale the selected wavelet function by one unit, and then repeat Steps 1–4.

Step 5. Repeat Steps 1–4 for all expansion scales.

The selection of the mother wavelet type and decomposition level are the two most important problems in wavelet analysis. In this study, the db5 wavelet was used to decompose the experimental sequence for the following two reasons:

(1) The db wavelets are more suitable for relatively stable sequences;
(2) db5 is also one of the most commonly used wavelets in the db wavelet family, which is suitable for smoother datasets.

Because Jinjiang water quality data has obvious smoothing characteristics, the db5 wavelet analysis was the most suitable method for this study.

The maximum decomposition levels of wavelet can be calculated by the following Equation (5):

$$L = \ln(n_d/(lw - 1)) \tag{5}$$

where lw is the length of the wavelet decomposition low-pass filter and n_d is the data length.

In this study, lw = 23 and n_d = 443 were selected, and L was calculated such that the number of wavelet decomposition layers was 3.

2.6. Basic Principle of LSTM

RNN was first proposed in the 1980s. As a popular algorithm in deep learning, compared with deep learning network (DNN), its circular network structure allows it to take full advantage of the sequence information in the sequence data itself. Therefore, it has many advantages in dealing with time series. Moreover, the ability to correct errors is achieved through back-propagation and a gradient descent algorithm. However, there are also many problems: as the time series grows, researchers have found that RNNs are weak for long time series, which means that the long-term memory of RNNs is poor. At the same time, as the length of the sequence increases, the depth of the model increases, and the problem of gradient disappearance and gradient explosion cannot be avoided when calculating the gradient. Therefore, Hochreiter et al. [69] proposed LSTM. The structure of LSTM is shown in Figure 6 [70].

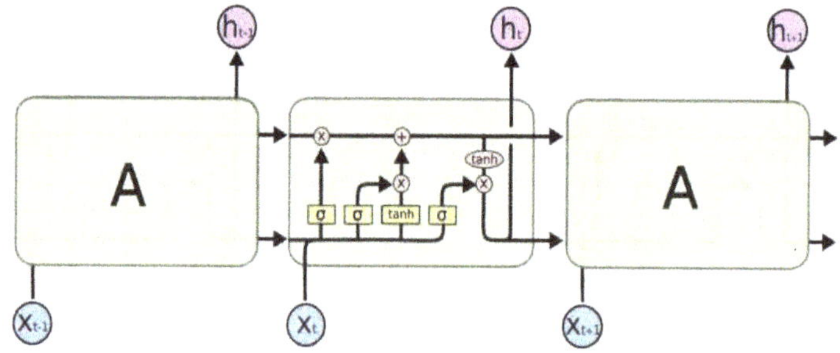

Figure 6. Long short-term memory network topology diagram.

The long-short term memory network is different from the traditional recurrent neural network in rewriting memory at each time step. LSTM will save the important features it has learned as long-term memory, and selectively retain, update, or forget the saved long-term memory according to the learning. However, the features with small weight in multiple iterations will be regarded as short-term memory and eventually forgotten by the network. This mechanism allows the important feature information to be transmitted all the time with the iteration so that the network has better performance in the classification task with a long-time dependence of samples. LSTM has been widely applied in flood sensitivity prediction [71], the prediction of key parameters of nuclear power plants [72], wind speed prediction [73,74], financial price trends [75], language processing [76], etc. In recent years, the LSTM model has made a series of improvements on the basis of RNN neurons. These include the addition of a transmission unit state in the RNN hidden layer controlled by three gating units: the forgetting gate, input gate, and output gate. Forgetting gates are used to control the forgetting of information and the extent to which it is retained. The calculation formula is:

$$F_t = \sigma(W_F \cdot [h_{t-1}, X_t] + b_F) \tag{6}$$

where X_t is the current input information, h_{t-1} is the data information in the previous hidden state, and the range of F_t is 0 to 1. When F_t = 1, it means that the information is completely retained, and when F_t = 0, it means that the information is completely abandoned.

The input gate is used to control how much input information at the current time is saved to the unit state. The expression is written as:

$$I_t = \sigma(W_i \cdot [h_{t-1}, x_t] + b_i) \tag{7}$$

where W_i is the weight matrix, b_i is the offset term, and I_t is the input layer vector value.

The input unit status C_t is represented as:

$$C_t = F_i \odot C_{t-1} + I_t \odot \widetilde{C}_t \tag{8}$$

$$\widetilde{C}_t = \tanh(W_c \cdot [h_{t-1}, x_t] + b_c) \tag{9}$$

where W_c is the weight matrix and b_c is the offset term.

The output calculation formula of the output gate O_t is shown as:

$$O_t = \sigma(W_o \cdot [h_{t-1}, X_t] + b_o) \tag{10}$$

where b_o is the offset value, W_o is the judgment matrix, and h_{t-1} is the hidden layer state at time $(t-1)$.

$$h_t = O_t \odot \tanh(C_t) \tag{11}$$

In Equation (11), $\odot$ is the Hadamard product and h_t is the hidden layer state at time t.

2.7. Evaluation Index

In this study, mean square error (MSE), root mean square error (RMSE), mean absolute error (MAE), and mean percentage error (MAPE) were selected as the basis for judging the prediction effect of the model. The calculation formulae are as follows:

$$\text{MSE} = \frac{1}{N}\sum_{t=1}^{N}(y_t - \overline{y_t})^2 \tag{12}$$

$$\text{RMSE} = \sqrt{\frac{1}{N}\sum_{t=1}^{N}(y_t - \overline{y_t})^2} \tag{13}$$

$$\text{MAE} = \frac{1}{N}\sum_{t=1}^{N}|(y_t - \overline{y_t})| \tag{14}$$

$$\text{MAPE} = \frac{1}{N}\sum_{t=1}^{N}\left|\frac{y_t - \overline{y_t}}{y_t}\right| \tag{15}$$

where N represents the total data volume, y_t represents the real value, and $\overline{y_t}$ represents the predicted value.

MAE is used to measure the mean absolute error between the predicted and actual values, RMSE is used to measure the deviation between the predicted and actual values (which is sensitive to outliers), and MAPE is used to measure the average relative error between the predicted and actual values.

3. Results

3.1. Artificial Neural Network Interpolation

In this study, we used MATLAB to construct a multilayer perceptron neural network to fill in the missing data in each dataset. The parameters of MLP were set as follows: the number of implied layers was two, the optimization algorithm was the conjugate scalar gradient algorithm, and the minimum relative change in the training error rate was 0.001. The artificial neural network training images for the DO, CODMn, and NH_3-N datasets are shown in the Figure 7.

As shown in Figure 7, the coefficient of determination was 0.99488 for the DO dataset, 0.99317 for the CODMn dataset, and 0.99525 for the NH_3-N dataset. It is clear that the fit of each dataset was good. Therefore, the model could be used to estimate the missing values in the DO, CODMn, and NH_3-N datasets.

3.2. Results of Wavelet Transform Model

Figure 8 shows the original images of DO, CODMn, TP and NH_3-N and the images after db5 wavelet decomposition.

However, after summing the data signals of each frequency band after wavelet reconstruction into the original signal, there is a certain error between this reduced data and the original data that is determined by the characteristics of the computer and is an error that cannot be eliminated. If this error value is too large, the experimental results will not be credible. Therefore, in this research, in order to verify the experimental accuracy of the model, the difference between the reduced data and the original data after wavelet reconstruction of the four indicators in the experiment was calculated. The errors are shown in Table 4.

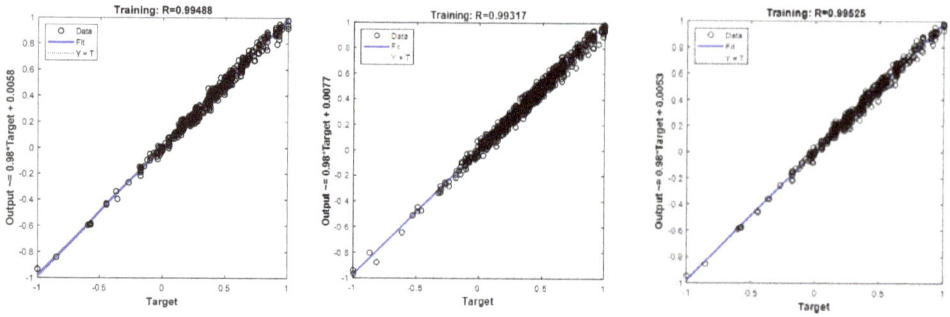

Figure 7. Residual images of ANN training on DO, CODMn, and NH_3-N datasets.

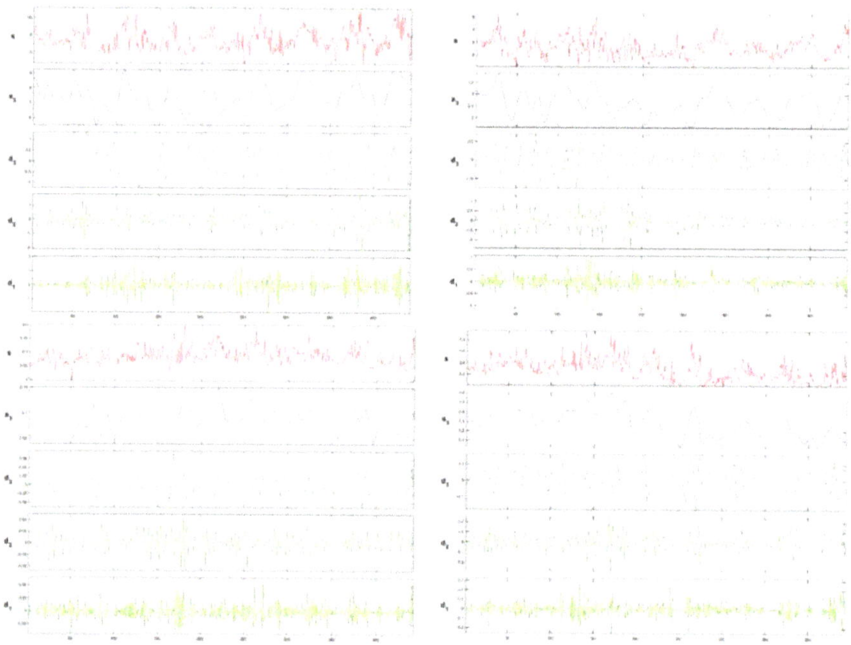

Figure 8. Original images and three-layer decomposition images of DO, CODMn, TP, and NH_3-N datasets.

From Table 4, it can be found that the error of CODMn was the largest (6.75×10^{-16}) and the error of DO was the smallest (9.1×10^{-16}). An error value in this range interval has a negligible effect on the experimental results, which proves the validity of the experiment.

3.3. Model Result Output

After wavelet transform of DO, CODMn, TP, and NH_3-N data, the decomposed low-frequency wavelet coefficient *cA* and high-frequency wavelet coefficient *cD* were used as the input of LSTM. Meanwhile, in order to verify that the ANN-WT-LSTM model had a higher prediction accuracy than other models, we selected ANN-LSTM, ARIMA, and NAR neural network models for comparison. The parameter settings of the other models are shown in Table 5.

Table 4. Wavelet reconstruction error table for each parameter.

Parameters	Error Value
DO	9.1×10^{-16}
CODMn	6.75×10^{-16}
NH_3-N	8.13×10^{-16}
TP	8.35×10^{-16}

Table 5. Parameter settings of ANN-WT-LSTM and traditional LSTM models.

Model Parameters	ANN-WT-LSTM		ANN-LSTM
Data Type	*cA*	*cD*	Data Interpolated by Artificial Neural Network
Number of hidden layer units [a]	300	300	300
Learning rate (%) [b]	0.003	0.003	0.001
Forgetting rate (%) [c]	0.2	0.2	0.2
Gradient threshold [d]	1	1	1
Number of iterations [e]	250	250	250
Batch size [f]	32	32	32

[a] NAR neural network parameter settings: autoregressive coefficient lag = 3, number of hidden layer units = 300. [b] Set to reduce the learning rate by multiplying by a factor of 0.2 after 125 rounds of training. [c] BPNN parameter settings: number of hidden layer units = 300 and number of iterations = 250; [d] SSA-LSTM parameters: dim is 4, the range of learning rate is (0.001,1), the range of the maximum number of iterations is (10,500), number of sparrows = 5, warning value ST = 0.6, proportion of discoverers PD = 0.6, proportion of sparrows aware of danger SD = 0.2, and population size = 5; [e] ISSA-BPNN: based on the fact that the sparrow search algorithm tends to fall into a local optimum, a tent mapping was used to initialize the population for the sparrow search algorithm. [f] The parameters were set as follows: lower boundary (lb) of the weight threshold is −5 and upper boundary (ub) of the weight queue is 5.

The results of the ANN-WT-LSTM model are shown in Figures 9–12.

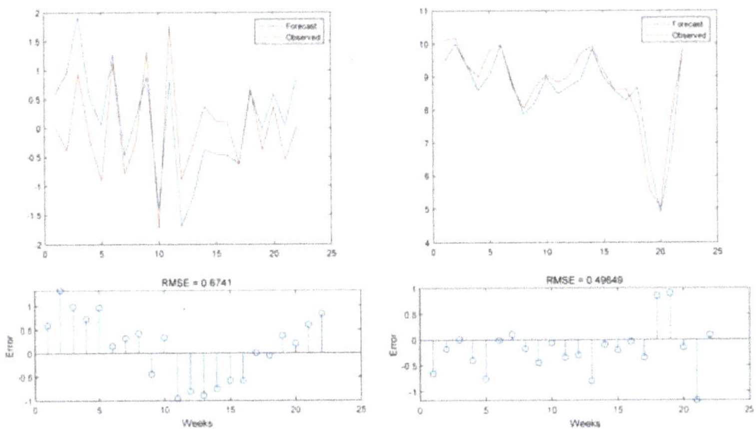

Figure 9. Prediction results and error images of the ANN-WT-LSTM model for high-frequency and low-frequency parts of the Do verification set.

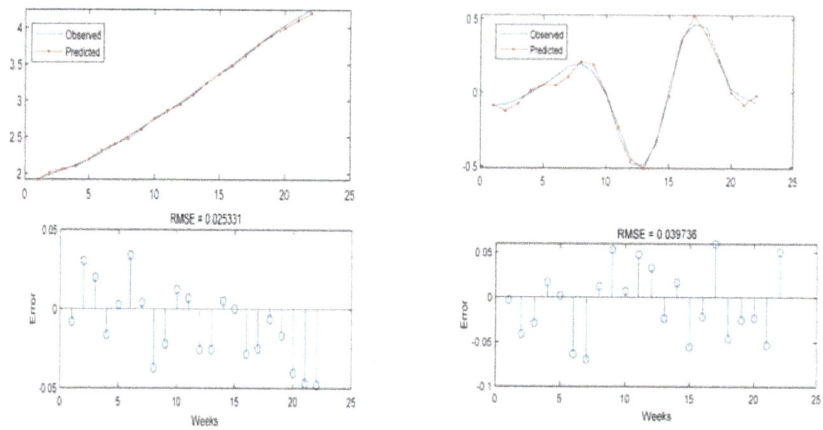

Figure 10. Prediction results and error images of the ANN-WT-LSTM model for high-frequency and low-frequency parts of the CODMN verification set.

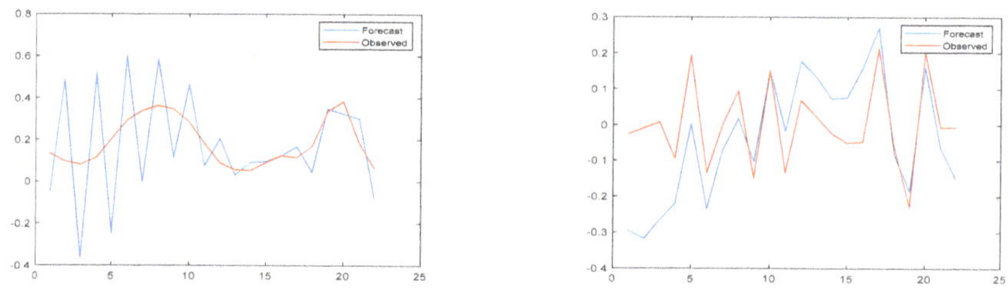

Figure 11. Prediction results and error images of the ANN-WT-LSTM model for high-frequency and low-frequency parts of the TP verification set.

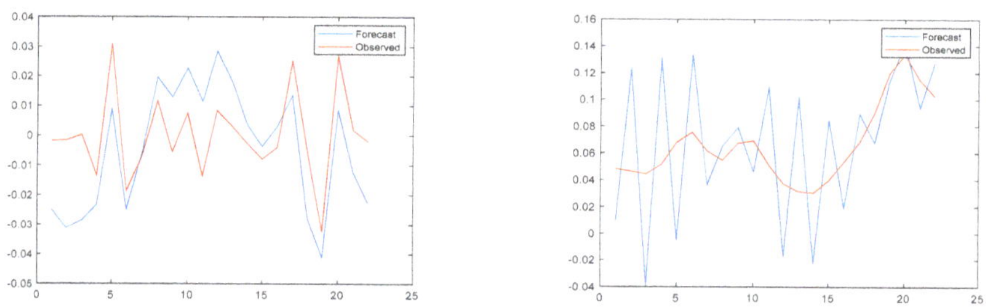

Figure 12. Prediction results and error images of the ANN-WT-LSTM model for high-frequency and low-frequency parts of the NH3-N verification set.

The prediction error images of the ANN-LSTM model on the test set data and the comparisons with the original images are shown in Figures 13–16.

3.4. Comparison with Other Models

We aimed to compare the prediction results of the ANN-WT-LSTM model with other existing models, compare the prediction accuracy of the different models, and analyze the prediction efficiency of the models. The comparison results of the model predictions are shown in Table 6.

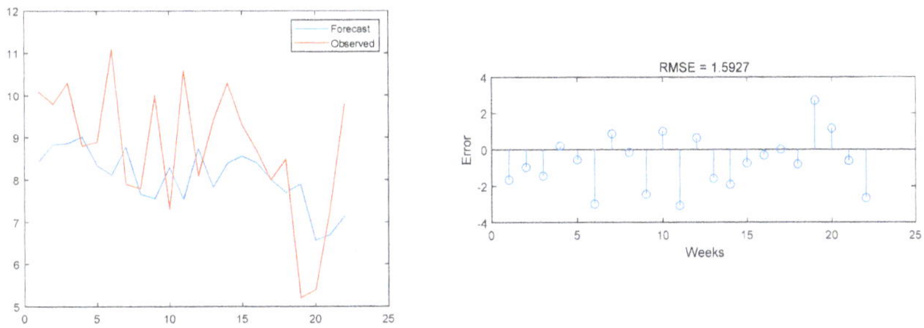

Figure 13. Prediction results of the ANN-LSTM model on the DO test set and comparison with the original image.

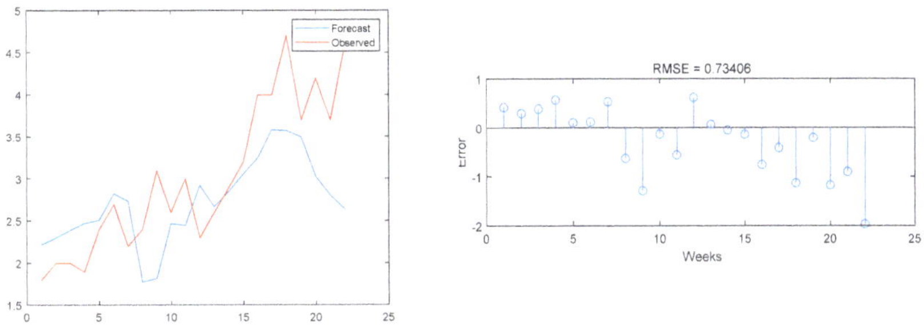

Figure 14. Prediction results of the ANN-LSTM model on the CODMn test set and comparison with the original image.

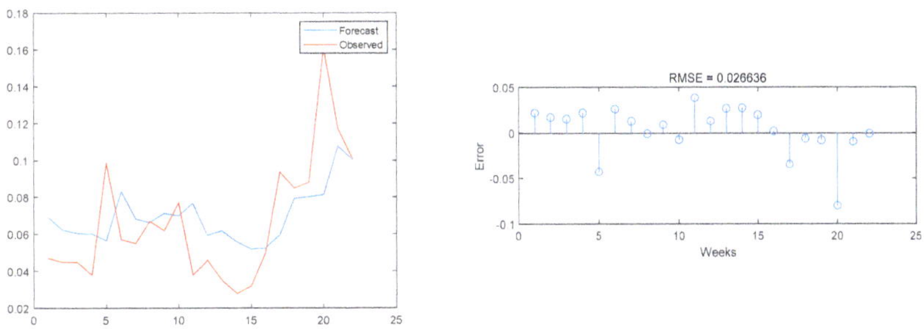

Figure 15. Prediction results of the ANN-LSTM model on the TP test set and comparison with the original image.

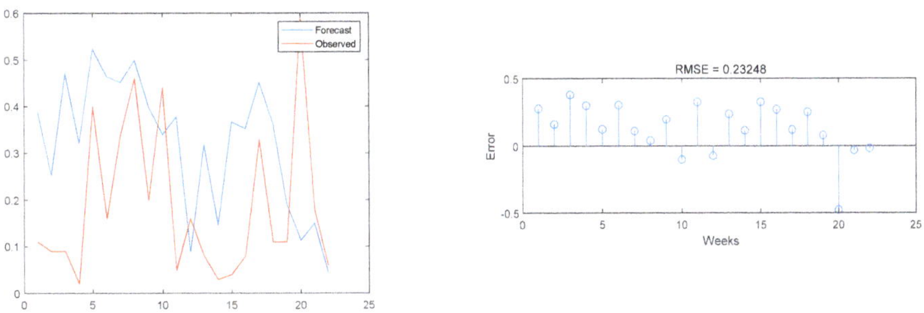

Figure 16. Prediction results of the ANN-LSTM model on the NH3-N test set and comparison with the original image.

Table 6. Analysis and comparison of ANN-WT-LSTM model and other models on dataset.

Model	Parameter	DO	CODMn	TP	NH_3-N
ANN-W-LSTM	MSE	8.3×10^{-25}	0.0006	0.00068	0.006
	RMSE	9.13×10^{-13}	0.024	0.026	0.0776
	MAPE	5.65×10^{-13}	0.021	0.243	0.0232
	MAE	4.39×10^{-12}	0.014	0.014	0.011
ANN-LSTM	MSE	2.536	0.539	0.0009	0.03
	RMSE	1.5927	0.7341	0.03	0.232
	MAPE	0.106	0.178	0.355	2.88
	MAE	0.94	0.52	0.02	0.181
NAR	MSE	2.2889	1.1353	0.0012	1.945
	RMSE	1.5129	1.0655	0.0345	1.395
	MAPE	0.1489	0.2593	0.563	13.639
	MAE	1.2417	0.813	0.026	1.149
ARIMA	MSE	3.1659	0.9277	0.0013	0.0297
	RMSE	1.7793	0.9632	0.0359	0.1723
	MAPE	0.1711	0.1959	0.6434	2.1502
	MAE	1.4999	0.7009	0.0306	0.1601
MLPNN	MSE	2.7947	0.728	0.0011	0.015
	RMSE	1.67	0.853	0.033	0.122
	MAPE	1.403	0.698	0.29	0.11
	MAE	0.159	0.267	0.589	1.27
CNN-LSTM	MSE	2.35	0.25	0.018	0.008
	RMSE	1.53	0.5	0.134	0.09
	MAPE	0.015	0.15	0.96	0.28
	MAE	1.17	0.4	0.11	0.02
BPNN	MSE	0.27	0.126	0.2	0.07
	RMSE	0.52	0.35	0.45	0.26
	MAPE	3.87	1.59	1.22	1.11
	MAE	0.42	0.29	0.41	0.22
SSA-LSTM	MSE	1.4	0.27	0.02	0;009
	RMSE	1.18	0.52	0.14	0.095
	MAPE	0.14	0.16	0.12	0.44
	MAE	1.16	0.42	0.11	0.024
ISSA-BPNN	MSE	0.14	0.062	0.12	0.05
	RMSE	0.37	0.25	0.35	0.22
	MAPE	5.13	1.3	0.92	2
	MAE	0.29	0.2	0.31	0.18
SSA-BPNN	MSE	0.15	0.063	0.13	0.1
	RMSE	0.38	0.251	0.36	0.32
	MAPE	1.73	0.77	0.98	1.86
	MAE	0.29	0.2	0.32	0.19

Table 6. *Cont.*

Model	Parameter	DO	CODMn	TP	NH_3-N
DWT-CNN-LSTM	MSE	0.24	0.04	0.002	0.03
	RMSE	0.49	0.2	0.045	0.173
	MAPE	0.05	0.05	0.5	1.62
	MAE	0.39	0.15	0.043	0.13
EMD-LSTM	MSE	1.67	0.07	0.035	0.022
	RMSE	1.3	0.26	0.19	0.15
	MAPE	0.14	0.06	0.62	1.96
	MAE	1.1	0.2	0.13	0.12
EEMD-LSTM	MSE	1.12	0.07	0.037	0.016
	RMSE	1.06	0.26	0.192	0.126
	MAPE	0.11	0.07	0.62	1.64
	MAE	0.89	0.20	0.13	0.105

4. Discussion

Water pollution is one of the biggest important environmental problems facing mankind, and the harm caused by it is largely due to the lack of prediction and early warning and emergency disposal capabilities. Therefore, the construction of an effective monitoring and early warning system to achieve intelligent decision making and the management of water quality is a key scientific and technological issue that needs to be addressed urgently. However, because water quality indicators usually have the characteristics of nonlinearity and non-smoothness, conventional statistical models often have difficulties making accurate predictions [77]. In recent years, there has been a rapid development of deep learning technology and wireless sensing technology. The model proposed in this study can be applied in the following aspects:

(1) Existing monitoring systems cannot achieve online high-frequency monitoring of all important pollutants, so the model proposed in this study can be used for soft computing to improve the timeliness, coverage and frequency of online monitoring and to form an effective early warning system for water quality management.

(2) According to real-time monitoring data for water quality change trend prediction and water quality risk judgment. When the prediction results show that the water quality situation has a deteriorating trend, the relevant management departments can make the corresponding measures of pollution prevention and control at the first time, so as to minimize the water quality losses caused by pollution incidents.

From the above results and error images, it can be seen that the accuracy of the ANN-WT-LSTM model prediction on the DO dataset was substantially improved compared with the MLPNN model, ANN-LSTM model, NAR neural network model, CNN-LSTM model, SSA-LSTM, SSA-BPNN model, and ISSA-BPNN model. For the CODMn dataset, the MAPE of the ANN-WT-LSTM model was 0.021, which was 0.157, 0.2383, 0.1749, 0.677, 0.129, 1.569, 0.139, 1.279, 0.749, 0.029, and 0.039 lower compared to the ANN-LSTM model, NAR neural network model, ARIMA model, MLPNN model, CNN-LSTM model, BPNN model, SSA-LSTM model, ISSA-BPNN model, SSA-BPNN model, DWT-CNN-LSTM model, and EMD-LSTM model, respectively. For the TP dataset, the RMSE of the ANN-WT-LSTM model was 0.026, which decreased by 0.004, 0.0085, 0.0099, 0.007, 0.108, 0.424, 0.114, 0.324, 0.334, 0.019, and 0.164, respectively, compared to the other models. For the NH_3-N dataset, the MSE of the ANN-WT-LSTM model was 0.006, which decreased by 0.024, 1.939, 0.0237, 0.009, 0.002, 0.064, 0.003, 0.044, 0.094, 0.024, and 0.016, respectively, when compared with the other models.

It is known that water quality prediction methods are divided into two main categories: mechanistic and non-mechanistic predictions. Mechanistic water quality models are derived using system structure data based on constraints in the underlying physical, biological, and chemical processes of the water environment system. A variety of water quality models have been developed, such as QUAL, WASP, MIKE, EFDC, SWAT, SMS, BASINS, etc., and have been widely used. However, these mechanistic water quality models are very complex and require a large amount of basic data information (such as simulation parameters, source and sink terms, etc.) to establish and solve the water quality control equations. This makes the complexity of building water quality models high and the parameters more difficult to determine, leading to limitations in the application of the models in many water bodies [78,79]. Moreover, for many aquatic environmental systems, the detailed mechanisms are not fully explained, and the evolutionary development of water quality is influenced and disturbed by many variables, such as physics, chemistry, biology, meteorology, and hydraulics, with strong non-linear characteristics. The existing water quality prediction models based on mathematical expressions are unable to take the influence of these factors into account, and it is difficult to accurately describe the migration and dispersion of the water environment using mechanistic modelling; hence, the predictions made on this basis have a "natural" bias. Furthermore, typical basin hydrological models, such as SWAT, HSPF and MIKE, have different scenarios that are able to simulate the hydrological processes and the evolution of point and non-point sources of pollution in large scale basins over long periods of time; however, they are not suitable for predicting water quality in larger water bodies, such as lakes and reservoirs. Water quality models such as CE-QUAL-W2, WASP, and EPD-RIV1 address the hydrodynamics and water quality of larger water bodies, but not the hydrological problems that occur in the basin.

In contrast, the ANN-WT-LSTM model proposed in this paper is based on the idea of neural networks to analyze historical water quality data to predict future water quality changes, and is one of the non-mechanical water quality prediction methods. Non-mechanical forecasting methods use the idea of statistics, through the water quality related to the historical time series data mining analysis, to find its data behind the law of change, and then deduce the trend of water quality changes. Compared with the mechanistic water quality prediction methods, the advantages are obvious. First of all, the modelling cost is lower as the modelling data requirements are not high. Therefore, the method can be applied to water quality prediction in areas where a large amount of hydrological data is missing. Secondly, the model prediction reliability is good, because the ANN-WT-LSTM model has good applicability to the analysis and prediction of non-linear problems in uncertain environment; thus, the water quality prediction accuracy has been improved a great deal compared with previous models (Table 6). In addition, the ANN-WT-LSTM model has good applicability. The model itself is a "black box" model analysis, which does not need the hydrological data of pollution sources for analysis. Whether the study area is the river basin environment or lakes, reservoirs or other large water bodies, it has wide applicability and universality. In summary, our view is that the ANN-WT-LSTM model proposed in this paper is not the only choice in water quality prediction models, but it still has great potential for application compared to other competing methods (including 1D, 2D, and 3D numerical models) due to its reliability, efficiency, and accuracy.

The ANN-DWT-LSTM model proposed in this study still has several aspects that can be improved.

(1) The model proposed in this paper only considers the historical data of water quality indicators in the Jinjiang River basin, while changes in the external environment have a greater impact on river water quality, which can interfere with the neural network training process, thus affecting the accuracy of the model. There is still room for further research into how to reduce the interference of external factors or consider the influence of water quality factors in the model.

(2) In this study, LSTM was used to predict water quality; however, there are numerous improved versions of the LSTM model, including the Bi-LSTM (bi-directional long short-term memory network) and the adaptive neuro-fuzzy inference system (ANFIS). These methods can be used to compare with the model proposed in this study.

Based on the powerful parallel data processing capability and non-linear processing ability of neural networks, we believe that the model proposed in this study can be combined with big data technologies, such as IoT, which can process large-scale data quickly and accurately and can meet the requirements of multi-sensor data fusion well.

5. Conclusions

To improve the accuracy of water quality prediction data, this study proposed the ANN-WT-LSTM model based on an artificial neural network, wavelet transform, and long short-term memory network, using the water quality data of the Jinjiang River basin in China as the research object for prediction analysis. For missing water quality data caused by instrument failure, this study used an artificial neural network to fill in the missing values based on the time-series information of water quality data. Then, we used wavelet transform to decompose and reconstruct the water quality time series, in order to remove the impact of short-term random disturbance noise, improve the prediction accuracy of the model on out-of-sample data, and the ability to predict future dynamic trends, so that it can more effectively predict the short-term as well as long-term dynamic trends in water quality time-series data. Subsequently, compared with the ANN-LSTM model and the NAR neural network model, the results show that the ANN-WT-LSTM proposed in this study is better than other models in all evaluation indexes, and the model effectively improves the accuracy of water quality prediction, which is significant for water environment protection. The study not only provides vital data support for water quality safety management decisions, but also has important theoretical and practical significance for safeguarding the sustainable development of the riverine areas and water environmental protection in the reservoir area.

This study predicts the possible future situation of reservoir water quality through the study of time series. However, due to the limitation of monitoring conditions, it can only predict the water quality at one point of the reservoir, which cannot reflect the overall spatial change of water quality. Therefore, in order to establish a more perfect reservoir early warning system, we suggest that water quality monitoring points be set up in many places to monitor the water quality in different directions of the reservoir to combine water quality prediction with GIS technology. In this way, we not only study the development trend of water quality in time, but also study the change of water quality in space, so as to combine time and space prediction and lay a good foundation for establishing a perfect water quality early warning system.

Author Contributions: Conceptualization, J.W.; methodology, J.W. and Z.W.; supervision, J.W.; investigation, J.W.; visualization, J.W.; writing—original draft, J.W. and Z.W.; writing—review and editing, J.W. and Z.W.; validation, J.W. and Z.W.; diagram and flowchart preparation, Z.W. All authors have read and agreed to the published version of the manuscript.

Funding: This research was supported by the Open Research Fund of the State Key Laboratory of Simulation and Regulation of Water Cycle in River Basin, China Institute of Water Resources and Hydropower Research (grant No. IWHR-SKL-201905).

Institutional Review Board Statement: Not applicable.

Informed Consent Statement: Not applicable.

Data Availability Statement: The data presented in this study are available on request from the corresponding authors. The data are not publicly available due to the continuation of a follow-up study by the authors.

Acknowledgments: We would like to thank the computing science center of Shanghai Ocean University for its support in the scientific research.

Conflicts of Interest: The authors declare no conflict of interest.

References

1. Wang, Q.; Hao, D.; Li, F.; Guan, X.; Chen, P. Development of a new framework to identify pathways from socioeconomic development to environmental pollution. *J. Clean Prod.* **2020**, *253*, 119962. [CrossRef]
2. Ministry of Water Resources. *Water Resources Assessment in China*; Water and Hydropower Publishing: Beijing, China, 2018; pp. 154–196.
3. Shi, X. The Safety of Drinking Water in China: Current Status and Future Prospects. *China CDC Wkly.* **2020**, *2*, 210–215. [CrossRef] [PubMed]
4. Hara, J.; Mamun, M.; An, K. Ecological river health assessments using chemical parameter model and the index of biological integrity model. *Water* **2019**, *11*, 1729. [CrossRef]
5. Chen, X.; Strokal, M.; Van, V.; Michelle, T.; Stuiver, J.; Wang, M.; Bai, Z.; Kroeze, C.; Ma, L. Multi-scale Modeling of Nutrient Pollution in the Rivers of China. *Environ. Sci. Technol.* **2019**, *53*, 9614–9625. [CrossRef] [PubMed]
6. Huang, J.; Zhang, Y.; Bing, H.; Peng, J.; Arhonditsis, G. Characterizing the river water quality in china: Recent progress and on-going challenges. *Water Res.* **2021**, *201*, 117309. [CrossRef] [PubMed]
7. Sarpong, K.; Xu, W.; Mensah-Akoto, J.; Neequaye, J.; Dadzie, A.; Frimpong, O. Waterscape, State and Situation of China's Water Resources. *J. Geosci. Environ. Prot.* **2020**, *8*, 26–51. [CrossRef]
8. O'Connor, D. Oxygen Balance of an Estuary. *Tran. Am. Soc. Civ. Eng.* **1961**, *126*, 556–575. [CrossRef]
9. Thomann, R. Mathematical Model for Dissolved Oxygen. *J. Sanit. Eng. Div.* **1963**, *89*, 1–32. [CrossRef]
10. Dobbins, W. BOD and Oxygen Relationship in Streams. *J. Sani. Eng. Div.* **1964**, *90*, 53–78. [CrossRef]
11. Ma, W.; Chao, Z. The numerical simulation of water quality of Suzhou Creek based on GIS. *Acta Geogr. Sin.* **1998**, *53*, 66–75.
12. Zhang, Y.; Xia, J.; Shao, Q.; Zhai, X. Water quantity and quality simulation by improved SWAT in highly regulated Huai River Basin of China. *Stoch. Env. Res. Risk Assess.* **2013**, *27*, 11–27. [CrossRef]
13. Peng, S.; Fu, G.; Zhao, X.; Moore, B. Integration of Environmental Fluid Dynamics Code (EFDC) Model with Geographical Information System (GIS) Platform and Its Applications. *J. Environ. Inform.* **2011**, *17*, 75–82. [CrossRef]
14. Huang, M.; Tian, Y. An Integrated Graphic Modeling System for Three-Dimensional Hydrodynamic and Water Quality Simulation in Lakes. *ISPRS Int. J. Geo-Inf.* **2019**, *8*, 18. [CrossRef]
15. Lee, I.; Hwang, H.; Lee, J.; Yu, N.; Yun, J.; Kim, H. Modeling approach to evaluation of environmental impacts on river water quality: A case study with Galing River, Kuantan, Pahang, Malaysia. *Ecol. Model.* **2017**, *353*, 167–173. [CrossRef]
16. Deus, R.; Brito, D.; Mateus, M.; Kenov, I.; Fornaro, A.; Neves, R.; Alves, C. Impact evaluation of a pisciculture in the Tucuruí reservoir (Pará, Brazil) using a two-dimensional water quality model. *J. Hydrol.* **2013**, *487*, 1–12. [CrossRef]
17. Al-Zubaidi, H.A.M.; Wells, S.A. Analytical and field verification of a 3D hydrodynamic and water quality numerical scheme based on the 2D formulation in CE-QUAL-W2. *J. Hydraul. Res.* **2018**, *58*, 152–171. [CrossRef]
18. Yang, L.K.; Peng, S.; Zhao, X.H.; Li, X. Development of a two-dimensional eutrophication model in an urban lake (China) and the application of uncertainty analysis. *Ecol. Model.* **2017**, *345*, 63–74.
19. Colton, B.; Tassiane, J.; Alice, D.; Bas, V. Mass-Balance Modeling of Metal Loading Rates in the Great Lakes. *Environ. Res.* **2022**, *205*, 112557.
20. Wang, X.; Jia, J.; Su, T.; Zhao, Z.; Xu, J.; Li, W. A fusion water quality soft-sensing method based on wasp model and its application in water eutrophication evaluation. *J. Chem.* **2018**, *2018*, 9616841. [CrossRef]
21. Yang, H.; Wang, J.; Li, J.; Zhou, H.; Liu, Z. Modelling impacts of water diversion on water quality in an urban artificial lake. *Environ. Pollut.* **2021**, *276*, 116694. [CrossRef]
22. Najafzadeh, M.; Noori, R.; Afroozi, D.; Ghiasi, B.; Hosseini-Moghari, S.M.; Mirchi, A.; Haghighi, A.T.; Kløve, B. A comprehensive uncertainty analysis of model-estimated longitudinal and lateral dispersion coefficients in open channels. *J. Hydrol.* **2021**, *603*, 126850. [CrossRef]
23. Song, C.; Yao, L.; Hua, C.; Ni, Q. A novel hybrid model for water quality prediction based on synchrosqueezed wavelet transform technique and improved long short-term memory. *J. Hydrol.* **2021**, *603*, 126879. [CrossRef]
24. Noori, R.; Safavi, S.; Shahrokni, S.A.N. A reduced-order adaptive neuro-fuzzy inference system model as a software sensor for rapid estimation of five-day biochemical oxygen demand. *J. Hydrol.* **2013**, *495*, 175–185. [CrossRef]
25. Noori, R.; Karbassi, A.R.; Ashrafi, K.; Ardestani, M.; Mehrdadi, N. Development and application of reduced-order neural network model based on proper orthogonal decomposition for BOD_5 monitoring: Active and online prediction. *Environ. Prog. Sustain. Energy* **2013**, *32*, 120–127. [CrossRef]
26. Noori, R.; Yeh, H.D.; Abbasi, M.; Kachoosangi, F.T.; Moazami, S. Uncertainty analysis of support vector machine for online prediction of five-day biochemical oxygen demand. *J. Hydrol.* **2015**, *527*, 833–843. [CrossRef]
27. Ahmed, U.; Mumtaz, R.; Anwar, H.; Shah, A.; García-Nieto, J. Efficient Water Quality Prediction Using Supervised Machine Learning. *Water* **2019**, *11*, 2210. [CrossRef]
28. Liu, P.; Wang, J.; Sangaiah, A.; Xie, Y.; Yin, X. Analysis and Prediction of Water Quality Using LSTM Deep Neural Networks in IoT Environment. *Sustainability* **2019**, *11*, 2058. [CrossRef]
29. Hu, Z.; Zhang, Y.; Zhao, Y.; Xie, M.; Zhong, J.; Tu, Z.; Liu, J. A water quality prediction method based on the deep LSTM network considering correlation in smart mariculture. *Sensors* **2019**, *19*, 1420. [CrossRef]

30. Eze, E.; Halse, S.; Ajmal, T. Developing a Novel Water Quality Prediction Model for a South African Aquaculture Farm. *Water* **2021**, *13*, 1782. [CrossRef]
31. Solanki, A.; Agrawal, H.; Khare, K. Predictive Analysis of Water Quality Parameters using Deep Learning. *Int. J. Comput. Appl.* **2015**, *125*, 29–34. [CrossRef]
32. Wang, J.; Zhang, L.; Zhang, W.; Wang, X. Reliable Model of Reservoir Water Quality Prediction Based on Improved ARIMA Method. *Environ. Eng. Sci.* **2019**, *36*, 1041–1048. [CrossRef]
33. Abdel-Fattah, M.; Mokhtar, A.; Abdo, A. Application of neural network and time series modeling to study the suitability of drain water quality for irrigation: A case study from Egypt. *Environ. Sci. Pollut. R.* **2021**, *28*, 898–914. [CrossRef] [PubMed]
34. Shi, B.; Peng, W.; Jiang, J.; Liu, R. Applying high-frequency surrogate measurements and a wavelet-ANN model to provide early warnings of rapid surface water quality anomalies. *Sci. Total Environ.* **2018**, *1390*, 610–611. [CrossRef] [PubMed]
35. Li, X.; Cheng, Z.; Yu, Q.; Bai, Y.; Li, C. Water-Quality Prediction Using Multimodal Support Vector Regression: Case Study of Jialing River, China. *J. Environ. Eng.* **2017**, *143*, 4017070. [CrossRef]
36. Ewaid, S.; Abed, S.; Kadhum, S. Prediction the Tigris river water quality within Baghdad, Iraq by using water quality index and regression analysis. *Environ. Technol. Innov.* **2018**, *11*, 390–398. [CrossRef]
37. Xu, L.; Liu, S. Study of short-term water quality prediction model based on wavelet neural network. *Mat. Comput. Model.* **2013**, *58*, 801–807. [CrossRef]
38. Long, S. Study of Short-term Water Quality Prediction Model Based on PSO-WSVR. *J. Zhengzhou Univ.* **2013**, *58*, 807–813.
39. Tizro, A.; Ghashghaie, M.; Georgiou, P.; Voudouris, K. Time series analysis of water quality parameters. *J. Appl. Res. Water Wastewater* **2014**, *1*, 43–52.
40. Faruk, D. A hybrid neural network and ARIMA model for water quality time series prediction. *Eng. Appl. Artif. Intel.* **2010**, *23*, 586–594. [CrossRef]
41. Zhang, L.; Zhang, G.; Li, R. Water Quality Analysis and Prediction Using Hybrid Time Series and Neural Network Models. *J. Agric. Sci. Tech. IRAN* **2016**, *18*, 975–983.
42. Than, N.; Ly, C.; Tat, P. The performance of classification and forecasting Dong Nai River water quality for sustainable water resources management using neural network techniques. *J. Hydrol.* **2021**, *596*, 126099. [CrossRef]
43. Zhou, J.; Wang, Y.; Xiao, F.; Wang, Y.; Sun, L. Water Quality Prediction Method Based on IGRA and LSTM. *Water* **2018**, *10*, 1148. [CrossRef]
44. Hameed, M.; Sharqi, S.; Yaseen, Z.; Afan, H.; Hussain, A.; Elshafie, A. Application of artificial intelligence (AI) techniques in water quality index prediction: A case study in tropical region, Malaysia. *Neural Comput. Appl.* **2017**, *28*, 893–905. [CrossRef]
45. Deng, J.; Guo, P.; Zhang, X.; Shen, X.; Su, H.; Zhang, Y.; Wu, Y.; Xu, C. An evaluation on the bioavailability of heavy metals in the sediments from a restored mangrove forest in the Jinjiang Estuary, Fujian, China. *Ecotox. Environ. Safe.* **2019**, *180*, 501–508. [CrossRef] [PubMed]
46. Liu, X.; Jiang, J.; Yan, Y.; Dai, Y.; Deng, B.; Ding, S.; Su, S.; Sun, W.; Li, Z.; Gan, Z. Distribution and risk assessment of metals in water, sediments, and wild fish from Jinjiang River in Chengdu, China. *Chemosphere* **2018**, *196*, 45–52. [CrossRef]
47. Chen, P. Effects of normalization on the entropy-based TOPSIS method. *Expert Syst. Appl.* **2019**, *136*, 33–41. [CrossRef]
48. Alkhodari, M.; Fraiwan, L. Convolutional and recurrent neural networks for the detection of valvular heart diseases in phonocardiogram recordings. *Comput. Methods Prog. Bio.* **2021**, *200*, 105940. [CrossRef]
49. Newman, D. Missing data: Five practical guidelines. *Organ. Res. Methods* **2014**, *17*, 372–411. [CrossRef]
50. Little, R.; Rubin, D. *Statistical Analysis with Missing Data*, 2nd ed.; John Wiley & Sons: Hoboken, NJ, USA, 2014.
51. Little, R.; Rubin, D. The analysis of social science data with missing values. *Sociol. Method Res.* **1989**, *18*, 292–326. [CrossRef]
52. Shan, Y.; Deng, G. Kernel PCA regression for missing data estimation in DNA microarray analysis. In Proceedings of the 2009 IEEE International Symposium on Circuits and Systems, Taipei, Taiwan, 24–27 May 2009; pp. 1477–1480.
53. Wang, Z.; Wu, X.; Wang, H.; Wu, T. Prediction and analysis of domestic water consumption based on optimized grey and Markov model. *Water Supply* **2021**, *21*, 3887–3899. [CrossRef]
54. Wu, J.; Wang, Z.; Dong, L. Prediction and analysis of water resources demand in Taiyuan City based on principal component analysis and BP neural network. *J. Water Supply Res. Tech.* **2021**, *70*, 1272–1286. [CrossRef]
55. Wu, X.; Wang, Z.; Wu, T.; Bao, X. Solving the Family Traveling Salesperson Problem in the Adleman-Lipton model based on DNA computing. *IEEE Trans. Nanobiosci.* **2022**, *21*, 75–85. [CrossRef] [PubMed]
56. Noori, R.; Karbassi, A.R.; Mehdizadeh, H.; Vesali-Naseh, M.; Sabahi, M.S. A framework development for predicting the longitudinal dispersion coefficient in natural streams using an artificial neural network. *Environ. Prog. Sustain.* **2011**, *30*, 439–449. [CrossRef]
57. Wang, Z.; Bao, X.; Wu, T. A Parallel Bioinspired Algorithm for Chinese Postman Problem Based on Molecular Computing. *Comput. Intel. Neurosc.* **2021**, *2021*, 8814947. [CrossRef]
58. Austin, C.; White, R.; Buuren, V. Missing Data in Clinical Research: A Tutorial on Multiple Imputation. *Can. J. Cardiol.* **2020**, *37*, 1322–1331. [CrossRef] [PubMed]
59. Chang, C.; Deng, Y.; Jiang, X.; Long, Q. Multiple imputation for analysis of incomplete data in distributed health data networks. *Nat. Commun.* **2020**, *11*, 5467. [CrossRef]
60. Goudarzi, M.A.; Cocard, M.; Santerre, R.; Woldai, T. GPS interactive time series analysis software. *GPS Solut.* **2013**, *17*, 595–603. [CrossRef]

61. Bao, Z.; Chang, G.; Zhang, L.; Chen, G.; Zhang, S. Filling missing values of multi-station GNSS coordinate time series based on matrix completion. *Measurement* **2021**, *183*, 109862. [CrossRef]
62. Gia, B.; Te, A.; Bksa, C. Evaluating missing value imputation methods for food composition databases. *Food Chem. Toxicol.* **2020**, *141*, 111368.
63. Tutz, G.; Ramzan, S. Improved methods for the imputation of missing data by nearest neighbor methods. *Comput. Stat. Data Anal.* **2015**, *90*, 84–99. [CrossRef]
64. Deng, A.; Wang, Z.; Liu, H.; Wu, T. A bio-inspired algorithm for a classical water resources allocation problem based on Adleman-Lipton model. *Desalin. Water Treat.* **2020**, *185*, 168–174. [CrossRef]
65. Li, R.; Chang, Y.; Wang, Z. Study on optimal allocation of water resources in Dujiangyan irrigation district of China based on improved genetic algorithm. *Water Supply* **2021**, *21*, 2989–2999. [CrossRef]
66. Ji, Z.; Wang, Z.; Deng, X.; Huang, W.; Wu, T. A new parallel algorithm to solve one classic water resources optimal allocation problem based on inspired computational model. *Desalin. Water Treat.* **2019**, *160*, 214–218. [CrossRef]
67. Yu, M. Short-term wind speed forecasting based on random forest model combining ensemble empirical mode decomposition and improved harmony search algorithm. *Int. J. Green Energy* **2020**, *17*, 332–348. [CrossRef]
68. Arvanaghi, R.; Danishvar, S.; Danishvar, M. Classification cardiac beats using arterial blood pressure signal based on discrete wavelet transform and deep convolutional neural network. *Biomed. Signal. Proces. Control.* **2022**, *71*, 103131. [CrossRef]
69. Hochreiter, S.; Schmidhuber, J. Long short-term memory. *Neural Comput.* **1997**, *9*, 1735–1780. [CrossRef]
70. Yan, S. Understanding LSTM and Its Diagrams. Available online: https://blog.mlreview.com/understanding-lstm-and-its-diagrams-37e2f46f1714 (accessed on 8 February 2021).
71. Fang, Z.; Wang, Y.; Peng, L.; Hong, H. Predicting flood susceptibility using LSTM neural networks. *J. Hydrol.* **2020**, *594*, 125734. [CrossRef]
72. Chen, Y.; Lin, M.; Yu, R.; Wang, T. Research on simulation and state prediction of nuclear power system based on LSTM neural network. *Sci. Technol. Nucl. Ins.* **2021**, *2021*, 8839867. [CrossRef]
73. Rui, F.; Suzuki, H.; Kitajima, T.; Kuwahara, A.; Yasuno, T. Wind Speed Prediction Model Using LSTM and 1D-CNN. *J. Signal. Process.* **2018**, *22*, 207–210.
74. Ehsan, M.; Shahirinia, A.; Zhang, N.; Oladunni, T. Wind Speed Prediction and Visualization Using Long Short-Term Memory Networks (LSTM). In Proceedings of the 10th International Conference on Information Science and Technology (ICIST), Bath/London/Plymouth, UK, 9–15 September 2020.
75. Troiano, L.; Villa, E.; Loia, V. Replicating a Trading Strategy by Means of LSTM for Financial Industry Applications. *IEEE T. Ind. Inform.* **2018**, *14*, 3226–3234. [CrossRef]
76. Sundermeyer, M.; Schlüter, R.; Ney, H. LSTM Neural Networks for Language Modeling. In Proceedings of the 13th Annual Conference of the International Speech Communication Association 2012 (INTERSPEECH 2012), Portland, OR, USA, 9–13 September 2012; pp. 194–197.
77. Liao, Z.; Li, Y.; Xiong, W.; Wang, X.; Liu, D.; Zhang, Y.; Li, C. An in-depth assessment of water resource responses to regional development policies using hydrological variation analysis and system dynamics modeling. *Sustainability* **2020**, *12*, 5814. [CrossRef]
78. Horn, A.L.; Rueda, F.J.; Hormann, G.; Fohrer, N. Implementing river water quality modelling issues in mesoscale watershed models for water policy demands–an overview on current concepts, deficits, and future tasks. *Phys. Chem. Earth.* **2004**, *29*, 725–737. [CrossRef]
79. Li, X.; Huang, M.; Wang, R. Numerical simulation of Donghu Lake hydrodynamics and water quality based on remote sensing and MIKE 21. *ISPRS Int. J. Geo-Inf.* **2020**, *9*, 94. [CrossRef]

water

MDPI

Article

Classification and Prediction of Fecal Coliform in Stream Waters Using Decision Trees (DTs) for Upper Green River Watershed, Kentucky, USA

Abdul Hannan [1] and Jagadeesh Anmala [2,*]

1 Department of Mechanical Engineering, Birla Institute of Technology and Science, Pilani, Hyderabad Campus, Jawahar Nagar, Kapra (Mandal), Medchal District, Hyderabad 500078, Telangana, India; f20170732@hyderabad.bits-pilani.ac.in
2 Department of Civil Engineering, Birla Institute of Technology and Science, Pilani, Hyderabad Campus, Jawahar Nagar, Kapra (Mandal), Medchal District, Hyderabad 500078, Telangana, India
* Correspondence: jagadeesh@hyderabad.bits-pilani.ac.in

Abstract: The classification of stream waters using parameters such as fecal coliforms into the classes of body contact and recreation, fishing and boating, domestic utilization, and danger itself is a significant practical problem of water quality prediction worldwide. Various statistical and causal approaches are used routinely to solve the problem from a causal modeling perspective. However, a transparent process in the form of Decision Trees is used to shed more light on the structure of input variables such as climate and land use in predicting the stream water quality in the current paper. The Decision Tree algorithms such as classification and regression tree (CART), iterative dichotomiser (ID3), random forest (RF), and ensemble methods such as bagging and boosting are applied to predict and classify the unknown stream water quality behavior from the input variables. The variants of bagging and boosting have also been looked at for more effective modeling results. Although the Random Forest, Gradient Boosting, and Extremely Randomized Tree models have been found to yield consistent classification results, DTs with Adaptive Boosting and Bagging gave the best testing accuracies out of all the attempted modeling approaches for the classification of Fecal Coliforms in the Upper Green River watershed, Kentucky, USA. Separately, a discussion of the Decision Support System (DSS) that uses Decision Tree Classifier (DTC) is provided.

Keywords: stream water quality; CART; ID3; random forest; bagging; boosting; extremely random trees; Gradient Boosting; land use factor; Decision Support System

Citation: Hannan, A.; Anmala, J. Classification and Prediction of Fecal Coliform in Stream Waters Using Decision Trees (DTs) for Upper Green River Watershed, Kentucky, USA. *Water* **2021**, *13*, 2790. https://doi.org/10.3390/w13192790

Academic Editors: Nigel W.T. Quinn, Ariel Dinar, Iddo Kan and Vamsi Krishna Sridharan

Received: 24 August 2021
Accepted: 5 October 2021
Published: 8 October 2021

1. Introduction

Water is a life-sustaining element that determines the establishment and survival of civilizations. The availability of innocuous and valuable quality water for household and industrial activities is essential for economic prosperity. Organizations worldwide, such as the World Health Organization (WHO), have specified quality standards for each natural source of this vital element. In the United States of America, the United States Environmental Protection Agency (USEPA) establishes standards for water quality and undertakes quality control measures. Due to the sporadic nature of rainfall globally, several countries are dependent on their rivers to be a primary source of water. Massive dams are constructed, and the water held is used for variegated purposes, including irrigation, electricity generation, domestic activities, etc. Over the past decade, industrial and human fecal waste deposition into the rivers, owing to burgeoning urbanization, had substantially exacerbated water contamination levels. Human interferences in nature have caused imbalances in nature, which, if not regulated, are deleterious and threaten humankind's very existence. Therefore, it is essential to monitor the damages that we have caused.

Water quality assessment is an integral part of environmental engineering. It includes the evaluation of the chemical, biological, and physical characteristics of water. Factors that determine water quality are (i) Physical: temperature, turbidity, suspended solids, color, taste, (ii) Chemical: pH, dissolved oxygen, Total Dissolved Solids (TDS), alkalinity, hardness, and (iii) Biological/Microbiological: Pathogens, coliforms. Fecal coliform is a bacterium that originates from the intestines of humans and other warm-blooded animals. The presence of this bacteria itself in a water body poses no direct harm, but it acts as an indicator of the existence of pathogens that may be harmful. Hence, tests and experiments are performed to measure fecal coliform concentration, which helps us determine the water sample's quality. Water quality determination is a tedious, lab-intensive process.

Moreover, current monitoring methods cannot provide real-time results because of testing time requirements. There is a need for practical, cost, and labor-efficient methods to indicate bacterial concentration on a real-time basis. The fecal coliform presence is measured by the number of colonies per 100 mL of sampled water. In general, the sources of fecal coliform loads in freshwater systems are due to wastewater treatment plant effluents, failed septic systems, human and animal manure [1]. High loads of bacterial contamination are found in rural or agricultural watersheds and urban watersheds streams. The farm cattle waste and failed septic systems of rural watersheds are replaced by domestic pets manure and failed sanitary sewers in urban watersheds. The current USEPA stipulations are as follows for four classes of freshwater systems: (i) Less than one colony/100 mL for drinking water standards, (ii) fewer than 200 colonies/100 mL for body contact recreation, (iii) fewer than 1000 colonies/100 mL for fishing and boating, and (iv) fewer than 2000 colonies/100 mL for domestic water supply. The fecal coliform present in the human and animal waste goes down the drain in houses and businesses from septic systems, overland plane areas through illegal and leaky sanitary sewer pipes to freshwater streams and rivers. It gets transported within the streams due to advection, diffusion, adsorption, and dispersion further down to the outlets. Due to the high affinity of bacteria to the soil, high sediment loads also contain high concentrations or loads of bacteria. High runoff events or storm runoff also known to contain higher levels of bacterial concentrations for the above reasons. The management actions usually include steps such as (i) routine maintenance of septic tanks, (ii) repair of broken field lines, (iii) elimination of straight pipes and failing septic systems, and (iv) isolation of cattle from streams [1].

In recent years, the Artificial Intelligence has been extensively used in the field of environmental engineering across several applications. Researchers have implemented various Machine Learning algorithms for water quality assessment.D'Agostino [2], Gaus [3], and Arslan [4] used Geographical Information System (GIS) to assess water quality parameters. Ahn & Chon [5] used thematic maps of pH, electrical conductivity, nitrate, and sulfate to create maps to utilize water for drinking purposes. Bae [6] used Classification and Regression Trees (CART) for prediction of indicator bacterial concentration in coastal Californian waters. Dissolved oxygen was found to be the most important parameter for the prediction of total and fecal coliforms, while the turbidity was found to be important for enterococci (ENT) using CART decision tree analysis. The pH, temperature, and streamflow were found to be less important for prediction of indicator bacteria. It was possible to predict the indicator bacterial concentrations in real time using CART, saving huge monitoring costs for the state of California. Liao & Sun [7] analyzed the water quality of Chao Lake in China using Improved Decision Tree Learning (IDTL) models that use the feedforward neural network model for preclassification. This model was found to be comparably successful with that of pure neural network models or pure decision tree models such as C4.5. This model was recommended for practitioners as it is faster and uses fewer decision rules. Nikoo [8] developed an integrated water quantity and quality model using the M5P decision tree algorithm for Total Dissolved Solids (TDS) as the water quality indicator. A comparison was made between optimization, support vector regression (SVR), and M5P models. It was found that the M5P model yields explicit relationships between inputs and output such as TDS, which are useful for a decision maker or a practitioner. Azam [9] classified the water quality data using Decision Trees (DTs), Logistic Regression (LR), and Linear Discriminant Analysis (LDA) for two cities. Maier and Keller [10] developed the Random

Forest (RF), Multivariate Adaptive Regression Splines (MARS), Extreme Gradient Boosting (XGB) regression models in conjunction with hyperspectral data to estimate water quality parameters for inland waters. Jerves Cobo [11] used Decision Tree models for assessment of microbial pollution in rivers by studying the presence of macroinvertebrates as indicators. This was needed to set up the pathogen pollution standards and to review the aquatic ecosystem health. Geetha Jenifel and Jemila Rose [12] have used recursive partitioning with decision trees and regression trees to predict water quality parameters and have found better results than the other models such as linear and support vector machine (SVM) models. They have found the decision tree models to be more accurate, practical, reasonable, and acceptable. Ho [13] used the ID3 decision tree model to predict the Water Quality Index (WQI) class for one of Malaysia's most polluted rivers. Sepahvand [14] compared the performances of the M5P model tree and its bagging, Random Forest (RF), and group method for data handling (GMDH) in the estimation of sodium absorption ratio (SAR). Out of all the models they have tried out, bagging M5P model was found to be the most accurate in estimating SAR. This was based on the indices such as correlation coefficient (CC), root mean square error (RMSE), and mean absolute error (MAE). The uncertainty analysis also revealed the accuracy of bagging M5P model compared to other models. Lu and Ma [15] used various hybrid Decision Tree methods and ensemble techniques such as Random Forest (RF), CEEMDAN, XGBoost, LSSVM, RBFNN, LSTM, etc and their combinations for the prediction of water quality parameters of the Tualatin river of Oregon, USA. Shin [16] predicted cholorophyll-a concentrations in the Nakdong River, Korea using various machine learning (ML) models and found the best results using recurrent neural networks (RNNs). The RNNs performed best when time-lagged memory terms are built into the model for predictions. Mosavi [17] compared the performance of two ensemble decision tree models- boosted regression trees (BRT) and random forest (RF) to predict hardness of groundwater quality. More recently, Naloufi [18] used six machine learning (ML) models, including Decision Trees, to predict *E. Coli* concentrations in Marne River in France. They found the Random Forest model to be the most accurate compared to other models. Using the results, they were able to come up with the best ML model for sampling optimization. The other recent discussions and applications of Decision Trees in the context of water quality modeling in rivers include that of studies [19–27].

In light of the above literature, the objectives of the current study are: (i) to study the potential and applicability of various Decision Tree (DTs) algorithms in prediction of Fecal Coliform from causal parameters such as climate (precipitation, and temperature), and land use parameters, (ii) to apply CART, ID3, RF and ensemble methods such as bagging and boosting specifically, and (iii) to suggest a Decision Support System (DSS) based on Decision Tree Classifier (DTC) for water quality management. The paper has been organized as follows. Section 2 describes the study area and the data considered for the present work. Section 3 details the methodology including GIS land use analysis, briefing of Decision Trees for classification, measures of attribute selection and their relation with the Decision Trees, and ensemble methods of bagging and boosting. The results are provided and discussed in the next Section 4. In the same section, a detailed discussion of Decision Tree Classifier based Decision Support System (DTCDSS) is provided. In the last section, i.e., Section 5, a concise summary of findings is provided.

2. Description of Study Area and Data

2.1. Study Area

The watershed under study is situated in the lower southwestern part of Kentucky in the USA. The watershed is called the Upper Green River watershed as the main stem of Green River, and its tributaries flow through it. The watershed consists of Russell, Adair, Taylor, Metcalfe, Green, Barren, Hart, and Edmonson counties. Although the watershed contains various land uses such as deciduous, evergreen, mixed forest, transitional, low and high intensity residential, commercial, industrial/transportation, open water, pasture or hay, emergent herbaceous wetlands, row crops, woody wetlands, they can be primarily grouped into three types of land using: urban, forest, and agricultural. The Green River flows through Edmonson, Hart, Green, and Taylor counties and joins the Ohio river downstream of the watershed. Extensive underground karst formations and springs exist between the tributaries of the main stem Green River. The watershed primarily consists of flat-lying limestones, sandstones, and shales forming the karst topography that passes all the surface water through caves and smaller underground passages below the ground surface [28]. The watershed is rated as the fourth most crucial watershed in the United States by the Nature Conservancy and the Natural Heritage Program. It is also the most critical watershed in Kentucky to protect fish and mussel species.

In this watershed, many rural households are not connected to wastewater treatment plants, and the untreated wastewater is directly discharged to streams and creeks, onto overland plains and soils, or into empty spaces of underground. This form of release is known as "straight pipe" discharge. Due to such discharges and failed septic systems, increased fecal coliform bacteria concentrations are found in various portions of the watershed and have essentially impacted the water quality of the Upper Green River basin [28]. The increased concentration of bacteria is so high that the streams are unsafe for fishing, swimming, and body contact. The storage and carefully timed application of animal manure as a fertilizer has been shown to reduce bacteria entering ground water and reduce the need for expensive chemical fertilizers.

2.2. Data

The minimum and maximum elevation levels of the watershed are at 123.14 m and 497.74 m. The minimum and maximum temperatures in a year are around 10.3 °C and 28.9 °C. The average annual precipitation is found to vary between 1041 mm to 1346 mm. The stream water quality sampling stations are located between the latitudes 36.94° and 37.43° and longitudes −86.04° and −85.16°. The water quality data is obtained by measuring samples collected from nearly 42 locations along the Green river monthly from May 2002 to October 2002. The stream network with sampling stations is shown in Figure 1, and the land use map of the watershed is shown in Figure 2. The climate parameters precipitation and temperature are obtained from Kentucky Climate Center. The two-day cumulative precipitation at each sampling location is computed by inverse distance weighted average procedure. The two-day cumulative precipitation, temperature at each sampling station are used along with the land use as inputs into the Decision Tree Classifier model. The six-month period includes a few rainy months and a few non-rainy months.

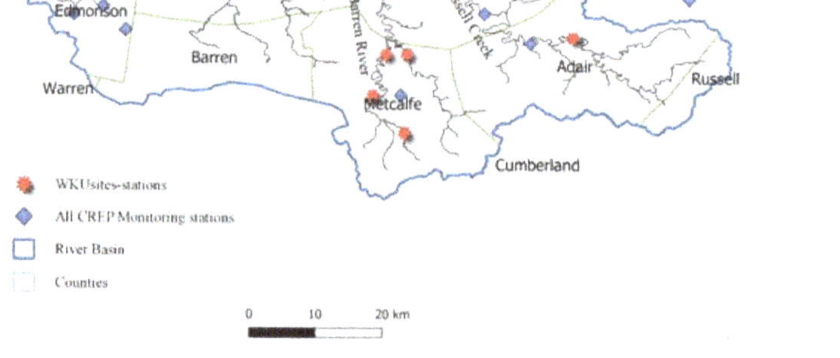

Figure 1. The stream networkand the sampling stations in the Upper Green River watershed.

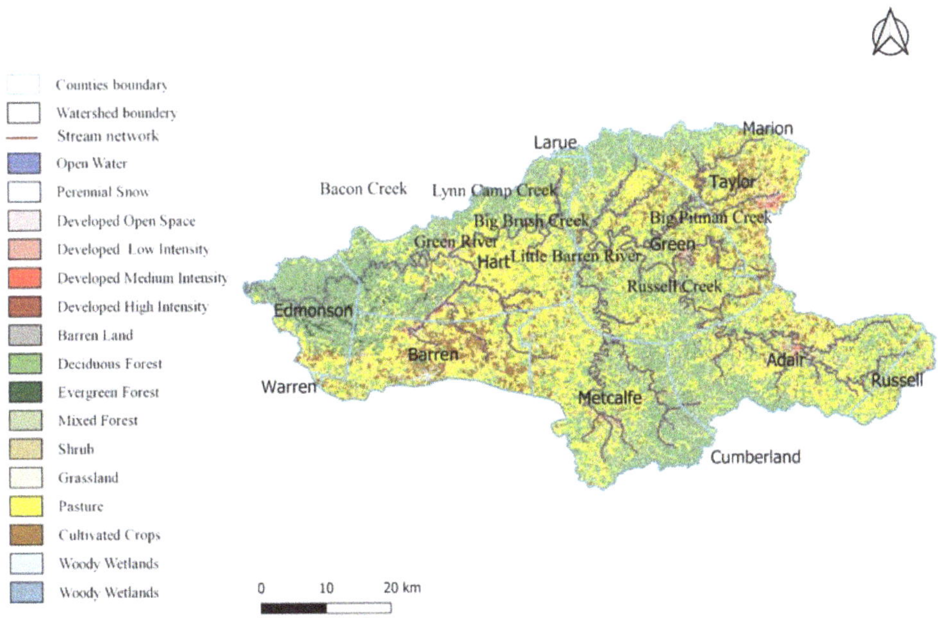

Figure 2. Land Use Map of Upper Green River Watershed.

3. Methodology

3.1. GIS Landuse Analysis

The land use factors developed using GIS analysis [29] essentially indicate the percentage contribution of each land use to the total catchment area or watershed area. The three dominant land use factors are ULUF (urban land use factor), FLUF (forest land use factor), and ALUF (agricultural land use factor). They are given by

$$ULUF = sin^{-1}\sqrt{urban\ area/\ watershed\ area} \quad (1)$$

$$FLUF = sin^{-1}\sqrt{forest\ area/\ watershed\ area} \quad (2)$$

$$ALUF = sin^{-1}\sqrt{agricultural\ area/\ watershed\ area} \quad (3)$$

Moreover, the above definitions have several advantages as compared to the land uses by simple fractions. The arcsine transformation helps in making the land use distribution near-normal or Gaussian. This transformation gives the values in radians proportional to the angle subtended at the center on a pie chart of land uses [25]. The transformation is known to stabilize the variance and scales the proportional data [30–33].

3.2. Decision Trees

The decision tree algorithm falls in the category of supervised learning methods [34]. The goal of a decision tree algorithm is to use the appropriate property of the input data (Information Gain, Gain Ratio, or Gini Index) depending on the type of decision tree selected with learning rules to cause splits in the input parameters and give output values as close to the target as possible. Various decision tree models have been devised, which differ in their data splitting methods. Each subsequent model is an improvement over the previous one. A decision tree classifies the input data based on these rules in a top-down fashion. The top node that is assigned all the data is called a root node. The leaf node contains data points that belong to one class or have one specific value, which demarcates the end of a decision tree branch. Further division of the data according to the rules, as mentioned earlier will not be possible. Suppose a node contains heterogeneous data belonging to two or more classes or has two different output values. In that case, that node can be further split to classify data in their respective categories. Such nodes that can be further split are called decision nodes. It is the decision node that takes decisions and propagates the tree to give relevant outputs.

Tree-structured regression methods form a hierarchical structure with the input data being split at the nodes creating smaller subsets till maximum homogeneity is obtained. For example, Y can first be split into $\{Y \mid y1 < 35\}$ and $\{Y \mid y1 \geq 35\}$ engender two nodes with greater homogeneity than the first node. These two nodes can further be split into $\{Y \mid y_4 \geq 27\}$ and $\{Y \mid y_4 < 27\}$ as shown in Figure 3.

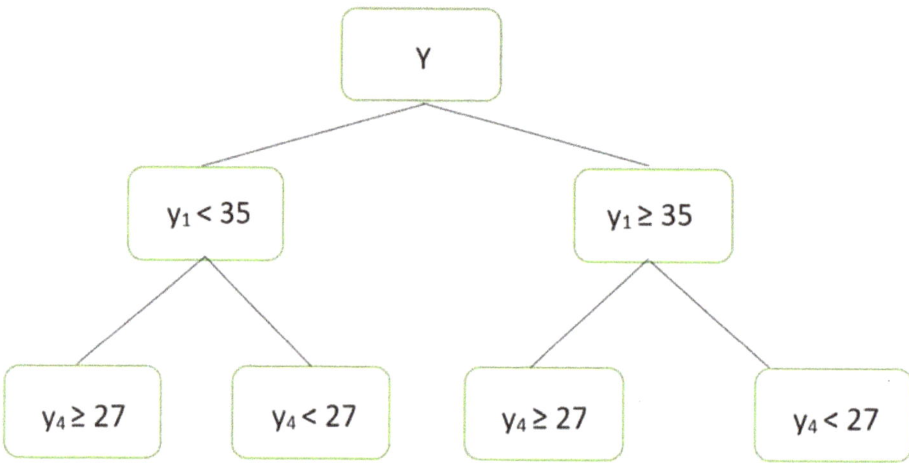

Figure 3. A simple tree at classification.

Consequently, similar splitting at each node will result in a tree structure with multiple branches. Selecting a particular attribute and the value of that particular attribute that determines the splitting of data is the basis for nuances in different decision tree models. Classification and Regression Trees (CART), CHAID, ID3, C4.5, and Random Forest are a few to name. For training and testing purposes, the six-month data set covering forty-two locations spread over the stream network is randomly divided into 70% for training and 30% for testing using the scikitlearn library of python programming language. The same training set and testing set are used to examine their classification accuracy for all the decision tree models.

3.3. *Attribute Selection Measures*

3.3.1. Entropy

Entropy indicates the degree of randomness in a given input set. A branch with 0 entropy is chosen to be the leaf node. If the entropy is not equal to zero, the branch is further split. The Entropy, E(S), measured in "bits" is given by

$$E(S) = \sum_{i=1}^{K} -p_i log_2 p_i \tag{4}$$

where p_i is the percentage of class i in the node or the probability, and index i runs from 1 to K number of classes or attributes. The process of splitting an attribute is continued until the entropy of resulting subsets is less than the previous input or training set, eventually leading to leaf nodes of zero entropy. Minimization of entropy is desired as it reduces the number of rules of the Decision Trees. The lowering of entropy essentially leads to Decision

Trees with fewer branches. The entropy is defined as an information-theoretic measure of the 'uncertainty' present in the dataset due to multiple classes [35].

3.3.2. Information Gain

The information gain of an input attribute gives its relation with the target output. A higher information gain suggests that the parameter can separate the training data following the target output to a greater extent. Information gain is inversely proportional to the entropy. The higher the randomness in a set of inputs, the lower will be the information gain. The ID3 model [36] uses Information Gain as the method for classification.

$$Information\ Gain = Entropy(before) - \sum_{j=1}^{K} Entropy(j, after) \quad (5)$$

where the index j runs from 1 to K possible classes. Maximizing the information gain essentially leads to minimization of entropy for that particular attribute. In the above Equation (5), the first term on the right-hand side is fixed or it is the entropy at the beginning. The attribute is selected for splitting first for which we obtain the minimal second term on the right-hand side resulting in maximization of information gain for that particular attribute.

3.3.3. Gini Index

It is obtained by the sum of squares of individual probabilities of each class from one. A Higher Gini index value indicates higher homogeneity. The CART algorithm uses the Gini Index to create splits in data [37]. The equation gives Gini index at a node

$$Gini = 1 - \sum_{i=1}^{K} (P_i)^2 \quad (6)$$

where p_i is the percentage of class i in the node, and the index i runs from 1 to K number of classes. It measures the "impurity" of a dataset. It takes a minimal value of zero to a maximal value of (1-1/K). In the attribute selection process of Decision Tree modeling, that particular attribute is selected for which there is a largest reduction in the value of Gini index. It turns out the reduction of Gini index essentially is accompanied by lowering of entropy.

3.3.4. Gain Ratio

C4.5, an improvised version of ID3, uses gain ratio to create splits in the input data. The Gain ratio removes the bias that exists in calculating the information gain of an input parameter. Information gain prefers the parameter with a large number of input values. To neutralize this, the Gain Ratio divides the Information Gain by the number of branches that would result from the split.

$$Gain\ Ratio = \frac{Information\ Gain}{Split\ Information} = \frac{Entropy(before) - \sum_{j=1}^{K} Entropy(j, after)}{\sum_{j=1}^{K} -w_j log_2 w_j} \quad (7)$$

where the index j runs from 1 to K possible number of nonempty classes, and wj is the percentage of class i in the node or the probability. The lower the Split Information, the higher the value of Gain Ratio. The Information Gain is essentially modified by the diversity and distribution of attribute values into the quantity known as Gain Ratio.

3.4. *Bagging and Boosting*

Bagging, or Bootstrap aggregation, proposed by [38], is a technique used with regression methods to decrease the variance and improve prediction accuracy. It is a simple technique where several bootstrap samples are drawn from the input data, and prediction is made following the same prediction method for each sample. The results are merged by averaging (regression) or simple voting (classification) that adumbrate the input data results subjected to the same prediction method as the bootstrap samples but with reduced variance. All the bootstrap samples have the same size as the original data. The sampling is done with replacement, because of which, a few instances/samples are repeated, and a few are omitted. The stability of base classifier of each bootstrap sample essentially determines the performance of bagging. Since all the samples are equally likely to get aggregated, bagging does not suffer from issues of overfitting and works well with noisy data. Thus, the focus on a specific sample of training data is removed.

Boosting, similar to bagging, is a sample-based approach to improve classification and regression models' accuracy; however, unlike bagging, which uses a direct averaging of individual sample results, boosting uses a weighted average method to reduce the overall prediction variance. All the samples are initialized with equal weights, then the weights are updated with each boosting classification round. The weights of samples that are harder to classify are increased, and the weights are decreased for the samples that are correctly classified. This ensures the boosting algorithm to focus on samples that are harder to classify with increase in iterations. All the base classifiers of each boosting round are aggregated to obtain the final ensemble boosting classification. The fundamentals of bagging and boosting could be found in [39].

4. Results and Discussion

4.1. Overview

Results obtained using decision trees are discussed in this section. A preliminary statistical analysis is first performed to study the statistics of the experimental data. The results from the initial statistical analysis are given in Table 1.

Table 1. Statistics of Water Quality Parameter and inputs.

Water Quality Parameter	Variable	Sum	Average	Standard Deviation	Input/Output
Precipitation (in cm)	P	817.0	3.6	4.21	Input
Temperature (°C)	T	4250.7	18.9	4.36	Input
Urban land-use factor	U	35.3	0.16	0.12	Input
Forest land-use factor	F	165.2	0.73	0.12	Input
Agricultural land-use factor	A	178.0	0.79	0.11	Input
Fecal Coliform (#colonies/100 mL)	FC	453,889	2017.3	3454	Output

The correlation structure of all the input water sample parameters and the fecal coliform is given by the correlation heatmap shown in Figure 4.

The agriculture land-use factor (ALUF or a) is highly and negatively correlated with the forest land use factor (FLUF or f). FLUF also has a strong negative correlation with urban land use factor (ULUF or u). An exciting inference from the correlation heatmap is the heavy positive correlation between precipitation and fecal coliform (please see Figure 4). A few scatter plots that reveal the relation between the input variables and the fecal coliform individually essentially display the results of the correlation map, i.e., Figure 4, and are not reproduced here. This makes precipitation the most significant variable in determining our output using the decision tree method, and this is evident from the CART and ID3 diagrams given a little later. The precipitation values are obtained at each sampling locations by interpolating the two-day cumulative precipitations at all the gauges in the watershed. The rainfall causes the surface water flow over the watershed's overland planes consisting of land uses (predominantly urban, forest, and agricultural) and eventually joining the tributaries and mainstem of the Green River. The various parts of the overland planes of the watershed contribute as surface water flows, which enters tributaries first and then the main stem of Green River before reaching the watershed outlet tip. Based on watershed characteristics and time of concentration studies, the two-day cumulative and interpolated precipitation values are most suitable drivers of fecal coliform concentrations than other precipitation measures at all the sampling sites [29]. The positive correlation of microbial indicators such as fecal coliform bacteria and precipitation/rainfall/wet weather conditions is in agreement with the other studies such as that of [40–44] for rivers and bays and [45,46] for lakes. The DTs are best suited for large datasets; however, an attempt is made in the current study because of the multi-dimensional feature space (five independent variables) for monthly instances of data of forty-two locations collected for the six-month period. The DTs are expected to work better for shorter data sets and fewer features as the intrinsic data complexities are reduced. In the present scenario, data limitations on the time frame are offset by the multi-dimensional feature space for varied spatial locations. The reduction of input dimension has been looked into for the same dataset using principal component analysis (PCA), canonical correlation analysis (CCA), and artificial neural networks (ANNs) elsewhere [47]. The authors have found comprehensive predictions using all the spatial parameters such as land uses, and temporal climate parameters. The histogram of fecal coliform is given in Figure 5. The variability of fecal coliform is large as is evident from the high standard deviation and the histogram.

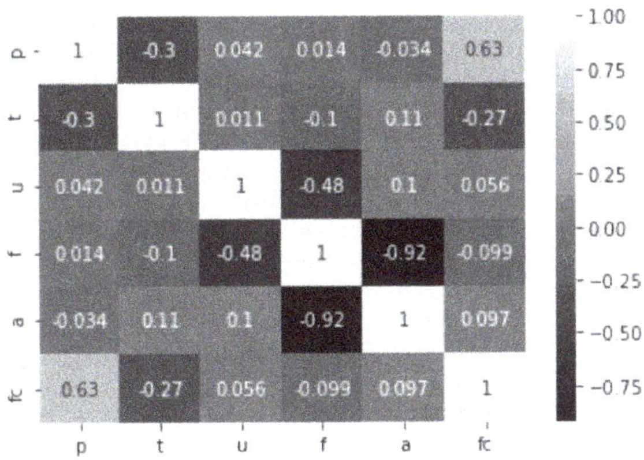

Figure 4. Correlation Heatmap.

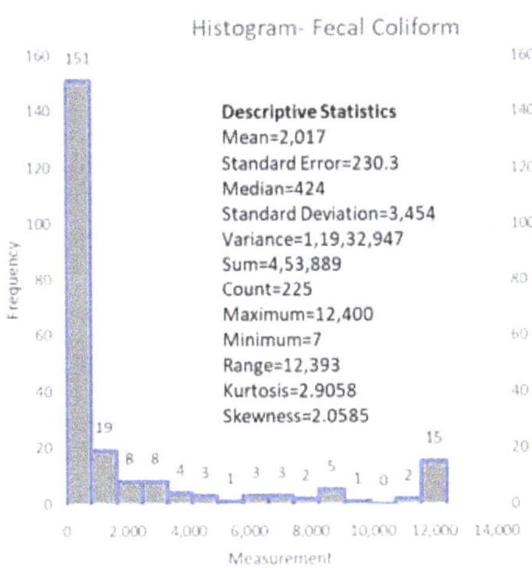

Figure 5. Histogram of Fecal Coliform (FC).

4.2. Results from Decision Tree Models

The precipitation, temperature, land use data, and experimentally measured fecal coliform values are used to formulate decision tree models. Several different decision tree models are developed for fecal coliform analysis. All input parameters (precipitation or P, temperature or T, ULUF or u, FLUF or f, ALUF or a) are used to create decision tree models with different data split methods. Precipitation was the single most crucial hydrological input parameter to determine fecal coliform. The correlation values obtained earlier also indicate similar results. 70% of the dataset is used as the training set, and the remaining 30% is used as the testing set. For CART, an accuracy of 63.05% was obtained in the training phase, and 60.29 was obtained in the testing phase. An accuracy of 62.22% was obtained on the entire data set. In CART decision tree modeling, the data is split based on the attribute with the lowest Gini index at the root node in a top-down process with the subsequent splitting of data of attributes of increasing Gini indices till we reach leaf nodes. In this way, the impurity or the uncertainty in the data is minimized with recursive portioning of the data [37]. Similar results were obtained for the ID3 model. Accuracies of 61.78%, 61.76%, and 61.77% were obtained in the training phase, testing phase, and the entire data set, respectively. In the ID3 model, the feature with maximum information gain or smallest entropy is used to split the data at the root node first and then the subsequent nodes till we reach leaf nodes. The least entropy corresponds to the features with the least uncertainty or randomness in the data [36]. Both DTs, CART, and ID3, belong to the family of Top-Down Induction Decision Trees (TDIDT). CART performs slightly better in training than ID3, and ID3 performs slightly better in testing than CART. However, the overall performance of CART is slightly better than ID3. CART and ID3 models were improved by augmenting the simple models with bagging and boosting methods. The highest test set accuracy was obtained for the CART model with adaptive boosting—the accuracy of 81.53%, 72.06%, and 78.67% was obtained in the training phase, testing phase, and entire dataset, respectively. The bagging and adaptive boosting of CART and ID3 perform much better than simple (without bagging and adaptive boosting) CART and ID3 models. Though bagging of ID3 results in largest training accuracy among simple and ensemble models of CART and ID3, the adaptive boosting of CART gives the largest testing accuracy among the same models. However, the overall accuracy of bagging of ID3 model is the highest among simple and ensemble models of CART and ID3 models. Apart from CART and ID3 models, Random Forest was also implemented on the experimental dataset to predict the fecal coliform density or concentration.

The Random Forest model gives an accuracy of 98.7% on the training set, 64.7% on the testing set, and 88.4% on the overall dataset. The Random Forest model is built by creating an ensemble of a large number of decision trees for classification and then predicting the mode or average/mean of all the individual decision tree classification results. The more uncorrelated the individual decision trees are, the better the final prediction or outcome [48]. The individual trees or sub-samples are drawn randomly from the original tree with replacement. The trees are grown to the largest extent possible for classification without pruning of the trees. The features selected in sub-sample trees need to be useful for the effectiveness of the Random Forest model than being pure, random guessing features in classification. The Random forest model outperforms decision trees such as CART, ID3 but its testing accuracy is slightly lower than gradient trees, extremely randomized trees, and DTs with bagging and adaptive boosting. A few other models, such as extremely randomized trees, were also implemented, and the accuracy results of all models are summarized in Table 2 below:

Table 2. Accuracies of various DT models in the prediction of FC.

Model	Training (%)	Testing (%)	Overall (%)
CART (Adaptive Boosting)	81.53	72.06	78.67
ID3-Bagg	86.62	70.58	81.78
ID3-AB	85.98	70.58	81.33
CART (Bagging)	80.25	70.58	77.33
Gradient Boosting (GBM)	98.19	69.12	89.33
Extremely Randomized Trees (ERT)	98.72	66.17	88.89
Random Forest (RF)	98.70	64.70	88.40
ID3	61.78	61.76	61.77
CART	63.05	60.29	62.22

Where ID3-Bagg is ID3 with Bagging, and ID3-AB is ID3 with adaptive boosting. The extremely randomized trees (also known as "Extra Trees") give a fourth-best testing accuracy of 66.17% and a second-best overall accuracy of 88.89% of all the models. While the Random Forest model uses subsamples with replacement, the extremely randomized trees use the whole input sample. Also, while Random Forest opts for optimum split to select cut points, the extremely randomized trees go for random cut points. The extremely randomized trees are faster and have both features of reducing bias and variance due to usage of original input sample and random split points [49]. Gradient boosting model (GBM) gives third-best testing accuracy of 69.12% and best overall accuracy of 89.33% of all the models. In the Gradient Boosting model, a loss function such as mean square error is minimized with the help of gradient descent principle and an ensemble of weak learners to eventually make correct predictions and become a strong learner tree [50]. The GBM, ERT, and RF perform better than simple and ensemble models of CART, and ID3 in training, and overall accuracies; the ensemble models of CART, and ID3 are slightly better in testing accuracies. This could also be due to possible overtraining of the Decision Tree models in the case of GBM, ERT, and RF. Further, optimal cutting down of trees may result in higher Training, Testing, and Overall accuracies of GBM, ERT, and RF than simple and ensemble models of CART, and ID3. The accuracy of a decision tree model is given by the number of correct predictions made divided by the total number of predictions. Here, the prediction is the class to which the water sample belongs. These results are presented in Figure 6.

We have used various decision tree algorithms to classify data rather than regression on our current data set. For classification, the target variable, i.e., fecal coliform (FC), was divided into four classes following the United States Environmental Protection Agency (USEPA) recommendations (given in Table 3).

Table 3. Fecal Coliform and its Class.

Class	Fecal Coliform (FC) Range (cfu/100 mL)
Body contact and recreation (BCR)	$0 < FC \leq 200$
Fishing and boating	$200 < FC \leq 1000$
Domestic utilization	$1000 < FC \leq 2000$
Dangerous	$FC > 2000$

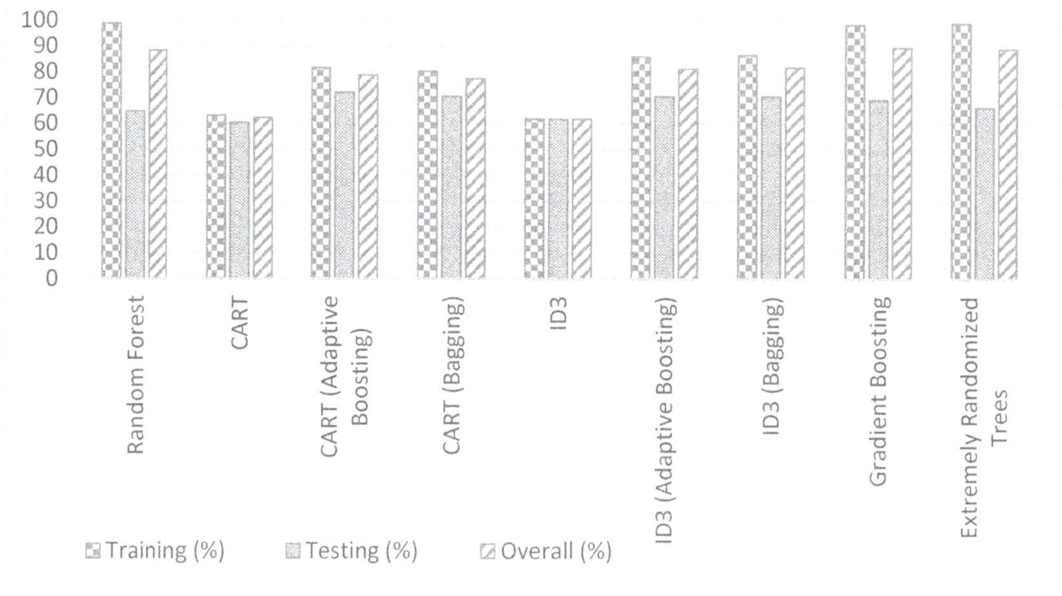

Figure 6. Bar graph of training, testing, and overall accuracies of different DT models.

The classified results into four classes or categories, namely- body contact and recreation, fishing and boating, domestic utilization, dangerous for all decision tree algorithms, are presented in Figure 7. The number of samples in each class is also shown in the same Figure. From Figure 6, we can see that high values of precision, recall, and F1-score are obtained for Random Forest, CART with adaptive boosting, ID3 with adaptive boosting, and Extremely randomized trees for the supports shown for all of the four classes. The definitions of accuracy, precision, recall, F1-score are given by:

$$Accuracy = \frac{TP + TN}{TP + FP + TN + FN} \tag{8}$$

$$Precision = \frac{TP}{TP + FP} \tag{9}$$

$$Recall = \frac{TP}{TP + FN} \tag{10}$$

$$F1_{score} = 2 \times \frac{Precision \times Recall}{Precision + Recall} \tag{11}$$

where TP = True Positive, TN = True Negative, FP = False Positive, FN = False Negative, and support is the number of occurrences of each class in ground truth (correct) target values.

This means that the above four algorithms can correctly classify the positive samples from negative samples for each of the respective class, able to recall all of its positive samples and that both of these abilities are equally important in the classification. Then the best values of precision, recall, and F1-score are obtained for CART with bagging and ID3 with bagging algorithms. Simple CART and ID3 yielded lower precision, recall, and F1-score than the rest of the above algorithms discussed in their respective classifications for each support class. Although not presented here, accuracies were also highest for Extremely randomized trees and Random Forest algorithms, slightly lower for CART and ID3 with bagging and boosting, and lowest for simple CART and ID3 algorithms.

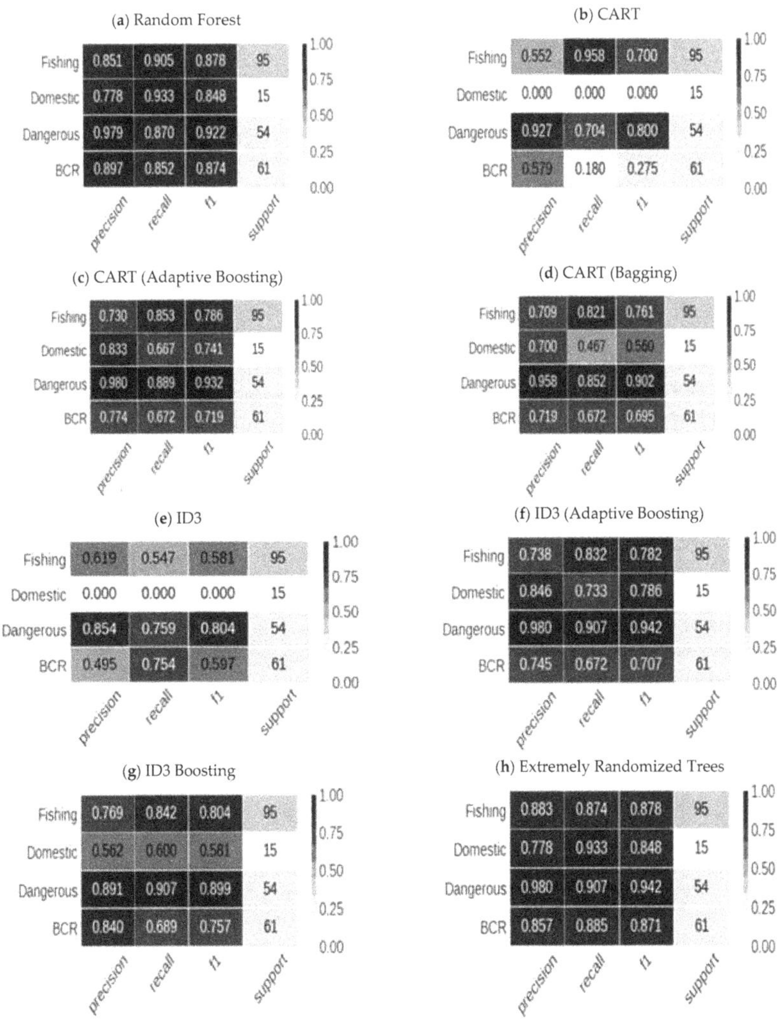

Figure 7. Classification reports of the entire dataset using different DT models.

4.3. CART with Bagging and Adaptive Boosting

The CART algorithm uses Gini impurity to split data and form a binary classification tree. Implementation of the CART algorithm on our data set results in a tree with four levels (please see Figure 8).

Level 1 has the root node. Level 2 and level 3 contain the decision nodes, and level 4 has the leaf nodes with data split into classes. The tree's root node containing all 225 data points of our data set has been split based on the precipitation input variable (p) since p is the variable of the highest significance. The decision nodes at level 2 are split using the forest land use factor (f) and the temperature (t) input variables since these nodes represent points of highest Gini impurity. At level 3, we get our first leaf node of class "Dangerous" with eight samples falling in this category ($p \leq 0.308$, $f \geq 0.763$). The final level contains the leaf nodes that classify the data into the specified classes. The class with the highest number of samples is fishing (156 samples), followed by BCR (51 samples). The classes "Domestic" and "Dangerous" have 5 and 13 samples, respectively, giving us a total of

225 samples in our data set. The CART algorithm is a moderately accurate method to classify our data set, giving an accuracy of 63.05% on the training set and 62.22% on the entire data set. However, improving our model using bagging and boosting methods yield even higher accuracies. From Table 2, CART with adaptive boosting gives the best testing accuracy out of all the decision trees. The adaptive boosting method enables to combine several weak classifiers into a strong classifier through an iterative decision tree modeling. The weak classifiers are weighted highly and trained with a few low-weighted strong classifiers to produce a strong ensemble classifier at the end [51]. From Table 2, we can also see that CART with bagging yields second-best testing accuracy. Bagging simply means "bootstrap aggregating." The implementation of CART with bagging results in creating many random sub-samples with replacement and training CART model on each sample. Then the average prediction is made on all the samples [38]. We can see that the ensemble predictions of bagging and boosting of the CART model are better than the simple CART model results.

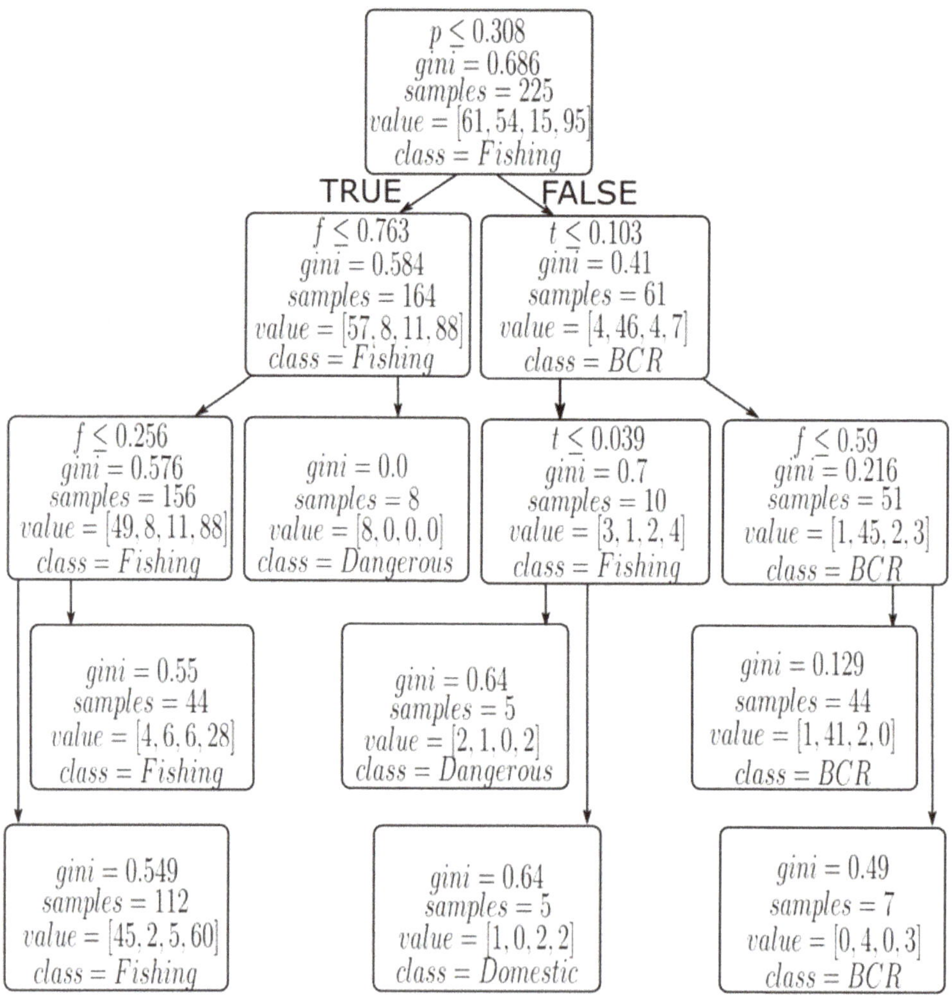

Figure 8. The Decision Tree developed using the CART algorithm.

4.4. ID3 with Bagging and Adaptive Boosting

Like the CART algorithm that uses Gini impurity to form splits in the data set, the ID3 decision tree utilizes the information gain and entropy. Implementation of the ID3 algorithm on our data set also yields a tree with four levels (please see Figure 9).

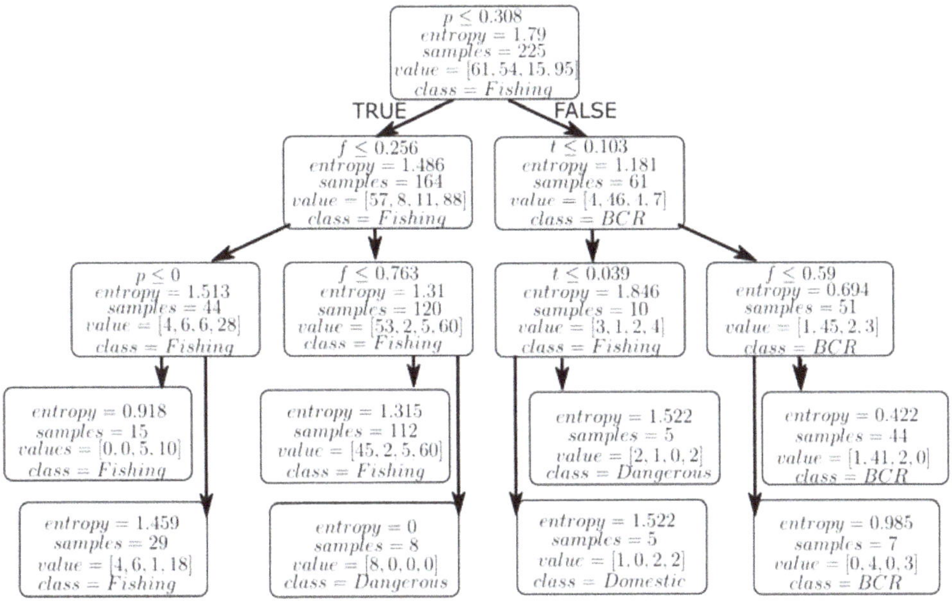

Figure 9. The Decision Tree developed using the ID3 algorithm.

Again, since "p" is the most significant variable, the root node is split using the "Precipitation" input variable. The decision nodes in level 3 and level 4 are split to maintain lower Information Gain and Entropy uncertainty. Level 4 of the ID3 decision tree has the dataset classified into four pre-defined classes. Similar to the CART algorithm results, the class with the highest number of samples is "Fishing" followed by "BCR," "Dangerous" and "Domestic." The accuracies obtained for the training data set and overall dataset are 61.78% and 61.77% respectively, which is slightly lower than that obtained by the CART algorithm. From Table 2, we can see that ID3 with adaptive boosting comparable results to that of CART with adaptive boosting. Adaboost is an iterative procedure with no replacement. It generates a strong ensemble classifier by putting high weights on the mis-classifiers and low weights on the correctly classified trees to reduce bias and variance in the model. For this reason, it is called the "best out-of-the-box classifier" usually. The second-best testing accuracy is obtained using this method with ID3 in all attempted decision tree models. ID3 with bagging results can also be seen from Table 2, which creates many independent bootstrap aggregation models and associates weak learner with each model to finally aggregate them to produce average or ensemble prediction that has a lower variance. This method also yields the second best testing accuracy of all the decision tree models. We can see that the bagging and boosting ensemble methods improve the testing accuracy compared to simple ID3.

4.5. Decision Tree Classifier Based Decision Support System (DTCDSS)

A Decision Support System (DSS) can be built from the above results of decision tree models, as shown in Figure 10.

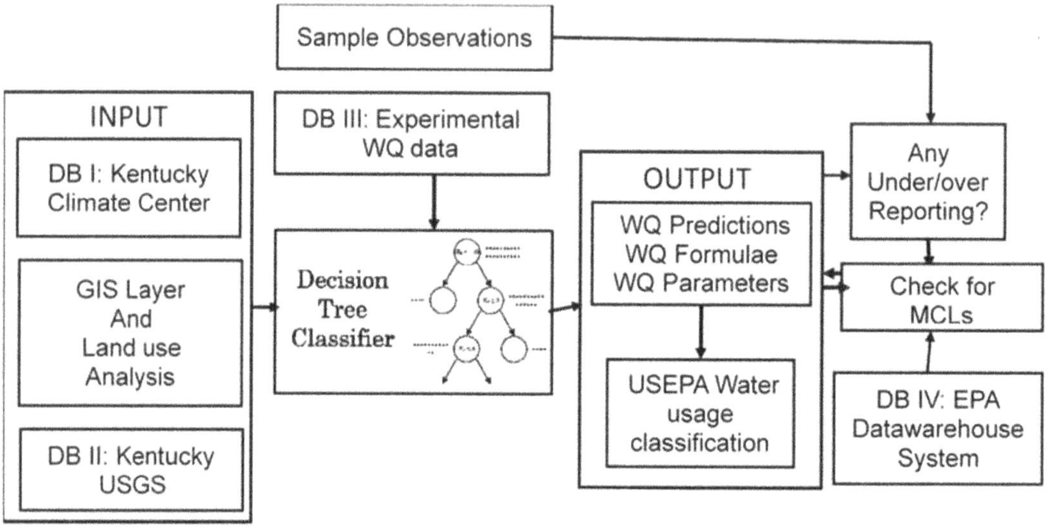

Figure 10. Decision Tree Classifier based Decision Support System (DTCDSS).

The input parameters and the sample water quality observations come from the input databases, as shown in the decision tree classifier (DTC). The above-discussed decision tree models help classify the water quality data based on input parameters and respective decision tree algorithms. The results are shown in the output database in "water quality predictions/formulae/bounds". The outputs are put into and as per classes of US EPA water usage classification. From the output classifications, one could figure out the under/over reporting issues, and check for maximum contamination levels (MCLs). In this process, the US EPA data warehouse system will help compare the output water quality parameters with stipulated MCLs. The inner working of any of the above-discussed decision tree classifiers is shown in the flow chart Figure 11.

Firstly, the DTC classifier receives the input data from the input databases such as climate, land use data, and water quality data. The entropy or information gain, gain ratio, and Gini index are computed based on the particular model chosen of the decision tree in the DTC classifier. For example, if the decision tree in the DTC classifier is ID3, then entropy or information gain is computed. If the decision tree in the DTC classifier is CART, then the gain ratio is computed. Similarly, if the decision tree is C4.5, then the Gini index is computed. The input parameters and the output parameters of a data sample are presented at the root node first. Then the tree is split based on the decision of the "if-else" statement minimizing the heterogeneity of data or increasing the homogeneity. The tree branches keep increasing with the addition of new data samples at the root node, and slowly, the leaf nodes get formed. At every level node of the tree, the entropy or information gain, gain ratio, Gini index are computed so that the data could be split easily and the data get traversed to the leaf nodes. Thus, the DTC classifier helps us classify the data or make a decision into four classes of output, namely, body contact and recreation, fishing and boating, domestic utilization, and dangerous at the leaf nodes. The model performance can be computed using the metrics such as accuracy, precision, recall, and F1-score. The particular DT model in the classifier can be any one of the CART, ID3, C4.5 with bagging and boosting variants, Random Forest, and Extremely Randomized Trees. The output performance of the DSS can be specific to the DT model chosen and could also be data sensitive. The best DT model for the DSS can be fixed only by experimenting with above-stated decision tree models for the data of several watersheds consisting of stream networks of varied conditions. Suppose we replace the DTC classifier with Decsion

Tree Regressor (DTR). In that case, we will be able to predict the output parameter such as fecal coliform, from the input parameters such as climate and land use. The output parameter predictions can also be further generalized or recasted using the regression formulae obtained of DTR and the output parameter bounds. The current Decision Support System is an improvised version of the DSS discussed in [29]. The Artificial Neural Network (ANN) model is replaced by DTC to suit the current problem of classification stream waters into four classes. The performances of DSS using DTR, the comparison of DTR and ANN models are out of the scope of the current work and are pursued elsewhere as separate research studies. Comparing output parameter predictions with respective MCLs within the DSS framework ensures the suitability of stream waters broadly into "safe" or "unsafe," thus making a helpful decision.

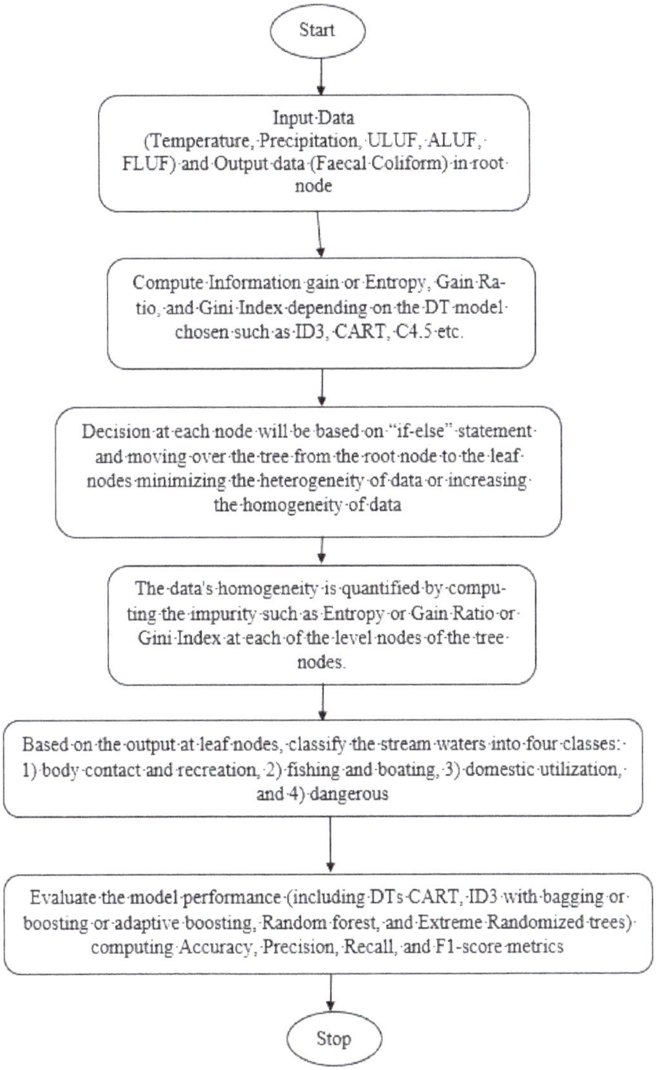

Figure 11. Flow Chart of a Decision Tree Classifier Model.

5. Conclusions

The classification abilities of Decision Trees such as CART, ID3, RF, and ensemble methods such as bagging and boosting are utilized for the classification and prediction of Fecal Coliform (FC) into four classes in this study. The variable with maximum information gain and gain ratio in the case of ID3 model, and the variable with maximum Gini Index in the case of CART model are selected at the root node, and such criteria used further down the tree till the leaf nodes, using DT algorithms for best classification in terms of maximum accuracy. The algorithms perform comparably well with each other, Random Forests being the most consistent in the classification of Fecal Coliform for the Upper Green River watershed overall. It performs better than CART and ID3 in all the phases, i.e., training, testing, and overall. Gradient Boosting and Extremely Randomized Trees are the other DT algorithms that show comparable accuracies as that of Random Forest in training and testing phases. The CART decision tree with Adaptive Boosting yielded the best testing accuracy. In contrast, the CART with Bagging and ID3 with Bagging and Adaptive Boosting yielded the comparable second-best testing accuracies respectively out of all the decision tree modeling attempts. There is no proof of exactly the same feature or attribute will be chosen for each node of the resulting tree for various DT algorithms. There is also no guarantee that accuracy of classification will be higher for the proposed classifier as it needs to be tested for a variety of water quality parameters of different watersheds under climate changes. Also, being greedy at each step/node may not ensure overall minimization of entropy or global optimization of the classification process. In the present work, the authors have focused on only the classification capabilities of the Decision Trees for this particular watershed/dataset. The present work explores the classification capabilities in training and testing phases only. The size of the data was one limitation because of which, the authors could not go for cross-validation. However, the depth of the successful trees is essentially governed by maximizing the information gain or minimizing the entropy, i.e., randomness at every level. It is generally found that the shorter trees are prone to better classification capabilities than the more extended trees [52] (Mitchell, 1997). This is due to lesser overtraining of the trees, leading to more successful generalization or predictions. From the above discussion of results, the following salient conclusions can be made as follows:

(i) The Decision Trees of Gradient Boosting (GB), Extremely Randomized Trees (ERT), and RF perform better than simple (without bagging and boosting) ID3, and CART models in training, testing, and overall.
(ii) The bagging and adaptive boosting Decision Trees of CART, and ID3 significantly improve the performance over simple (without bagging and boosting) CART, and ID3 models.
(iii) The performances of bagging and adaptive boosting Decision Trees of CART, and ID3 are slightly better than GB, ERT, and RF in testing. However, the training and overall accuracies of GB, ERT, and RF are better than all the models (including bagging and adaptive boosting) of CART and ID3.
(iv) The Decision Tree models of GB, ERT, and RF are more consistent than other models in training, testing, and overall accuracies.
(v) Overtraining the trees increases training accuracy at the expense of testing accuracy. A judicious choice need to be made in cutting down the trees, so that an optimal performance of training, testing, and overall accuracies is obtained.

Author Contributions: A.H.—Methodology, software, validation, writing-original draft preparation, visualization; J.A.—Conceptualization, methodology, validation, writing-review and editing, visualization, validation, software, formal analysis, investigation, supervision, project administration, funding acquisition; All authors have read and agreed to the published version of the manuscript.

Funding: Council of Scientific and Industrial Research (CSIR), India grant (No. 24 (0356)/19/EMR-II), project titled "Experimental and Computational studies of Surface Water Quality parameters from Morphometry and Spectral Characteristics".

Institutional Review Board Statement: Not applicable.

Informed Consent Statement: Not applicable.

Data Availability Statement: The data is owned by Upper Green Biological Preserve, Department of Biology, Western Kentucky University, Bowling Green, Kentucky, USA.

Acknowledgments: The corresponding author would like to express acknowledgments to the Council of Scientific and Industrial Research (CSIR), India grant (No. 24 (0356)/19/EMR-II) of the project titled "Experimental and Computational studies of Surface Water Quality parameters from Morphometry and Spectral Characteristics." The authors would like to thank Ouida Meier, Albert Meier, Stuart Foster, Tim Rink, and Jenna Harbaugh for providing us with the required data. The authors would like to thank Vamsi Krishna Sridharan, Institute of Marine Sciences, University of California, Santa Cruz, California, USA for a detailed review and comments of the manuscript. The authors would also like to thank M.M. Prakash Mohan and N.Satish, Research Scholars, Department of Civil Engineering, Birla Institute of Technology and Science, Pilani, Hyderabad Campus for helping out with a couple of figures.

Conflicts of Interest: The authors declare no conflict of interest.

References

1. Ormsbee, L.J.; Anmala, S.E.M. *Total Maximum Daily Load (TMDL) Development for Eagle Creek*; Report submitted to Kentucky Department for Environmental Protection, Division of Water, Frankfort; University of Kentucky: Lexington, Kentucky, 2002.
2. D'Agostino, V.; Greene, E.A.; Passarella, G.; Vurro, M. Spatial and temporal study of nitrate concentration in groundwater by means of coregionalization. *Environ. Earth Sci.* **1998**, *36*, 285–295. [CrossRef]
3. Gaus, I.; Kinniburgh, D.G.; Talbot, J.C.; Webster, R. Geostatistical analysis of arsenic concentration in groundwater in Bangladesh using disjunctive kriging. *Environ. Earth Sci.* **2003**, *44*, 939–948. [CrossRef]
4. Arslan, H. Spatial and temporal mapping of groundwater salinity using ordinary kriging and indicator kriging: The case of Bafra Plain, Turkey. *Agric. Water Manag.* **2012**, *113*, 57–63. [CrossRef]
5. Ahn, H.; Chon, H. Assessment of groundwater contamination using geographic information systems. *Environ. Geochem. Health* **1999**, *21*, 273–289. [CrossRef]
6. Bae, H.-K.; Olson, B.H.; Hsu, K.-L.; Sorooshian, S. Classification and regression tree (CART) analysis for indicator bacterial concentration prediction for a Californian coastal area. *Water Sci. Technol.* **2010**, *61*, 545–553. [CrossRef]
7. Liao, H.; Sun, W. Forecasting and Evaluating Water Quality of Chao Lake based on an Improved Decision Tree Method. *Procedia Environ. Sci.* **2010**, *2*, 970–979. [CrossRef]
8. Nikoo, M.R.; Karimi, A.; Kerachian, R.; Poorsepahy-Samian, H.; Daneshmand, F. Rules for Optimal Operation of Reservoir-River-Groundwater Systems Considering Water Quality Targets: Application of M5P Model. *Water Resour. Manag.* **2013**, *27*, 2771–2784. [CrossRef]
9. Azam, M.; Aslam, M.; Khan, K.; Mughal, A.; Inayat, A. Comparisons of decision tree methods using water data. *Commun. Stat. -Simul. Comput.* **2016**, *46*, 2924–2934. [CrossRef]
10. Maier, P.M.; Keller, S. Machine Learning Regression on Hyperspectral Data to Estimate Multiple Water Parameters. In *2018 9th Workshop on Hyperspectral Image and Signal Processing: Evolution in Remote Sensing (WHISPERS)*; IEEE: Piscataway, NJ, USA, 2018; pp. 1–5.
11. Jerves-Cobo, R.; Córdova-Vela, G.; Iñiguez-Vela, X.; Díaz-Granda, C.; Van Echelpoel, W.; Cisneros, F.; Nopens, I.; Goethals, P.L.M. Model-Based Analysis of the Potential of Macroinvertebrates as Indicators for Microbial Pathogens in Rivers. *Water* **2018**, *10*, 375. [CrossRef]
12. Geetha Jenifel, M.; Jemila Rose, R. Recursive partitioning algorithm in water quality prediction. *Int. J. Environ. Sci. Technol.* **2020**, *17*, 745–754. [CrossRef]
13. Ho, J.Y.; Afan, H.A.; El-Shafie, A.H.; Koting, S.B.; Mohd, N.S.; Jaafar, W.Z.B.; Sai, H.L.; Malek, M.A.; Ahmed, A.N.; Mohtar, W.H.M.W.; et al. Towards a time and cost effective approach to water quality index class prediction. *J. Hydrol.* **2019**, *575*, 148–165. [CrossRef]
14. Sepahvand, A.; Singh, B.; Sihag, P.; Samani, A.N.; Ahmadi, H.; Nia, S.F. Assessment of the various soft computing techniques to predict sodium absorption ratio (SAR). *ISH J. Hydraul. Eng.* **2019**, 1–12. [CrossRef]
15. Lu, H.; Ma, X. Hybrid decision tree-based machine learning models for short-term water quality prediction. *Chemosphere* **2020**, *249*, 126169. [CrossRef]
16. Shin, Y.; Kim, T.; Hong, S.; Lee, S.; Lee, E.; Hong, S.; Lee, C.; Kim, T.; Park, M.S.; Park, J.; et al. Prediction of Chlorophyll-*a* Concentrations in the Nakdong River Using Machine Learning Methods. *Water* **2020**, *12*, 1822. [CrossRef]
17. Mosavi, A.; Hosseini, F.S.; Choubin, B.; Abdolshahnejad, M.; Gharechaee, H.; Lahijanzadeh, A.; Dineva, A.A. Susceptibility Prediction of Groundwater Hardness Using Ensemble Machine Learning Models. *Water* **2020**, *12*, 2770. [CrossRef]

18. Naloufi, M.; Lucas, F.S.; Souihi, S.; Servais, P.; Janne, A.; De Abreu, T.W.M. Evaluating the Performance of Machine Learning Approaches to Predict the Microbial Quality of Surface Waters and to Optimize the Sampling Effort. *Water* **2021**, *13*, 2457. [CrossRef]
19. Chen, K.; Chen, H.; Zhou, C.; Huang, Y.; Qi, X.; Shen, R.; Liu, F.; Zuo, M.; Zou, X.; Wang, J.; et al. Comparative analysis of surface water quality prediction performance and identification of key water parameters using different machine learning models based on big data. *Water Res.* **2020**, *171*, 115454. [CrossRef]
20. Alizamir, M.; Heddam, S.; Kim, S.; Mehr, A.D. On the implementation of a novel data-intelligence model based on extreme learning machine optimized by bat algorithm for estimating daily chlorophyll-a concentration: Case studies of river and lake in USA. *J. Clean. Prod.* **2021**, *285*, 124868. [CrossRef]
21. Khullar, S.; Singh, N. Machine learning techniques in river water quality modelling: A research travelogue. *Water Supply* **2021**, *21*, 1–13. [CrossRef]
22. Asadollah, S.B.H.S.; Sharafati, A.; Motta, D.; Yaseen, Z.M. River water quality index prediction and uncertainty analysis: A comparative study of machine learning models. *J. Environ. Chem. Eng.* **2021**, *9*, 104599. [CrossRef]
23. Zounemat-Kermani, M.; Alizamir, M.; Fadaee, M.; Namboothiri, A.S.; Shiri, J. Online sequential extreme learning machine in river water quality (turbidity) prediction: A comparative study on different data mining approaches. *Water Environ. J.* **2021**, *35*, 335–348. [CrossRef]
24. Bui, D.T.; Khosravi, K.; Tiefenbacher, J.; Nguyen, H.; Kazakis, N. Improving prediction of water quality indices using novel hybrid machine-learning algorithms. *Sci. Total. Environ.* **2020**, *721*, 137612. [CrossRef]
25. Jeihouni, M.; Toomanian, A.; Mansourian, A. Decision Tree-Based Data Mining and Rule Induction for Identifying High Quality Groundwater Zones to Water Supply Management: A Novel Hybrid Use of Data Mining and GIS. *Water Resour. Manag.* **2020**, *34*, 139–154. [CrossRef]
26. Saghebian, S.M.; Sattari, M.T.; Mirabbasi, R.; Pal, M. Ground water quality classification by decision tree method in Ardebil region, Iran. *Arab. J. Geosci.* **2014**, *7*, 4767–4777. [CrossRef]
27. Anmala, J.; Turuganti, V. Comparison of the performance of decision tree (DT) algorithms and extreme learning machine (ELM) model in the prediction of water quality of the Upper Green River watershed. *Water Environ. Res.* **2021**. [CrossRef] [PubMed]
28. *Status Report, Green and Tradewater Basins*; Kentucky Division of Water: Frankfort, KY, USA, 2001.
29. Anmala, J.; Meier, O.W.; Meier, A.J.; Grubbs, S. A GIS and an artificial neural network based water quality model for a stream network in Upper Green River Basin, Kentucky, USA. *ASCE J. Environm. Eng.* **2015**, *141*, 04014082. [CrossRef]
30. Kaplan, C.; Pasternack, B.; Shah, H.; Gallo, G. Age-related incidence of sclerotic glomeruli in human kidneys. *Am. J. Pathol.* **1975**, *80*, 227–234.
31. Sokal, R.R.; Rohlf, J.F. *Biometry: The Principle and Practice of Statistics in Biological Research*, 2nd ed.; W.H. Freeman and Company: San Francisco, CA, USA, 1981.
32. Snedecor, G.W.; Cochran, W.G. *Statistical Methods*, 8th ed.; Iowa University Press: Ames, IA, USA, 1989. [CrossRef]
33. Rao, P.V. *Statistical Research Methods in the Life Sciences*; Duxbury Press: Austin, TX, USA, 1998; 889p.
34. Sutton, C.D. Classification and Regression Trees, Bagging, and Boosting. *Hand. Stat.* **2005**, *24*, 303–329. [CrossRef]
35. Bramer, M. *Principles of Data Mining*; Springer: London, UK, 2007; 343p, ISBN 978-1-84628-765-7.
36. Quinlan, J.R. Induction of Decision Trees. *Mach. Learn.* **1986**, *1*, 81–106. [CrossRef]
37. Breiman, L.; Firedman, J.H.; Olshen, R.A.; Stone, C.J. *Classification and Regression Trees*; Chapman & Hall/CRC: Boca Raton, FL, USA, 1984; ISBN 978-0-412-04841-8.
38. Breiman, L. Bagging predictors. *Mach. Learn.* **1996**, *24*, 123–140. [CrossRef]
39. Tan, P.-N.; Steinbach, M. *Introduction to Data Mining*; Pearson India Education Services Pvt. Ltd: Bengaluru, India, 2016; p. 760.
40. Shehane, S.; Harwood, V.; Whitlock, J.; Rose, J. The influence of rainfall on the incidence of microbial faecal indicators and the dominant sources of faecal pollution in a Florida river. *J. Appl. Microbiol.* **2005**, *98*, 1127–1136. [CrossRef]
41. Santiago-Rodriguez, T.M.; Tremblay, R.L.; Toledo-Hernandez, C.; Gonzalez-Nieves, J.E.; Ryu, H.; Domingo, J.W.S.; Toranzos, G.A. Microbial Quality of Tropical Inland Waters and Effects of Rainfall Events. *Appl. Environ. Microbiol.* **2012**, *78*, 5160–5169. [CrossRef]
42. Islam, M.M.M.; Hofstra, N.; Islam, A. The Impact of Environmental Variables on Faecal Indicator Bacteria in the Betna River Basin, Bangladesh. *Environ. Process.* **2017**, *4*, 319–332. [CrossRef]
43. Leight, A.K.; Crump, B.C.; Hood, R. Assessment of Fecal Indicator Bacteria and Potential Pathogen Co-Occurrence at a Shellfish Growing Area. *Front. Microbiol.* **2018**, *9*, 384. [CrossRef]
44. Seo, M.; Lee, H.; Kim, Y. Relationship between Coliform Bacteria and Water Quality Factors at Weir Stations in the Nakdong River, South Korea. *Water* **2019**, *11*, 1171. [CrossRef]
45. Kagalou, I.; Tsimarakis, G.; Bezirtzoglou, E. Inter-relationships between Bacteriological and Chemical Variations in Lake Pamvotis—Greece. *Microb. Ecol. Health Dis.* **2002**, *14*, 37–41. [CrossRef]
46. Jin, G.; Englande, A.; Bradford, H.; Jeng, H.-W. Comparison of E.Coli, Enterococci, and Fecal Coliform as Indicators for Brackish Water Quality Assessment. *Water Environ. Res.* **2004**, *76*, 245–255. [CrossRef] [PubMed]
47. Venkateswarlu, T.; Anmala, J.; Dharwa, M. PCA, CCA, and ANN modeling of climate and land-use effects on stream water quality of Karst watershed in Upper Green River, Kentucky, USA. *ASCE J. Hydrol. Eng.* **2020**, *25*, 05020008. [CrossRef]
48. Breiman, L. Random forests. *Mach. Learn.* **2001**, *45*, 5–32. [CrossRef]

49. Geurts, P.; Ernst, D.; Wehenkel, L. Extremely randomized trees. *Mach. Learn.* **2006**, *63*, 3–42. [CrossRef]
50. Breiman, L. *Arcing the Edge*; Technical Report 486, Ann. Prob; Statistics Department, University of California: Berkeley, CA, USA, 1997; Volume 26, pp. 1683–1702.
51. Freund, Y.; Schapire, R.E. A short introduction to Boosting. *J. Jpn. Soc. Artif. Intell.* **1999**, *14*, 771–780.
52. Mitchell, T.M. *Machine Learning*, 1st ed.; Mc-Graw Hill Education: New York, NY, USA, 1997.

water

MDPI

Article

Comparison of Deterministic and Statistical Models for Water Quality Compliance Forecasting in the San Joaquin River Basin, California

Nigel W. T. Quinn [1,2,3,*], Michael K. Tansey [2] and James Lu [2]

1 Lawrence Berkeley National Laboratory, Climate and Ecosystem Sciences Division, Berkeley, CA 94720, USA
2 US Bureau of Reclamation, Washington, DC 20240-0001, USA; mtansey@usbr.gov (M.K.T.); jlu@usbr.gov (J.L.)
3 School of Engineering, University of California, Merced, CA 90095, USA
* Correspondence: nwquinn@lbl.gov; Tel.: +1-510-612-8802

Citation: Quinn, N.W.T.; Tansey, M.K.; Lu, J. Comparison of Deterministic and Statistical Models for Water Quality Compliance Forecasting in the San Joaquin River Basin, California. *Water* **2021**, *13*, 2661. https://doi.org/10.3390/w13192661

Academic Editors: Francesco De Paola and Thomas Meixner

Received: 10 June 2021
Accepted: 22 September 2021
Published: 27 September 2021

Abstract: Model selection for water quality forecasting depends on many factors including analyst expertise and cost, stakeholder involvement and expected performance. Water quality forecasting in arid river basins is especially challenging given the importance of protecting beneficial uses in these environments and the livelihood of agricultural communities. In the agriculture-dominated San Joaquin River Basin of California, real-time salinity management (RTSM) is a state-sanctioned program that helps to maximize allowable salt export while protecting existing basin beneficial uses of water supply. The RTSM strategy supplants the federal total maximum daily load (TMDL) approach that could impose fines associated with exceedances of monthly and annual salt load allocations of up to $1 million per year based on average year hydrology and salt load export limits. The essential components of the current program include the establishment of telemetered sensor networks, a web-based information system for sharing data, a basin-scale salt load assimilative capacity forecasting model and institutional entities tasked with performing weekly forecasts of river salt assimilative capacity and scheduling west-side drainage export of salt loads. Web-based information portals have been developed to share model input data and salt assimilative capacity forecasts together with increasing stakeholder awareness and involvement in water quality resource management activities in the river basin. Two modeling approaches have been developed simultaneously. The first relies on a statistical analysis of the relationship between flow and salt concentration at three compliance monitoring sites and the use of these regression relationships for forecasting. The second salt load forecasting approach is a customized application of the Watershed Analysis Risk Management Framework (WARMF), a watershed water quality simulation model that has been configured to estimate daily river salt assimilative capacity and to provide decision support for real-time salinity management at the watershed level. Analysis of the results from both model-based forecasting approaches over a period of five years shows that the regression-based forecasting model, run daily Monday to Friday each week, provided marginally better performance. However, the regression-based forecasting model assumes the same general relationship between flow and salinity which breaks down during extreme weather events such as droughts when water allocation cutbacks among stakeholders are not evenly distributed across the basin. A recent test case shows the utility of both models in dealing with an exceedance event at one compliance monitoring site recently introduced in 2020.

Keywords: water quality forecasting; decision support; WARMF; regression model; salinity; irrigated agriculture; stakeholder involvement

1. Introduction

Water quality forecasting in arid river basins is especially challenging given the importance of protecting beneficial uses in these environments and the livelihood of agricultural communities. Model selection for water quality forecasting depends on many factors including analyst expertise and cost, stakeholder involvement and expected performance.

An American Society of Civil Engineers (ASCE) Task Committee was convened within the Environmental Water Research Institute (ASCE, 2021) to document the state of the practice in the use of water quality models that addresses selection, data collection and organization, calibration, and independent testing to define uncertainty and to envisage both the state of the art and future development. This paper draws on this effort focusing specifically on two distinctly different approaches to involving stakeholders in salinity management in a highly regulated river basin in California, dominated by agricultural and managed wetland return flows.

Management of salinity in the United States and around the world is typically performed through environmental regulation. In the United States, the federal Environmental Protection Agency (USEPA) uses the concept of a total maximum daily load (TMDL) to establish safe and sustainable pollutant concentrations in receiving waters and pollutant load assimilative capacity to help guide stakeholders in determining pollutant load reduction strategies. The TMDL goal for salt loading to impaired waterbody [1,2] can be defined as:

$$TMDL = \sum WLA + \sum LA + MOS + RC \quad (1)$$

where WLA is the waste load allocation for each point source of salt load, LA is the salt load allocation for non-point sources and MOS is the margin of safety selected that accounts for measurement and analytical uncertainty. The RC is a reserve capacity that is seldom used in California applications but that could be used to account for future anticipated loading from both point sources and non-point sources. Possible examples are future climate change, population growth, land use and land cover changes, sea-level rise and environmental policy initiatives.

Models are commonly used in the development of TMDLs and to assess their impact under a range of environmental conditions [2–8]. Models can range in complexity from simple salinity mass balances, that may use simple regression equations to relate salinity to flow and other water quality parameters, to comprehensive, physically-based hydrologic and water quality models [3,9] that attempt to simulate important processes. These models may be used at various phases of TMDL development and implementation including (a) the assessment of the level of impairment and the impacts of existing best management practices on the water quality; (b) the evaluation and comparison of load reduction strategies; (c) the computation of TMDL uncertainty and margin of safety [10]; (d) as decision support tools [11–13]; and (e) for real-time or near-real-time forecasting after implementing a TMDL [14–16]. This paper compares the performance of two modeling techniques used in near-real-time forecasting of compliance with salinity objectives in the San Joaquin River Basin in California.

2. Background

The San Joaquin River (SJR) drains approximately 8.7 million acres (4 million ha) of California's San Joaquin Valley including 1.4 million acres (0.64 million ha) of agricultural land (Figure 1). The San Joaquin River Basin (SJRB) watershed is bounded by the Sierra Nevada mountains on the east, the Coast Range mountains on the west, the Sacramento–San Joaquin Delta to the north, and the closed Tulare Lake Basin on the south. The Coast Range mountains are relatively recent in geologic history and formed of an uplifted seabed whose sedimentary constitution is naturally high in salinity including trace elements such as selenium, boron and molybdenum. [17–19]. Additional salt is imported to the basin from large state and federal water pumping facilities in the Sacramento–San Joaquin Delta. These facilities replace water supply that was diverted from the SJR to irrigate farmland in the southern part of the San Joaquin Valley in the 1960s [17] and constitute more than 47% of the salts imported to the basin. For this reason, as the main purveyor of irrigation water supply, the federal government is considered a stakeholder in actions to manage salinity impairments in the SJR.

Since the 1940s, mean annual salinity concentrations in the SJR measured at the Vernalis monitoring station, the most downstream station not impacted by tidal flows in the Delta, have more than doubled. West-side SJRB sources that include agricultural surface and subsurface drainage and surface drainage from seasonally managed wetlands that comprise the 140,000 acres (64,000 ha) Grasslands Ecological Area discharge through Mud and Salt Sloughs and accounted for more than 37% of the salt loading to the SJR for the period 2000–2009 (Figure 1). Several smaller, ephemeral streams including Hospital, Ingram, Del Puerto, Orestimba and Los Banos Creeks contribute an additional 30% to SJR salt loads [13]. The major tributaries to the SJR, the Stanislaus, Tuolumne, and Merced Rivers, drain the east side of the basin and are the major source of dilution flow and salt load assimilative capacity to the SJR (Figure 1).

Water quality data collected by the Central Valley Regional Water Quality Control Board (CVRWQCB) staff since 1985 indicate that the 30 day running average electrical conductivity (EC) water quality objectives of 1000 μS/cm in the non-irrigation season and 700 μS/cm in the irrigation season (1 April–31 August) have been routinely exceeded at the Vernalis compliance monitoring station, especially prior to 2005 [12,13]. The non-irrigation season salinity objective was exceeded 11 percent of the time and the irrigation season salinity objective was exceeded 49 percent of the time during the period 1986–1998 [13]. This rate of exceedance occurred even though releases were made from New Melones Reservoir on the Stanislaus River to help meet salinity objectives at Vernalis [12].

The SJR TMDL for salinity had several objectives, namely (a) to identify and quantify the sources of salt loading to the SJR; (b) determine the load reductions necessary to achieve attainment of applicable water quality objectives in order to protect beneficial uses of SJR water supply; and (c) to allocate salt loads to the various sources and source areas within the watershed which, once implemented, would result in attainment of applicable water quality objectives [13,14]. Figure 1 shows the seven source areas identified by the CVRWQCB that each were assigned annual and monthly salinity load objectives, modified to account for wet, normal, dry and critically dry water year classifications. However. realization of these objectives using a 10% low flow hydrology to account for critically low-flow conditions over a 73 year historical flow record, in lieu of the standard MOS, produced a TMDL where the base load allocations were overly conservative.

The TMDL already recognized a consumptive use allocation to account for irrigation evapotranspiration of applied water, a Delta Mendota Canal supply relaxation load for salt imported with water supply deliveries to the west side of the basin, a SJR supply water relaxation for salts diverted from the SJR and an allocation to the federal agency for actions related to mitigation of salts imported by the agency in irrigation water supply. The USBR was assigned responsibility for 47 percent of the salt load discharged to the SJR [13].

Analysis conducted by CVRWQCB staff showed that stakeholder adherence to these salt load limits would result in salt accumulation in the watershed and long term of both degradation of ground and surface waters. Continuation of existing drainage practices could result in average annual fines of over $300,000 in each of the subareas (Table 1) assuming a fine schedule of $5000 per day for each month the load allocation for each subarea was exceeded [20]. These fines would be borne largely by agricultural water district stakeholders, some of whom are those adversely impacted by elevated EC in the SJR primarily along Reach 83 (Figure 1). Agriculture is the primary beneficial use impaired by salinity in the SJR Basin recognized in the SJR Basin plan [13,14]. To overcome the constraints imposed by the conservative nature of the strict TMDL formulation, the CVRWQCB made provision for an additional real-time salt load allocation in lieu of the fixed base load allocation to maximize salt export from the LSJR Basin while still meeting water quality objectives [14]. The real-time load allocation would apply any time salt load assimilative capacity was available in the SJR.

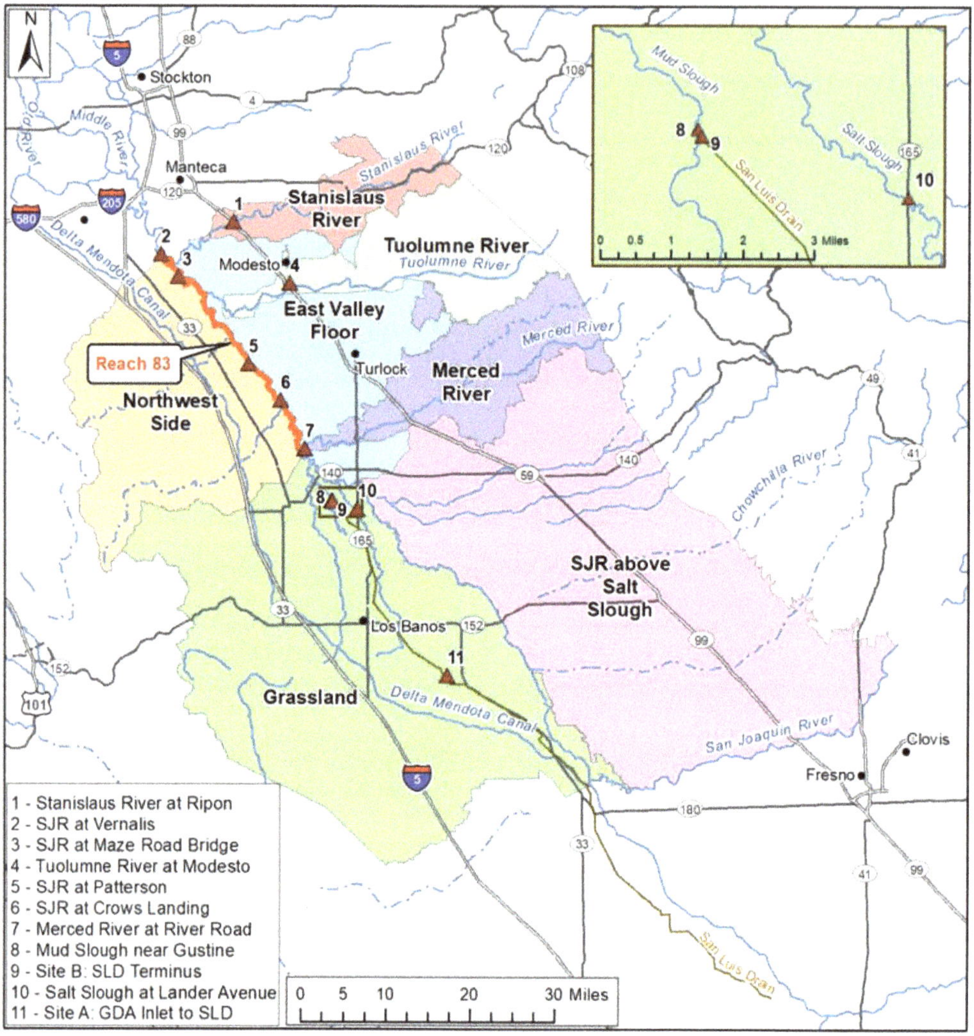

Figure 1. Major subareas within the SJR Basin that drain to the SJR as defined in the salinity TMDL [13]. Reach 83 shown in the figure is the reach for which water quality (salinity) is regulated through the recognition of three compliance monitoring stations at Crows Landing, Maze Road Bridge and Vernalis. The most salient feature of the SJR Basin is that drainage from sources to the west of the SJR are elevated in salinity by virtue of native salts in alluvial sediments deposited from the coastal range mountains west of the Valley floor and the importation of irrigation water supply from the Sacramento-San Joaquin Delta that is also salt impacted. Tributary inflow from land areas to the east of the SJR are of high quality, derived from snowmelt from the Sierra Nevada mountains. Real-time management is essentially a scheduling activity—coordinating salt load assimilative capacity consumed by west-side saline drainage with salt load assimilative capacity supplied by east-side reservoir releases along the major tributaries.

Table 1. Hypothetical SJR daily salt discharge exceedance fees by subarea (10-year period 2001–2012) using an assumed $5000/day fine for exceedance of the 30-day running average mean EC objective [20].

	LSJR Salt Discharge Exceedence Fees by TMDL Subarea for a 10 Year Period 2001–2012				
		Northwest Side	**Grasslands**	**Upstream San Joaquin River**	**East Valley Floor**
days exceeded by period	Oct	0	0	0	0
	Nov	90	60	0	0
	Dec	124	248	0	0
	Jan	186	0	310	0
	Feb	28	196	0	0
	Mar	0	279	0	0
	Apr	28	56	42	14
	VAMP	0	0	30	30
	May	0	0	51	17
	Jun	30	30	210	90
	Jul	0	0	248	91
	Aug	0	0	248	31
	Sep	0	0	0	0
Total days of exceedences		486	869	1139	273
$5000 per day penalty		$5000	$5000	$5000	$5000
Total penalties		$2,430,000	$4,345,000	$5,695,000	$1,365,000
Years calculated		8	10	10	3
Average penalty per year		$303,750	$434,500	$569,500	$455,000
Acres of agriculture		118,000	353,000	187,000	201,000
Average penalty per acre		$2.57	$1.23	$3.05	$2.26

3. Real-Time Salinity Management

The USBR provides water to west-side agricultural and wetland resource contractors via the Delta-Mendota Canal (DMC). The USBR's water rights under which the USBR delivers water to the SJR Basin were amended to require that the USBR meet the 1995 Bay Delta Plan Salinity objectives at Vernalis, which are equivalent to the numeric targets established by the salinity TMDL [13]. Upstream salinity objectives at the Crows Landing Bridge compliance monitoring site was ratified in 2017 to protect riparian diverters downstream of Crows Landing and upstream of the Vernalis compliance monitoring site [21]. The control program requires the USBR to meet DMC salt load allocations or provide dilution flows to create additional assimilative capacity for salt in the LSJR equivalent to DMC salt loads in excess of their allocation. Thec program includes an innovative provision that provides relief from the restrictive salinity load restrictions imposed by the salinity TMDL and codified in the Basin Water Quality Control Plan. This provision states that "Participation in a Regional Board approved real-time management program (RTMP) and attainment of salinity and boron water quality objectives will constitute compliance with this control program" [21]. Participation in the RTMP was designed to promote cooperation and data sharing between entities, effectively replacing a costly salt load-based regulatory program with a more cost-effective, stakeholder-driven program that permitted full use of the river's assimilative capacity for salt [14,15,21]. Participation in the RTMP also included the development and use of a water quality forecasting model to provide stakeholder decision support and allow stakeholders sufficient time to address anticipated violations of the 30 day running average EC at compliance monitoring stations along the SJR [14].

The WARMF model was chosen for this task [22–24]. Compliance became the collective responsibility of SJRB stakeholders including the USBR.

The RTMP strategy increases potential management flexibility for agricultural, wetland and municipal dischargers to the SJR and provides an opportunity to maximize salt load export from the basin without exceeding environmental objectives. However, it assumes a level of coordination and cooperation amongst stakeholders that does not currently exist. The core elements of this program have led to: (a) the development of a basin-scale, sensor network to collect real-time monitoring of flow and salinity data; (b) an information dissemination system for effective sharing of data among basin stakeholders; (c) a need for continual calibration of the WARMF hydrology and salinity model of in the SJR and its contributing watersheds to improve the accuracy of forecasting and daily assessment of river assimilative capacity; (d) the creation and funding of stakeholder institutional entities responsible for coordinating salinity management actions and ensuring compliance with SJR salinity objectives; and € continued oversight and sanction of the CVRWQCB [14–16].

3.1. WARMF Water Quality Simulation Model

The San Joaquin River Basin application of the public-domain, Watershed Analysis Risk Management Framework (WARMF) model [14,24] was developed in 2004 by Systech Water Resources Inc. as a TMDL decision support tool. The first application of the model was to assess options for control of dissolved oxygen sag in the SJR Deep Water Ship Channel [23–25]. The SJRB WARMF model application is a physically-based, data-intensive watershed model that simulates the hydrologic, chemical, and physical processes in the river and contributing waterbodies (Figure 2). The model was derived from the San Joaquin River Input–Output (SJRIO) model [26,27]. The model was updated and reconfigured as a salinity forecasting tool in 2014 [14,24] as the USBR's contribution to stakeholder-led real-time salinity management activities. The WARMF model application simulates flow and water quality in surface water diversions, groundwater pumping, and irrigation water supply, while keeping track of crop evapotranspiration, seepage, and irrigation surface and subsurface return flows [25]. Delineation of land catchments in WARMF conforms to both irrigation and drainage district boundaries and natural catchments, allowing the model to track salt loads from their points of diversion in delivery canals back to the river [25].

The data-intensive WARMF model is supplied with daily meteorology, diversion flows, and measured flow and electric conductivity (EC) at the upstream model boundaries [23,25]. The current upstream model boundaries are at gages where flow and EC are measured continuously in the SJR and along its major tributaries including the Merced River, Tuolumne River and Stanislaus River. Real-time data, tributary reservoir release forecasts, and meteorology forecasts are collected and imported into WARMF using an automated process consisting of custom scripts and web scraping tools that interact with agency web portals for hydrology and water quality monitoring [25,28]. WARMF model data acquisition accesses seven agency web portals and is accomplished as a separate data acquisition and pre-processing routine.

The combination of real-time monitoring, simulation modeling and forecasting of SJR assimilative capacity has the potential to optimize use of available river salt assimilative capacity, generated by releases of high quality Sierran water, which provides dilution to saline west-side agricultural and managed wetland return flows. However, there needs to be coordination and sufficient lead time to allow entities being asked to charge drainage practices or alter reservoir release patterns to be able to respond. Agricultural return flows and salt loads are highest during the summer irrigation season whereas return flows and salt loads from seasonally managed wetlands are highest during the spring months of March and April, when most seasonal wetland ponds are drained to promote establishment of moist soil plants and habitat for waterfowl [15]. These anticipated hydrologic patterns help to screen the array of practices on both the east and west sides of the basin that will be most effective at managing salinity.

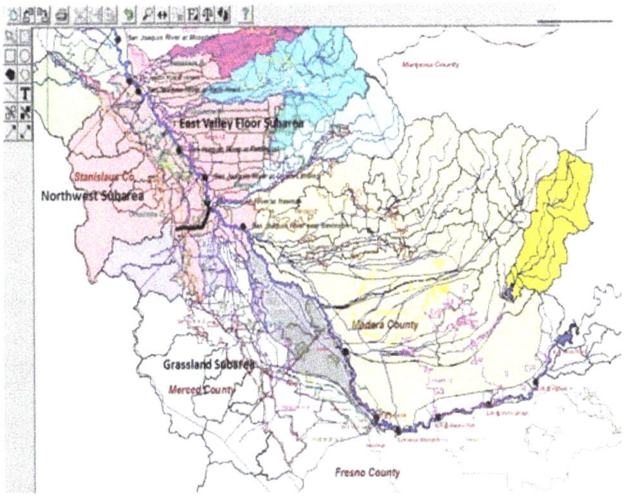

Figure 2. Map of the SJR Basin represented as major contributing watersheds within the WARMF model. The WARMF model allows further disaggregation of these watersheds into small contributing subareas and allows the substitution of available data at the major outlets of these subareas for model-derived flow and water quality estimates.

Given the uncertainty associated with estimates of salt assimilative capacity, the need for adequate lead time for stakeholders to adjust tributary inflow and drainage return flow schedules and the fact that most weather forecasts provided by news organizations rarely extend beyond two weeks—a two-week forecast period and a one-week hindcast period was chosen for the real-time salinity management program. The one-week hindcast refers to the technique of beginning the simulation one week in arrears so that the first week of the forecast can be compared to observed flow and electrical conductivity (EC) data [14,16,22]. Model parameters affecting river and tributary inflow and water quality such as the partitioning coefficients that allocate watershed runoff and deep percolation to groundwater can be adjusted to recalibrate the model during periods when model output and river observations diverge. This activity is infrequently performed due to the significant effort involved and the fact that the WARMF model has exhibited excellent performance for simulation of flow EC and EC along Reach 83 of the SJR. Simulated flow and EC are compared to measured data along the SJR for model calibration including drainage return flows from east- and west-side catchments and direct diversions from the SJR to riparian water districts. Although agricultural and managed wetland stakeholders have yet to fully embrace the model as a decision support tool both have concurred that the suggested two-week forecast and one-week hindcast periods are a good compromise balancing the utility and credibility of the forecasts with the time stakeholders might need to adjust water management and drainage discharge operations.

The SJR WARMF model has a number of customized output visualization options designed to enhance user understanding of salinity fate and transport in the SJR Basin and the use of salt load assimilative capacity by river mile along the mainstem of the SJR [28]. The output visualization also allows users to estimate if and when the salinity concentration at the compliance monitoring sites will approach or exceed objectives. The model is also capable of showing the impact of potential salinity management changes in the watershed designed to comply with regulatory limits (Figure 3). For example, the SJR WARMF model can simulate the effect of increased irrigation water diversions from the river into riparian water districts, lowering salt loading in the river, which may help to improve compliance with salinity concentration objectives [28].

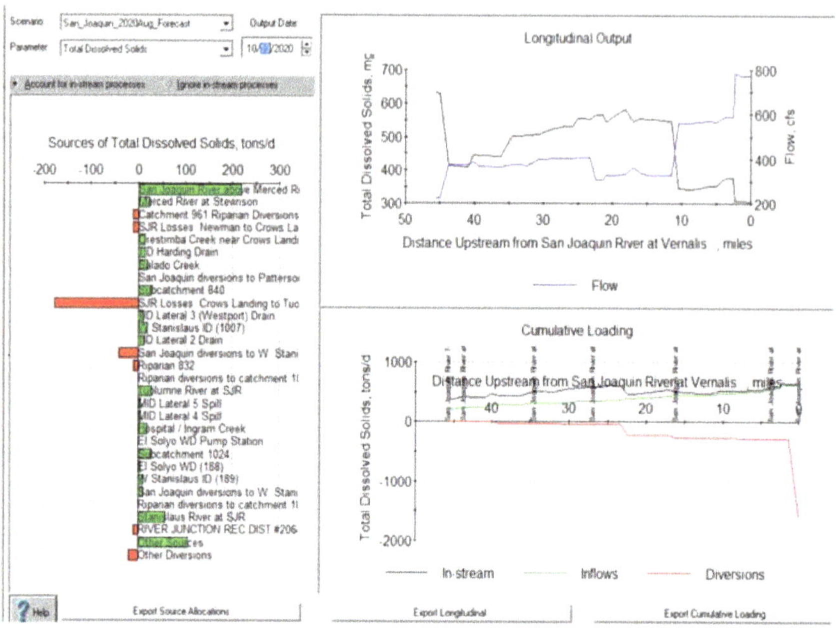

Figure 3. A unique feature of the WARMF model is the availability of customized model outputs such as the "Gowdy" output (named after its developer) shown here. This depicts a Lagrangian view of the SJR at any point in time showing the major inflow to and diversions from the river approximately every $\frac{1}{2}$ mile (800 m) along its main reach as well as the incremental flow and EC concentration from the origin at Lander Avenue to the EC compliance monitoring station at Vernalis [23,25].

The SJR WARMF model has been improved and customized over the past 15 years with the USBR and research grant support as a watershed-based simulation tool for flow and salinity forecasting in the SJR [14,25]. Updating time series data inputs and maintaining model calibration are expensive and time consuming. This constraint has restricted the stakeholders' pool and agency individuals able to run the model on a regular basis and has been an impediment for stakeholder entities such as the San Joaquin Valley Drainage Authority (SJDVA) to take over operation and maintenance of the model as a decision support tool. As a result, the USBR evaluated other approaches for providing flow and salinity forecasts of SJR at Vernalis and Crows Landing, the two salinity concentration compliance points for the TMDL. Although the WARMF model has been used for various decision support activities in the SJR for over 15 years, other less data-intensive and more easily understood approaches may be better received by stakeholders [28].

3.2. ANN-Based Statistical Models

The USBR developed a statistical approach as an alternative to the physically-based SJR WARMF model for flow and salinity forecasting in the SJR. This approach was limited to the Vernalis, Crows Landing and Maze Road Bridge compliance monitoring stations (Figure 1) [29]. Two Artificial Neural Networks (ANN) based models a Recurrent ANN and an Autoregressive ANN were identified as potential alternatives [30]. The most salient features of these ANN alternatives were that the underlying basis should be easy to understand and independent of having a deep understanding of basin hydrology [29,30]. ANN and regression-based approaches have the advantage of ready automation and have the advantage that daily flow forecasts are available online from the National Oceanic and Atmospheric Administration (NOAA) California River Forecast Center (RFC) providing the basis for river EC forecasts at the compliance monitoring stations. The significance of this work is that daily bulletins from dam operators along the three major tributaries to the SJR are recognized in these forecasts.

Under normal basin hydrologic conditions, there is sufficient salt load assimilative capacity in the river when defined as the 30 day running average EC. Only in rare circumstances such as a prolonged drought is action required to limit salt loading during certain months to the SJR. During these periods, the more comprehensive WARMF model could be called upon to assist stakeholder management entities to determine appropriate salt loading reduction by subarea within the basin to avoid fines.

Recurrent ANN models are statistical learning models that are used in machine learning, inspired by biological neural networks such as in the human brain [30]. A number of ANN and recurrent neural network architectures with both short- and long-term memory were developed and applied to the Vernalis compliance monitoring station using existing flow and salinity data resources. None of the ANN architectures or network hyperparameters performed sufficiently well due to time series water quality data limitations and the impact of random anthropogenic factors that can affect reservoir operations [29]. In conducting the analysis, less than 5000 observations were available, whereas most applications of this method typically require well over a million observations to be successful. An additional ANN-based model was investigated using the MATLAB machine learning toolbox using an embedded machine learning application called Autoregressive ANN that accommodated external inputs. Although the Autoregressive ANN approach performed better in salinity forecasts compared to the Recurrent ANN model, the model salinity forecast performance was unsatisfactory [29]. Future work in the application of neural networks to flow and EC time series forecasting on the SJR may find more success in the use of Bayesian neural networks for capturing water quality forecast uncertainty.

3.3. Simple Regression Model

Water agency analysts have long recognized the inverse relationship between flow and EC. This relationship was utilized for many years in applications of the previous USBR water supply allocation models for the federal service area within the San Joaquin Valley to estimate New Melones reservoir releases for water quality. However, the poor performance of these models for estimating EC at low-flow conditions, based on simple regression relationships, was one of the reasons a data-driven flow and salinity mass balance approach was adopted for the state-federal California (Water Allocation) Simulation Model (CALSIM) model that replaced the previous models. A re-examination of the flow–EC relationship [29] suggested a new approach using the rate of change of salinity that was found to be approximately proportional to the rate of change (or gradient) of the measured flow in the SJR. This new algorithm was not as susceptible to low-flow conditions as the prior approach.

The flow gradient was calculated as follows:

$$Q_{grad} = (Q_t - Q_{(t-1)})/Q_{(t-1)}$$

where Q_t is the flow at time t, and $Q_{(t-1)}$ is the flow at the previous time step.

The salinity gradient was calculated in a similar fashion. Further analysis of daily flow and salinity data of the SJR at Vernalis for the period 2000–2018 showed that a clear linear regression relationship exists between flow and salinity gradients. After removing one percent of the outliers from the plot of flow and salinity gradients using daily data for the 2000–2018 time period, the resulting regression equation of flow and salinity relationship at Vernalis became (Lu et al., 2019):

$$EC_{grad} = -0.5396 \times Q_{grad} + 0.0038$$

or

$$[\![(EC)_t - [\![EC_{(t-1)})/[\![EC)_{(t-1)}] = -0.5396 \times Q_t - Q_{(t-1)})/Q_{(t-1)} + 0.0038$$

Using this relationship, the salinity forecast (measured as EC) at time step t can be determined as follows:

$$[\![EC]\!]_t = [\![EC]\!]_{(t-1)} - [0.5396 \times (Q_t - Q_{(t-1)})/Q_{(t-1)} + 0.0038] \times [\![EC]\!]_{(t-1)}$$

This equation was initially applied to daily Vernalis flow and salinity data (Figure 4) for the period 2000–2018 to generate six-day model-based forecasts that were compared to historical data. The correlation coefficients for the relationship between the six-day forecasted salinity and observed flow ranged from 0.8780 to 0.9787. The same regression method was then applied to the upstream Crows Landing compliance monitoring station, resulting in the following equation for forecasting the SJR salinity concentration downstream of that location.

$$[\![EC]\!]_t = [\![EC]\!]_{(t-1)} + [-0.4413 \times (Q_t - Q_{(t-1)})/Q_{(t-1)} + 0.0036] \times [\![EC]\!]_{(t-1)}$$

The correlation coefficients of the relationship of observed flow and the six-day forecasted salinity concentration ranged from 0.9831 to 0.9154.

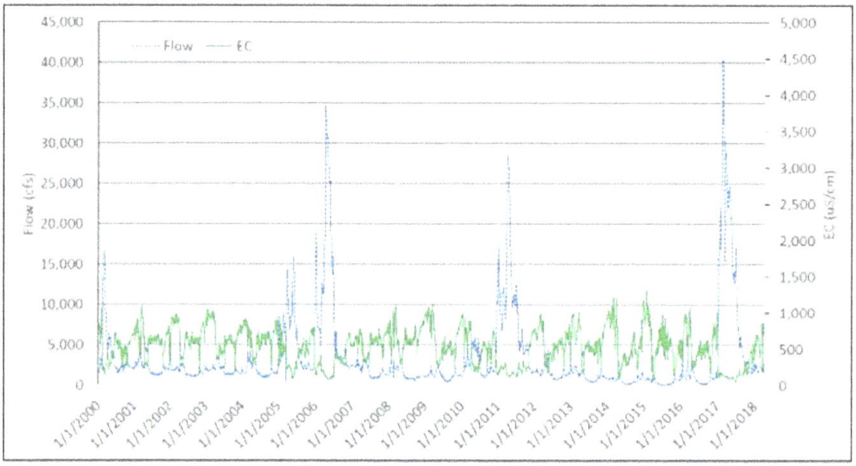

Figure 4. Flow and EC observations at Vernalis compliance monitoring station on the SJR for the period 2000–2018.

4. Comparison of the SJR WARMF and Regression Model Applications

A comparison of the SJR WARMF and Regression models was undertaken to evaluate the performance of the models for water quality forecasting. This evaluation initially compared differences between forecasted and observed water quality measured as EC at the monitoring station located at Vernalis (Figure 4). Similar analyses were performed in Excel using an algorithm that computed the difference (Δ) between the daily forecasted (FC) and observed (OBS) EC (Δ = FC − OBS) starting on the forecast day (FC Day + 0) and each consecutive day within the lead forecast time of 14 days (FC Day + 14). The analyses were conducted with observations and forecasts made between 22 February 2018 and 22 May 2020. During this period, a total of 820 EC observations were measured. However, for all the forecast lead times considerably fewer forecasts were actually made. In the case of the Regression model, the number of forecasts ranged from 399 for forecasts of less than 6 days (FC Day + 6) down to 347 forecasts for lead times of 7 days or more (FC + 7 to FC + 14). Forecasts were made only on regular workdays and were not conducted on certain days due to personnel availability and periods of downtime in the monitoring system. Forecasts for days 11 through 15 were simply repeats of the FC + 10 forecast given that the California River Forecast Center (RFC) does not extend its daily forecasts, used by the WARMF and Regression models, past 10 days.

In the case of the WARMF model, there were even fewer forecasts throughout the evaluation period. The greater personnel time commitment to make WARMF model forecasts limited the forecast frequency to once per week, usually on a Monday. There were 131 forecasts for lead times from FC + 0 to FC + 7 and fewer forecasts for greater lead times. Table 2 presents the frequency count and statistics (mean and standard deviation) for the observations and model forecasts in the initial comparison of results produced by the Regression and WARMF models. Table 2 confirms that the Regression model forecasts were made approximately 3 times more often than those for the WARMF model.

In general, the Regression model forecasts had mean EC predictions that are approximately equal to the mean EC of the observations but increased to above the observation's mean EC after FC Day + 5 through the end of the forecast period. The WARMF model had slightly lower mean forecast EC values until FC Day + 4 after which they increased throughout the remainder of the forecast period. The observed EC, Regression and WARMF forecast mean EC values were compared in Figure 5 at each of the forecast lead times.

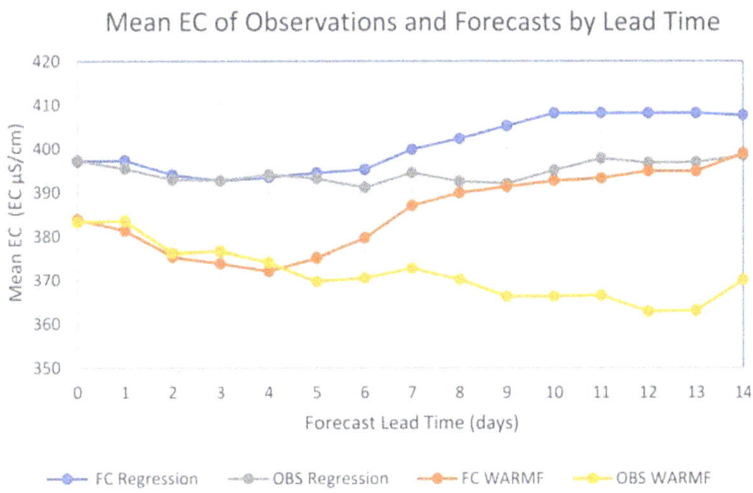

Figure 5. Means of the observed (OBS) EC and Forecast (FC) EC for the Regression and WARMF models for all forecast lead times between 22 February 2018 and 22 May 2020.

Table 2. Statistics of observed (OBS) and forecasted (FC) EC (μS/cm) for the Regression and WARMF models made between 22 February 2018 and 22 May 2020 by lead time.

	Regression Model EC Data				WARMF Model EC Data		
	Count	Mean	Std Dev		Count	Mean	Std Dev
OBS Day + 0	399	397	224	OBS Day + 0	131	401	235
FC Day + 0		397	223	FC Day + 0		384	192
OBS Day + 1	399	395	224	OBS Day + 1	131	383	214
FC Day + 1		397	225	FC Day + 1		381	182
OBS Day + 2	399	393	224	OBS Day + 2	131	376	211
FC Day + 2		394	225	FC Day + 2		375	178
OBS Day + 3	399	393	225	OBS Day + 3	131	377	208
FC Day + 3		393	225	FC Day + 3		374	182
OBS Day + 4	399	394	223	OBS Day + 4	131	374	209
FC Day + 4		393	226	FC Day + 4		372	183
OBS Day + 5	399	393	222	OBS Day + 5	131	370	207
FC Day + 5		394	224	FC Day + 5		375	187
OBS Day + 6	398	391	219	OBS Day + 6	131	371	201
FC Day + 6		395	222	FC Day + 6		380	190
OBS Day + 7	347	394	218	OBS Day + 7	131	373	204
FC Day + 7		400	218	FC Day + 7		387	194
OBS Day + 8	347	393	217	OBS Day + 8	129	370	203
FC Day + 8		402	220	FC Day + 8		390	200
OBS Day + 9	347	392	218	OBS Day + 9	129	366	202
FC Day + 9		405	223	FC Day + 9		391	204
OBS Day + 10	347	395	222	OBS Day + 10	128	366	204
FC Day + 10		408	225	FC Day + 10		393	208
OBS Day + 11	347	398	224	OBS Day + 11	128	366	207
FC Day + 11		408	225	FC Day + 11		393	211
OBS Day + 12	347	397	223	OBS Day + 12	126	363	203
FC Day + 12		408	225	FC Day + 12		395	213
OBS Day + 13	347	397	225	OBS Day + 13	126	363	204
FC Day + 13		408	225	FC Day + 13		395	214
OBS Day + 14	347	398	229	OBS Day + 14	124	370	209
FC Day + 14		408	224	FC Day + 14		399	214

A comparison of the mean of differences between forecasted EC and observed EC for both Regression and WARMF models is shown in Table 3. For both models, the mean of the differences between forecasted EC minus observed EC was computed for the period between 22 February 2018 and 22 May 2020. For the Regression model, the differences were small ($\leq$+5) for until Δ Day + 6. The mean EC differences increase to maximum of 15 μS/cm at Δ Day + 9. From Δ Day + 10 to the end of the forecast period, the mean EC differences decrease slightly to a value of 12 μS/cm. For the WARMF model, the mean of the EC differences were small, decreasing from +1 to −3 at Δ Day + 3. From Δ Day + 4 to Δ Day + 12, the mean of the EC differences increases consistently reaching a peak value of

+33 μS/cm at Δ Day + 12 after which there is a slight decrease to 30 μS/cm at the end of the forecast period. These results are illustrated in Figure 6.

Table 3. Comparison of mean differences (Δ) between forecasted EC and observed EC (μS/cm) for all model forecasts made between 22 February 2018 and 22 May 2020.

Regression Model EC Differences				WARMF Model EC Differences			
	Count	**Mean Δ**	**Std Dev Δ**		**Count**	**Mean Δ**	**Std Dev Δ**
Δ Day + 0	398	0	14	Δ Day + 0	131	1	78
Δ Day + 1	397	2	37	Δ Day + 1	131	−2	85
Δ Day + 2	396	2	48	Δ Day + 2	131	−1	92
Δ Day + 3	395	1	57	Δ Day + 3	131	−3	86
Δ Day + 4	394	1	69	Δ Day + 4	130	−2	100
Δ Day + 5	394	3	80	Δ Day + 5	130	5	105
Δ Day + 6	393	5	86	Δ Day + 6	130	9	108
Δ Day + 7	341	7	91	Δ Day + 7	130	14	115
Δ Day + 8	340	12	103	Δ Day + 8	128	20	122
Δ Day + 9	339	15	116	Δ Day + 9	128	25	134
Δ Day + 10	338	15	131	Δ Day + 10	127	27	142
Δ Day + 11	337	13	144	Δ Day + 11	126	27	151
Δ Day + 12	337	14	153	Δ Day + 12	124	33	164
Δ Day + 13	337	14	163	Δ Day + 13	124	33	173
Δ Day + 14	336	12	171	Δ Day + 14	122	30	179

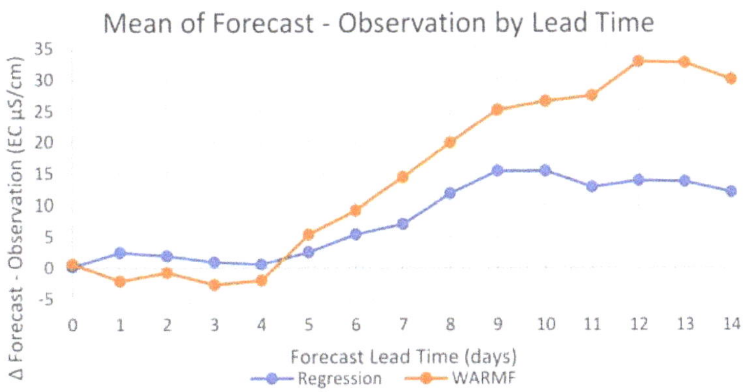

Figure 6. Comparison of mean differences in forecasted EC and observed EC for the Regression and WARMF models for the period between 22 February 2018 and 22 May 2020.

The forecast standard deviation is a measure of the dispersion of the forecast EC predictions around the mean EC value. Larger standard deviations imply a wider range of forecast predictions of EC and/or differences between forecasted EC values and observed EC. Figure 7 presents the standard deviations of the EC observations and EC forecasts for both models (Figure 7a) as well as the standard deviations of the EC differences between the forecasts minus observations (Figure 7b) over the forecast period. As illustrated, the standard deviations of the Regression model EC forecasts closely approximate the standard deviations of the EC observations at all lead times. In contrast, the standard deviations of the WARMF model EC forecasts are consistently less than standard deviations of the EC observations until lead time day 8 as shown in Figure 7a. The maximum difference (33 μS/cm) between forecast and observation standard deviations occurs at lead time day 2. In Figure 7b, the standard deviation of the differences between the EC forecasts minus EC observations for both models increase consistently with lead time indicating increasing uncertainty in the EC forecasts. Additionally, the WARMF model has consistently greater standard deviations in EC differences relative to the Regression model.

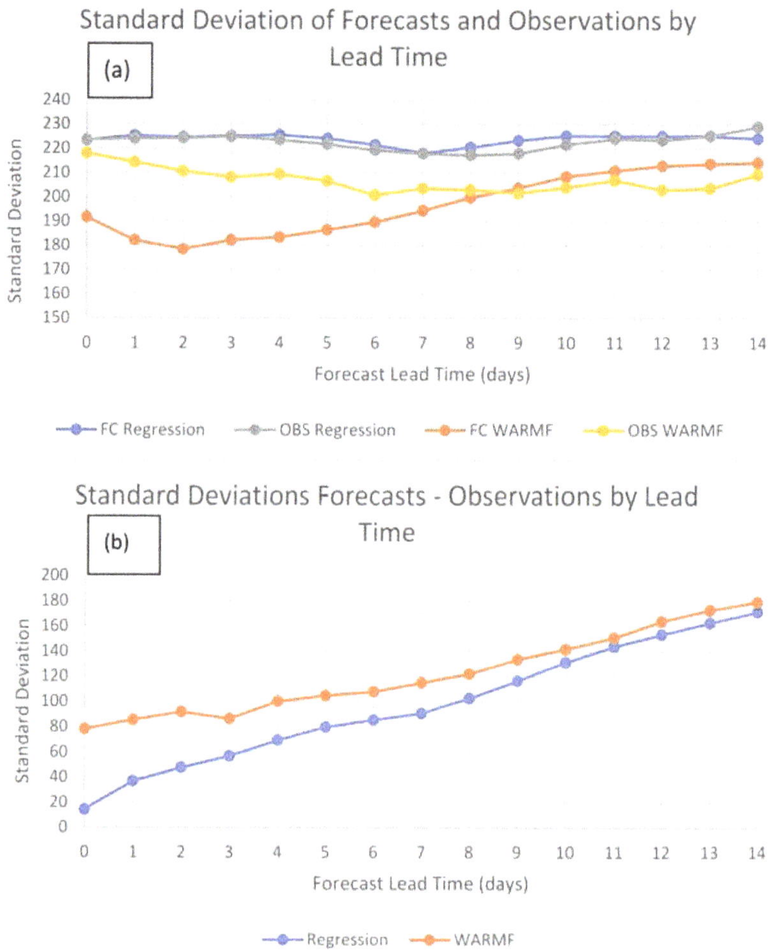

Figure 7. (**a**,**b**). Comparison of the standard deviations of forecasted EC and observed EC and standard deviations of differences between EC forecasts and EC observations for the Regression and WARMF models by lead time in the period between 22 February 2018 and 22 May 2020.

An additional evaluation was performed to determine the extent to which model bias affects the mean of differences between the forecasts and observations. For example, the models could forecast values significantly greater than the observations. However, a few large underestimates could potentially offset the positive bias and make the model appear to show better performance. In order to examine this effect, forecasts which were greater than the corresponding observations were examined separately from those in which the forecasts were less than the corresponding observations. After this sorting into positive (forecast >= observation) and negative (forecast < observation) bias groups, the means of the EC differences (forecast–observation) over the study period were calculated for each forecast lead time.

Figures 8 and 9 illustrate comparisons of the Regression and WARMF models for the positive and negative bias results, respectively. For the positive bias differences, the Regression model has lower differences at all lead times than the WARMF model.

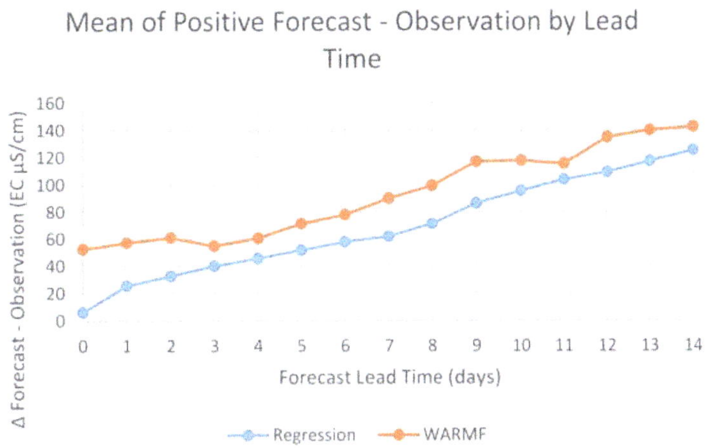

Figure 8. Comparison of means of forecasted EC and observed EC for the Regression and WARMF models for the period between 22 February 2018 and 22 May 2020. Data censored to include only over (positive) predictions.

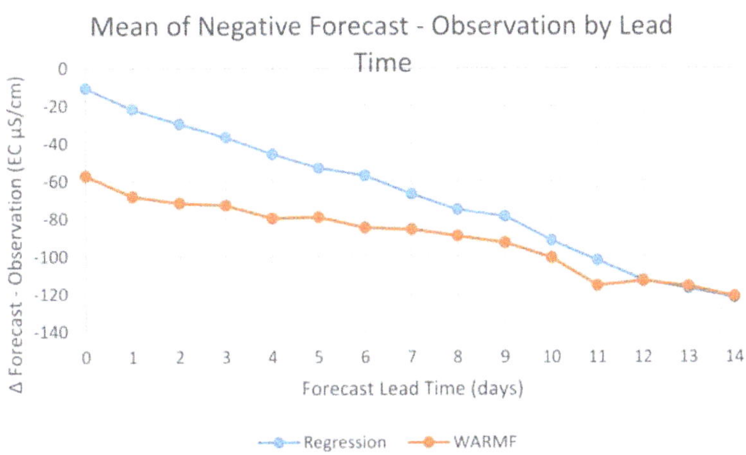

Figure 9. Comparison of means of forecasted and observed EC for the Regression and WARMF models for the period between 22 February 2018 and 22 May 2020. Data censored to include only under (negative)-predictions.

For the negative bias differences, the Regression model has lower negative mean differences than the WARMF model from Δ Day + 0 to Δ Day + 11 after which both models have nearly equal EC differences.

Another aspect of the potential bias introduced by these forecasting methods is how frequently do the overpredictions (positive) or underpredictions (negative) of mean differences in EC occur as a function of forecast lead times. For instance, the mean EC forecast bias could be overly influenced by a small number of very large EC discrepancies—either positive or negative. Figure 10 compares the percentages of positive bias differences in EC for both models.

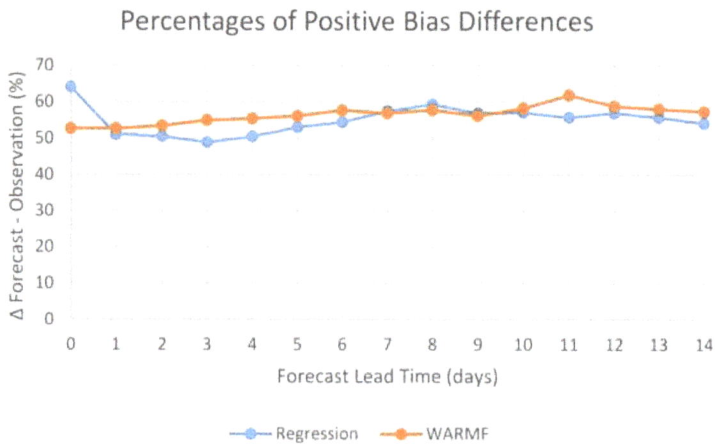

Figure 10. Comparison of the percentages of higher (positive bias) EC forecasts for the Regression and WARMF models for the period between 22 February 2018 and 22 May 2020.

As illustrated above, both the Regression and WARMF models exhibit a slight positive EC forecast bias. The Regression model exhibits a higher frequency (65%) of positive EC forecast bias differences on Δ Day + 0 for the period between 22 February 2018 and 22 May 2020. From Δ Day + 1 to Δ Day + 4, the Regression model has a neutral EC forecast bias frequency of approximately 50%. From Δ Day + 5 to Δ Day + 8, the Regression model EC forecast bias becomes increasingly positive reaching a maximum of 60% before declining gradually to 55% by Δ Day + 14. The WARMF model exhibits a gradually increasing positive EC forecast bias from 53% on Δ Day + 0 to 58% on Δ Day + 6. Subsequently, the EC forecast bias declines slightly to Δ Day + 9.

In summary, the results of the model comparison analyses indicate that the Regression model EC forecasts were closer to the overall mean of the EC observations than the WARMF model forecasted EC (Figure 5). As illustrated by Figure 6, the Regression model provided EC forecasts with mean differences of less than or equal to 5 μS/cm for the first 7 days (Δ Day + 0 to Δ Day + 6). In comparison, the WARMF model provided EC forecasts with mean differences of less than or equal to 5 μS/cm for only 5 days (Δ Day + 0 to Δ Day + 4). Based on these measures of performance, the Regression model provided EC forecasts with reduced error relative to the WARMF model especially for the period from Δ Day + 4 to Δ Day + 6.

The standard deviations of Regression model EC forecasts closely approximated the standard deviations EC observations at all lead times. In contrast, the standard deviations of the WARMF model EC forecasts were consistently less than the corresponding standard deviations of the EC observations at lead time less than day (Figure 7a). For both models, the standard deviation of EC forecast differences steadily increased with forecast lead time, as expected, while the WARMF model had higher standard deviations of EC than the Regression model throughout the forecast period (Figure 7b).

When the EC forecasts were separated into those with overestimate (positive) and underestimate (negative) biases, the mean differences between the EC forecasts and observations were seen to increase predictably with forecast lead times. For both the positive and negative forecast EC mean differences, the Regression model performed better than the WARMF model for lead times from Δ Day + 0 to Δ Day + 10. From Δ Day + 12 to Δ Day + 14, the performance of both models was approximately the same.

As illustrated in Figure 10, both models have slightly positive EC forecast biases. With the exception of the high overprediction (positive) bias (65%) for the Regression model EC on Δ Day + 0, the Regression model predictions were relatively unbiased between Δ Day + 1 to Δ Day + 4 and subsequently remained slightly positively biased throughout the remainder of the forecast period. The WARMF model made consistently greater overpredictions (positive biases in EC) than the Regression model.

It is also important to note that the Regression model EC and WARMF model EC results were originally based on different forecasted flows. Up until mid-2020, the WARMF model used prior water year operations forecast for the 14 day flow forecast along the three major east-side tributaries. From July 2020 onward, the WARMF model has been using the same flow forecasts as the Regression model which come directly from the NOAA California-Nevada River Forecast Center. The analyst who makes these daily forecasts is in regular communication with reservoir operators at Modesto Irrigation District, Merced Irrigation District and the USBR Central Valley Operations Office who control releases and provide regular bulletins of changes in release schedules. Hence, any differences between the models are no longer a function of the flow release forecasts but rather the WARMF model's watershed simulation and prior knowledge of diversions and drainage inflow along each tributary research and along the mainstem of the SJR.

5. Time Series Comparisons of the WARMF and Regression Models

The preceding analysis focused on comparisons of mean EC values and differences between model-predicted EC and observations for the Regression and WARMF models for various forecast lead times. In this section, time series comparisons of the EC predicted by each model compared to EC observations for the same time period were made for selected lead times. As shown in Figure 11, both models have relatively small mean EC differences at forecast lead times of less Δ Day + 4. From Δ Day + 5 to Δ Day + 8, mean differences increased. After Δ Day + 9, the EC predictions of both models reached a relatively constant plateau. Figure 12 also shows a comparison of Regression model EC forecasts and observations at Δ Day + 4, Δ Day + 8 and Δ Day + 12. As illustrated, there was a good match between observations and forecasts. However, as the forecast lead time increased the differences between model forecast of EC and observations also increased. This relationship between model EC forecasts and observations can be quantified using the root mean square error (RMSE) statistic which increases from 69.4 at Δ Day + 4 to 103 at Δ Day + 8 to 154 at Δ Day + 12. Figure 12 shows a similar relationship between model EC forecasts and observations for the WARMF model. In this case, the RMSE increases from 99.8 at Δ Day + 4 to 123 at Δ Day + 8 to 166 at Δ Day + 12. As illustrated by the figures and RMSE values, the Regression model performed somewhat better than the WARMF model in predicting EC for similar lead times.

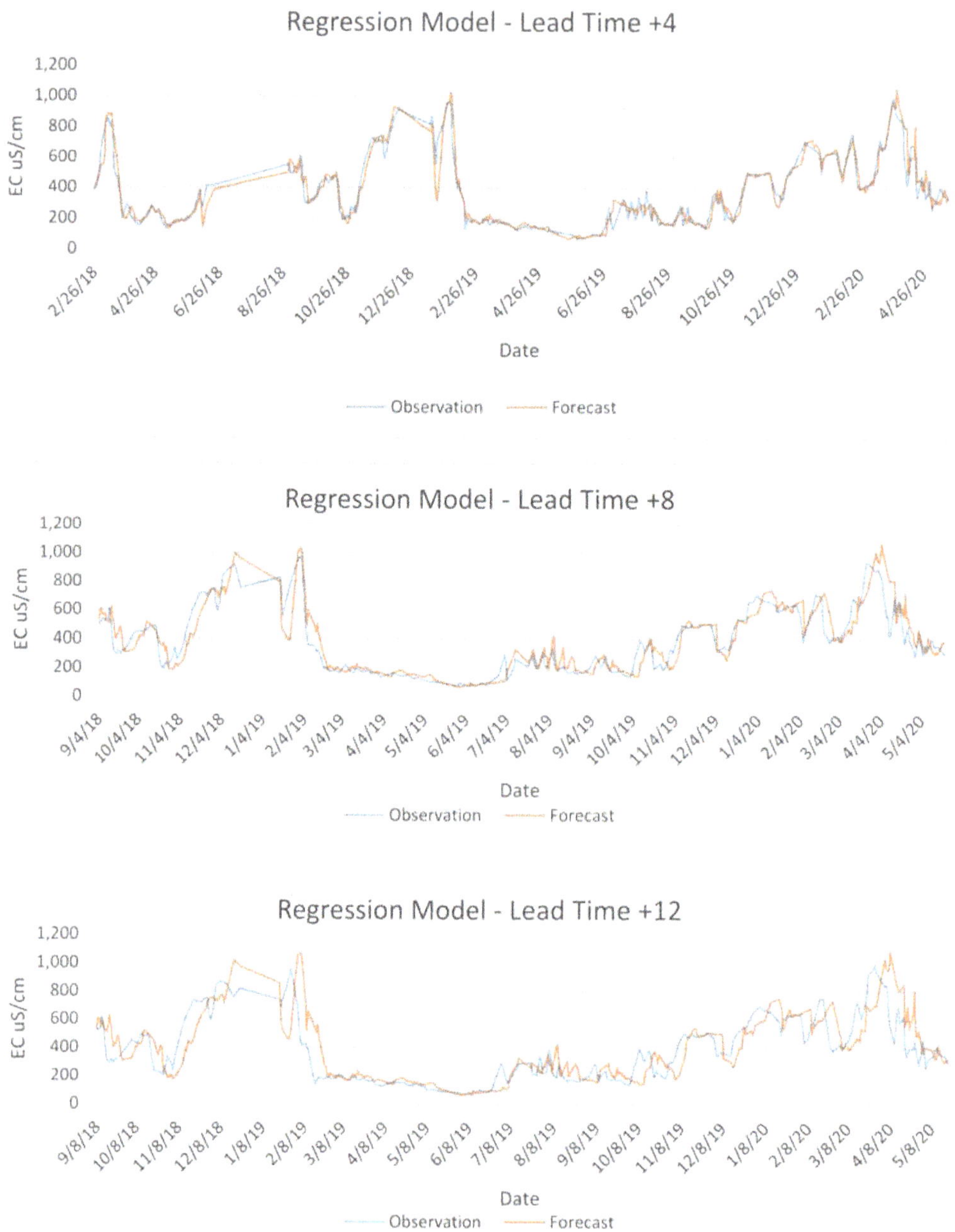

Figure 11. Comparison of Regression model forecasts and observations of EC at various lead times for the period between 22 February 2018 and 22 May 2020.

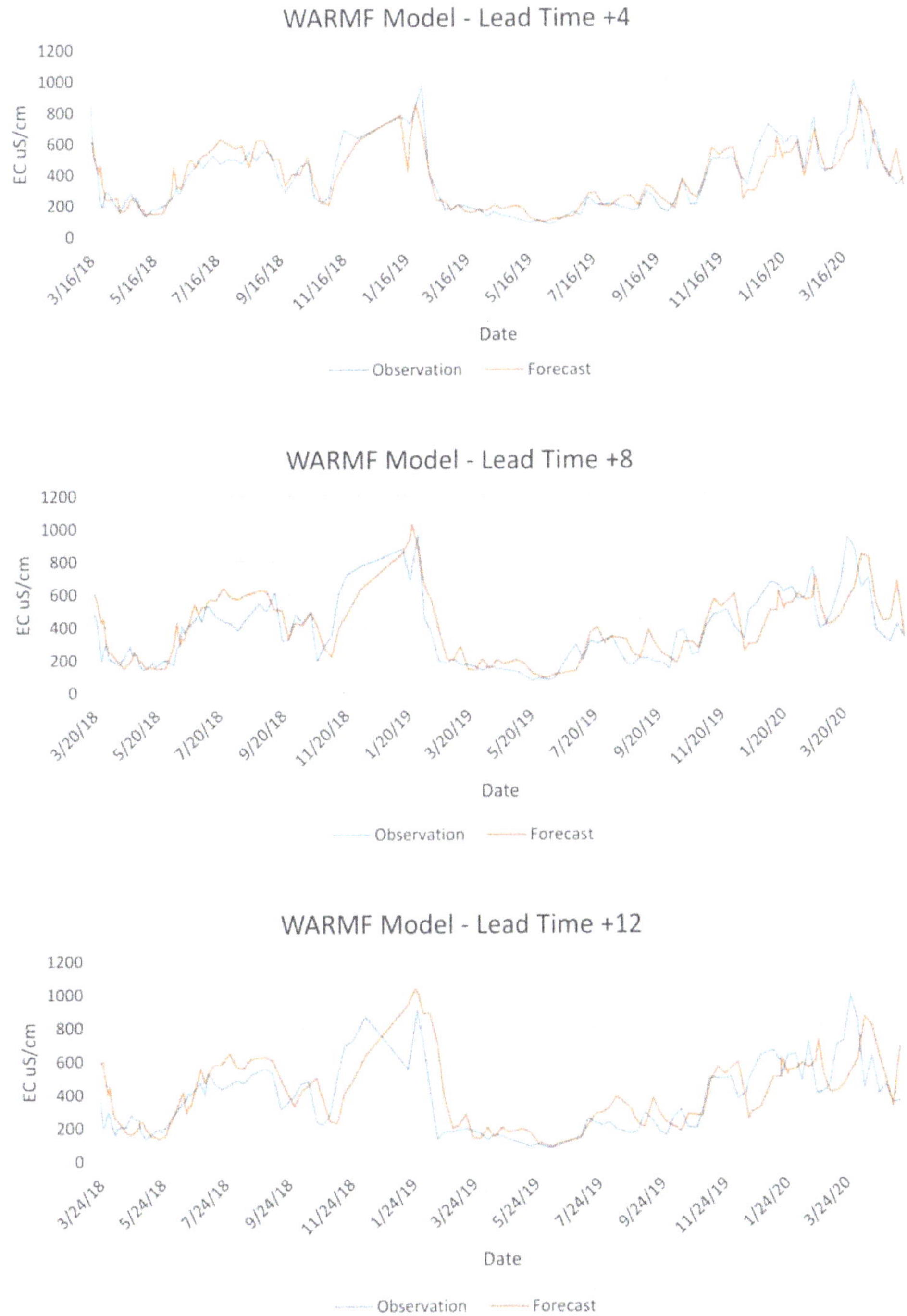

Figure 12. Comparison of WARMF model forecasts and observations of EC at various lead times for the period between 22 February 2018 and 22 May 2020.

6. Statistical Analyses

The prior analysis focused on comparisons between model EC forecasts for the regression and WARMF models and how the differences between model predictions and observations change over time for forecast lead times ranging from Δ Day + 0 to Δ Day + 14. Correspondence between model EC forecasts and EC observations exhibited significant variability. In general, as expected, the differences between model predictions of EC and observations increase with forecast lead time. Consequently, the question arises up to what lead time can the model forecasts of EC be considered reasonably reliable. In this section, a statistical approach to comparing the means of the observations and model EC forecasts is described.

The application of statistical testing methods for comparing the two models requires that careful consideration be given to the underlying assumptions made in the analysis. A preliminary decision is what statistical property should be tested. For the statistical analysis, a comparison of observation and forecast means was selected following the prior analysis based on the fact that mean salinity load, the product of the mean concentration (EC) and the mean flow, is the parameter of primary interest.

In general, most environmental data do not follow a normal distribution, as will be demonstrated for the observed EC monitoring data presented in this study. This fact has important impacts on the statistical tests that can be employed to test the equivalence of the observation and forecast EC mean values. The classical t-test statistic assumes the data are normally distributed. If they are not normally distributed, it might be possible to transform the data (e.g., using logarithmic transformations) so that, when plotted, they appear normally distributed. Such transformations can sometimes complicate the interpretation of the results. Non-parametric methods, that do not assume the data are normally distributed, are tests on the median values of the sampled data and therefore are not appropriate for this study. Another approach is the use of a permutation test. This method employs large numbers of stochastically generated realizations based on the underlying data to obtain a reasonably normal distribution of values. This is the statistical analysis approach chosen for this study.

The application of these methods was accomplished with the use of the R-commander software platform (R version 3.5.3). R is public domain software available under the "Great Truth" Copyright (C) 2019 The R Foundation for Statistical Computing. Additionally employed in the analysis were several R scripts developed by Practical Statistics Inc. and made available through their Applied Environmental Statistics courses. The statistical methods deployed in the analysis that follows were chosen based on their relative accessibility and the perception that these could be easily explained to program participants and interested stakeholders. Given the differences in the ways each of the models has been deployed for forecasting (one run daily and the other weekly), it was thought necessary to address these potential biases through the use of standard, well recognized methods. These included

1. Visual examination of the observed EC data and Regression and WARMF model EC forecasts at selected forecast lead times using boxplot graphical output.
2. Statistical testing of the normality of the observed EC data and model EC forecasts using the Shapiro–Wilks test at selected forecast lead times.
3. Statistical testing to determine whether the observed EC data and model EC forecasts have similar variances using the Fligner–Killeen test at selected forecast lead times.
4. Scatterplots of the output from the Regression and WARMF model forecasts data at selected forecast lead times.
5. Developing linear models using a forecast response variable and observation explanatory variable and computing the adjusted R-squared as an indicator of model goodness of fit at selected forecast lead times.
6. Matched pair permutation testing to evaluate the whether the means of the observed EC and model forecast EC are statistically significant at the selected forecast lead times.

The results of these analyses are presented for selected lead times of Δ Day + 12 representing the late forecast period. The boxplots showing the results of the Regression (Figure 13a) and WARMF (Figure 13b) model forecast EC comparisons with the observed data EC. Boxplots are visual tools that can be used to indicate whether the data are normally distributed. If the distribution is normal, the boxplot would be divided into equal (blue) areas by the median (black line) and the data range represented by the dashed line would have equal lengths on the top and bottom of the box. As illustrated, these conditions are not met by the EC observations and EC forecasts for either model. The Shapiro–Wilks test is a statistical test used to evaluate whether data are normally distributed. Commonly, a p-value of less than 0.05 is considered indicative of a non-normal distribution. As shown in Figure 13, the p-values are considerably less than 0.05 confirming the boxplot interpretation. At forecast lead time Δ Day + 12, the boxplots in Figure 13 suggest that neither the observed EC or model forecast EC are normally distributed but have similar variances.

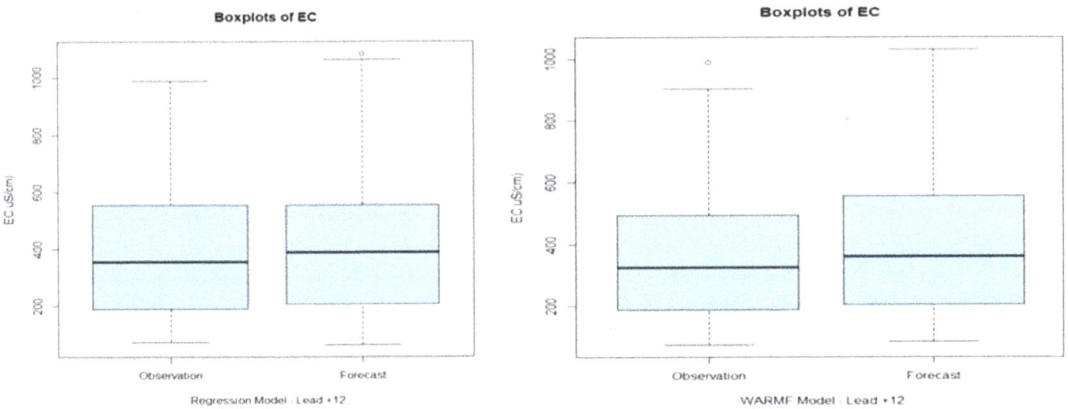

(a) (b)

Figure 13. (**a**,**b**). Boxplots of observed EC and forecast EC by the Regression (**a**) and WARMF (**b**) models are shown for forecast lead time Δ Day + 12. Fligner–Killeen variance p values are 0.6244 and 0.2703 for the Regression and WARMF models, respectively.

Scatterplots of observed EC data and both Regression and WARMF model models EC forecasts are shown in Figure 14a,b, respectively, with their linear regression plots superimposed. The Regression model EC forecasts shows slightly less scatter around the "best fit" regression line than the WARMF model EC forecasts. However, neither model shows a high R-squared coefficient indicating poor fit.

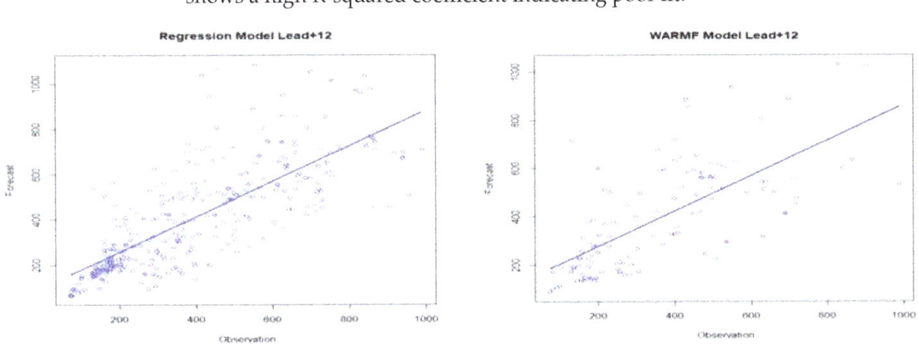

(a) (b)

Figure 14. (**a**,**b**). Calculated linear regression relationship (solid blue line) for the Regression (**a**) and WARMF (**b**) models together with a scatterplot of the underlying observed EC data and model forecast EC for lead time Δ Day + 12.

Figure 15 shows the histograms and *p*-values associated with the matched pair permutation test for both Regression (15a) and WARMF (15b) model EC forecasts for forecast lead time Δ Day + 12. The results of the matched pair permutation test indicate that neither the Regression model nor WARMF model EC forecasts are good representations of the observed EC values at lead day Δ Day + 12. The Regression model EC has a *p*-value of slightly greater than 0.05 (0.1021) while the WARMF model EC has a *p*-value is slightly less than 0.05 (0.0283).

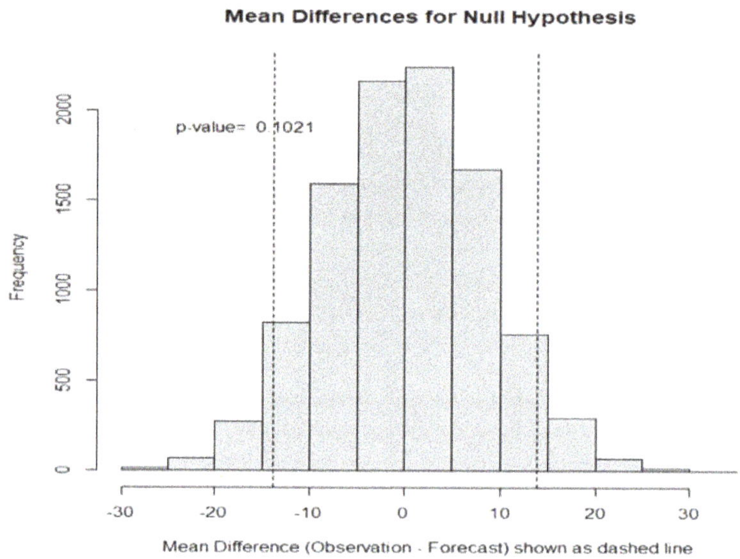

(**a**)

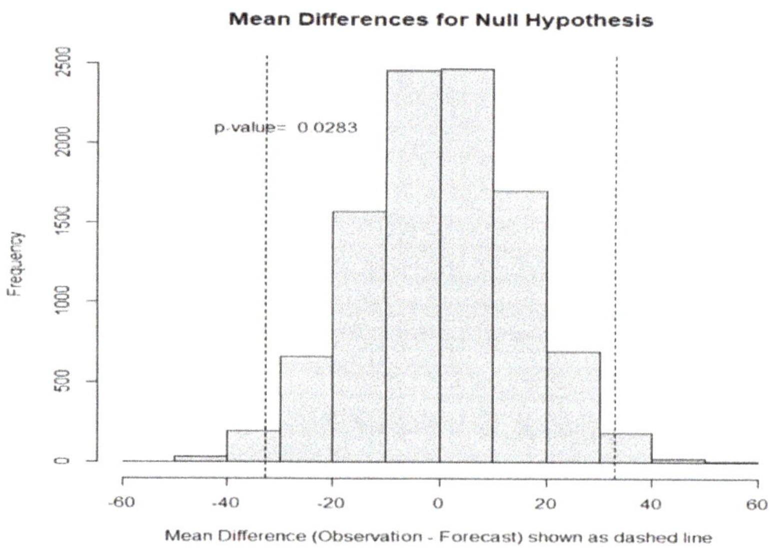

(**b**)

Figure 15. (**a**,**b**). Histograms of the mean differences between observed EC and model forecast EC for the Regression (**a**) and WARMF (**b**) models for model forecast lead time Δ Day + 12.

In addition to the selected lead times presented above, adjusted R-squared and matched pair permutation tests were computed for EC predictions from both Regression and WARMF models for all EC forecast lead times from Δ Day + 0 to Δ Day + 14. Figure 15a,b shows the adjusted R-squared values for both models. As illustrated, the Regression model has higher adjusted R-squared values than the WARMF model throughout the forecast period indicating a better goodness of fit. However, it is also worth noting that the adjusted R-squared values for both models decline progressively over the forecast period indicating a declining goodness of fit at longer lead times.

The results of the matched pair permutation tests comparing the mean of the observed EC and forecast EC for both regression and WARMF models are shown in Figure 16.

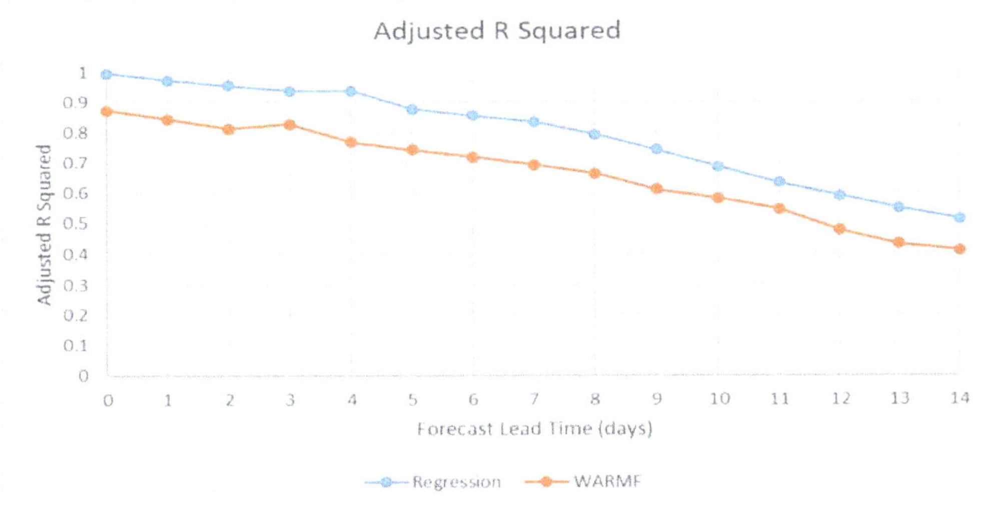

Figure 16. Adjusted R-squared values for the Regression and WARMF models for all EC forecast lead times.

The results of the statistical analyses are summarized as follows:

- Visual analysis and statistical tests indicate that although both observed EC and model forecast EC are not normally distributed their variance are sufficiently similar to validate the use of the matched pair permutation test to test whether the mean of the EC observations and model EC forecasts are statistically similar.
- The Regression model has consistently higher adjusted R-squared values than the WARMF model at all lead times indicating it has a relatively better goodness of fit.
- The matched pair permutation testing suggests that both models can make reasonably good EC forecasts out to approximately 7 days.

Discussion of Model Evaluations

Qualitative and quantitative comparisons of the performance of the WARMF and Regression models for forecasting EC at the compliance monitoring station at Vernalis were made to assess the utility of both models. The simple evaluation of the Regression and WARMF forecasting models comparing the differences between the observed salinity and the model-based forecasts of EC at the Vernalis compliance monitoring station between 22 February 2018 and 22 May 2020 suggested that the Regression model EC forecasts were generally closer to the overall mean of the observations than the WARMF model EC forecasts (previously shown in Figure 5). Although there were a total of 820 EC observations made at the Vernalis monitoring station fewer forecasts were made due to personnel availability and occasionally data validity issues. The WARMF model EC forecasts were made on the Monday of each week owing to the greater amount of time required to assemble model time series input data and complete each forecast and associated personnel constraints—hence forecasting frequency was roughly three times higher in the case of the Regression model (previously shown in Table 2).

The results of the model performance comparison as was shown in Figure 7, the Regression model provides EC forecasts with mean differences of less than or equal to 5 μS/cm for the first 7 days (Δ Day + 0 to Δ Day + 6). Alternately, the WARMF model provides EC forecasts with mean differences of less than or equal to 5 μS/cm for only 5 days (Δ Day + 0 to Δ Day + 4). Based on these measures of performance, the Regression model provides EC forecasts with reduced error relative to the WARMF model for the period from Δ Day + 4 to Δ Day + 6.

Forecast EC standard deviation, a measure of the dispersion of the EC forecasts or EC forecast differences around the mean EC value, showed that Regression model EC forecasts closely approximated of the EC observations at all lead times. The standard deviations of the WARMF model EC forecasts were consistently less than standard deviations of the EC observations until lead time day + 8. The standard deviation of forecast EC differences steadily increased with forecast lead time for both models with the WARMF model EC forecasts exhibiting greater values of standard deviation than the Regression model throughout the forecast period (previously shown in Figure 2).

To examine the effect where individual model bias affected the mean of differences between the observed EC and the model forecasted EC, EC forecast values that were higher than the measured EC were examined separately from those for which the EC forecast values were lower than the corresponding EC observations. Figures 8 and 9 showed comparisons of the positive and negative bias EC results for the Regression and WARMF models, respectively. For the positive bias differences in EC, the Regression model had smaller differences at all lead times than the WARMF model. For the negative bias differences in EC, the Regression model had smaller negative mean differences than the WARMF model. For both the positive and negative bias forecast mean differences in EC, the Regression model performed better than the WARMF model for lead times from Δ Day + 0 to Δ Day + 10. From Δ Day + 12 to Δ Day + 14, the performance of both model EC forecasts was approximately the same.

Visual inspection of the forecast EC time series results did not reveal any particular seasonal influence on the results. The RMSE between the observed EC data and model EC forecasts was also calculated as a function of forecast EC lead time. These results revealed that RMSE increased with EC forecast lead time indicating a decrease in the reliability of model forecasts. The Regression model showed consistently lower RMSE values compared to the WARMF model. The California Nevada River Forecast Center has typically run its published forecasts out only 10 days. As previously discussed, fourteen days has been considered by technical analysts associated with the real-time salinity management program to be a minimum period that would reasonably allow agricultural and wetland managers time to make adjustments to salt load export to the SJR.

Visual analysis and statistical tests suggested that neither the observed EC data or the model EC forecasts were normally distributed whereas the variances were sufficiently similar to validate the use of the matched pair permutation test, used to test whether the mean of the observed EC and model EC forecasts are statistically similar. The Regression model showed better goodness of fit relative to the WARMF model (Figure 16) as assessed by the R-squared coefficient The matched pair permutation tests indicated that both Regression and WARMF models provided reasonable forecasts extending out to approximately 7 days—not quite long enough to satisfy the goal of 14 days suggested for agricultural and wetlands stakeholder operations.

The prior analyses were based on using the full data set of all available daily observation EC data–model forecast EC paired values for both the Regression and WARMF models. However, the Regression model EC forecasts were made approximately three times more frequently than the WARMF model EC forecasts over the past 2 years (Table 3).

Comparisons of the concurrent day EC forecast results with those made with the full data set suggested that the results of the analysis were similar. For both cases, the WARMF model EC forecasts were consistently lower than those the Regression model and also lower than the observed data (comparing Figure 1 with Figure 17). The standard deviations of differences between forecasts and observations for the WARMF model EC forecasts for both the full and concurrent data sets were greater than those for the Regression model at all forecast lead times.

In general, the Regression model performed better than the WARMF model for forecasting EC for up to one week into the future.

7. Case Study: Forecasts of EC Exceedances during Spring 2021

During February 2021, an opportunity arose to compare the forecasting capability of both models in real-time during a time period where the trend in the 30 day running average EC at two of the three SJR compliance monitoring stations suggested potential future exceedance of EC objectives. California is in the second year of a severe drought and water shortages in the State's reservoirs have resulted in severe curtailment of surface deliveries to some farmers. Federal contractors with junior water rights in the SJR Basin, south of the Delta, may receive no surface water deliveries at all during the 2021 irrigation season. The central premise of the real-time salinity management program remains that coordinated actions on the part of stakeholders can optimize the use of SJR assimilative capacity preventing violations of water quality objectives.

The real-time water quality management program was initiated during a time when Vernalis was the only compliance monitoring station for salinity on the SJR. During 2020, two additional water quality stations were added for salinity management in the lower SJR—Reach 83. This action, that was subsequently introduced as an amendment to the Basin Water Quality Control Plan, ostensibly places limits on the degradation of water quality (EC) of riparian diversions into the Patterson and West Stanislaus Irrigation Districts. Although it is unclear what enforcement actions might follow non-compliance with the new 1550 μS/cm salinity objective for Reach 83, the current WARMF model and the USBR's Regression model were extended to supply 14 day forecasts of EC and salt load assimilative capacity at these stations. The basin Plan amendment provided some compliance relief for various sequences of wet, dry and critically dry years where the 30 day running average EC limit was raised using a weighting schema. Unfortunately, the formula does not provide any means to avoid the EC objective for the current water year.

The USBR's obligation under a Management Agency Agreement (MAA) signed with the CRWQCB (the State regulator) is to meet the 30 day rolling average EC objectives at the Vernalis, Crows Landing and Maze Road Bridge, the current compliance monitoring sites for EC. These objectives are ostensibly to provide suitable water quality for riparian agricultural diversions along the mainstem of the SJR and in the Delta. The premise was that stakeholders would help to sustain water quality improvements in the SJR with the help of the USBR-funded cyberinfrastructure by scheduling drainage salt loads from west-side sources to coincide with dilution flows generated from east-side sources so as not to exceed the salt load assimilative capacity of the SJR, estimated at each of these stations.

In late February 2021, as watershed inflow to the SJR subsided after a series of rainfall events, both the WARMF and Regression forecasting models suggested a slowly increasing trend in the daily and 30 day running average EC (Figures 17 and 18) that might exceed the various compliance monitoring station EC objectives at Crows Landing and Maze Road (EC 30 day running average objectives of 1550 μS/cm) and at Vernalis which was transitioning from the winter 30 day running average EC objective of 1000 uS/cm to the irrigation season objective of 700 μS/cm. Note that the irrigation season objective applies after April 30 (when 30 days have elapsed). The weekly WARMF model forecast (green background) suggested on 22 February 2021 that the 30 day running average EC threshold of 1550 μS/cm at Crows Landing could be exceeded on 6 March 2021 (Figure 18a) whereas the Vernalis site still showed salt load assimilative capacity (Figure 17a). The USBR had been making regular adjustments of New Melones reservoir releases to maintain compliance with EC objectives at Vernalis as required under the MAA. The Regression model (blue background) that was run on the same Monday February 22 (Figure 18b) suggested an occurrence of the same exceedance event although the date of the exceedance was predicted one day earlier. In order to lower the 30 day running average EC at Crows Landing, west-side return flows upstream of Crows Landing would need to fall below the 1550 µS/cm criterion.

WARMF and Regression model forecasts made on April 26 were much closer in their predictions (Figure 18c,d) and neither suggested that 30 day running average would drop below the zero line—indicating continuing exceedance and lack of SJR SLAC (Figure 18e,f). The forecasts made by the models on 1 June 2021 show that the daily mean EC dropped below the objective on 19 May 2021 and continued to drive the 30 day rolling average downward until it dropped below the 1550 µS/cm objective and transitioned into positive territory on 28 May 2021 (Figure 19a,b).

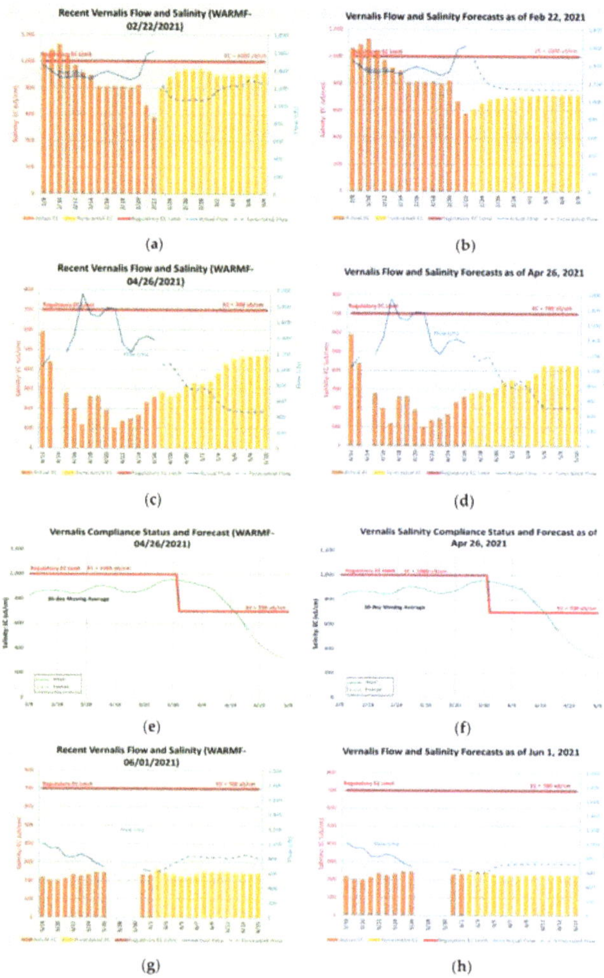

Figure 17. Comparison of daily WARMF and Regression model forecasts for EC at the Crows Landing compliance monitoring station on 22 February 2021 (**a,b**); 26 April 2021 (**c,d,e,f**); and 1 June 2021 (**g,h**). Graphs (**e,f**) show the 30-day running average EC forecast on 26 April 2021 relative to the the 30-day running average EC compliance objective. Conversion of flow in cfs to m^3/s: 100 cfs = 2.83 m^3/s.

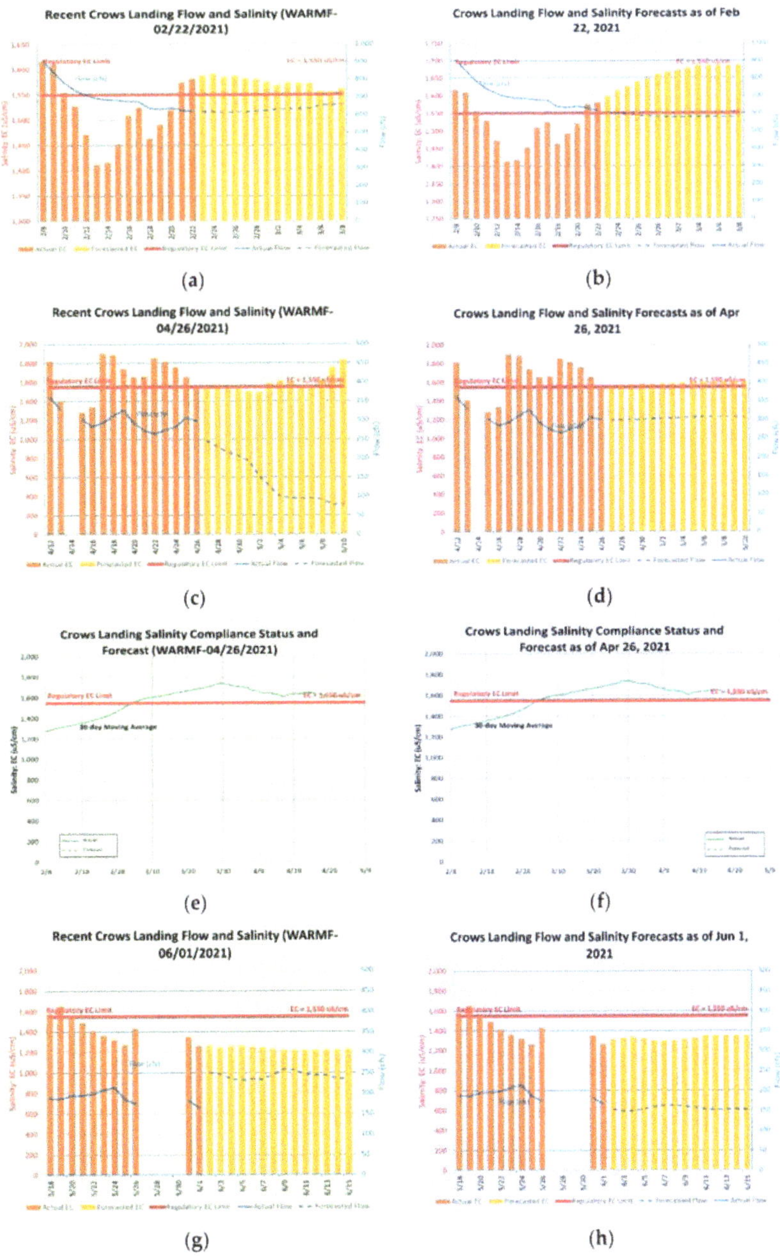

Figure 18. Comparison of daily WARMF and Regression model forecasts for EC at the Crows Land-ing compliance monitoring station on 22 February 2021 (**a**,**b**); 26 April 2021 (**c**,**d**,**e**,**f**); and 1 June 2021 (**g**,**h**). Graphs (**e**,**f**) show the 30-day running average EC forecast on 26 April 2021 relative to the the 30-day running average EC compliance objective. Conversion of flow in cfs to m3/s: 100 cfs = 2.83 m^3/s.

This transition is also shown in Figure 19. Figure 19c, which was produced on 1 June 2021, correctly predicted the transition to positive salt load assimilative capacity on 26 May 2021. This plot also shows the proportion of the salt load contributed by the combination of Mud and Salt Slough relative to the total salt load measured at the Crows Landing compliance monitoring station. At this time of year, the majority of the salt load in these Sloughs is seasonal wetland drainage, which typically has an EC in excess of 1500 μS/cm.

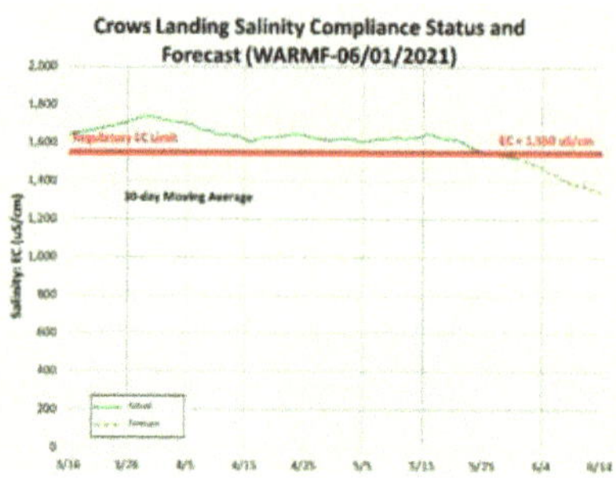

(a)

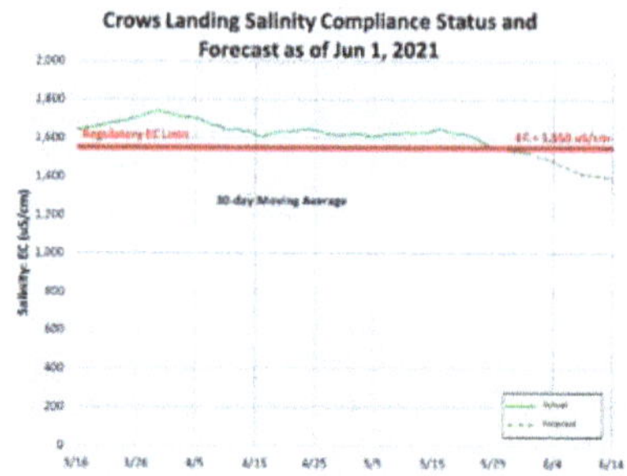

(b)

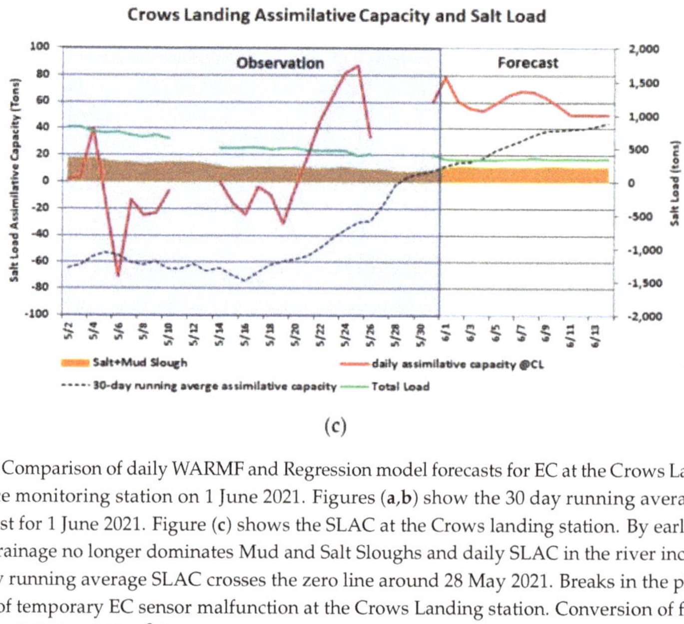

(c)

Figure 19. Comparison of daily WARMF and Regression model forecasts for EC at the Crows Landing compliance monitoring station on 1 June 2021. Figures (**a**,**b**) show the 30 day running average EC and forecast for 1 June 2021. Figure (**c**) shows the SLAC at the Crows landing station. By early May wetland drainage no longer dominates Mud and Salt Sloughs and daily SLAC in the river increases. The 30 day running average SLAC crosses the zero line around 28 May 2021. Breaks in the plot are the result of temporary EC sensor malfunction at the Crows Landing station. Conversion of flow in cfs to m^3/s: 100 cfs = 2.83 m^3/s.

8. Stakeholder Response and Coordination

As previously noted, this event has provided the USBR with an opportunity to demonstrate the agency's commitment to its obligations under the MAA, reminded stakeholders of their role in the real-time program and exposed deficiencies in real-time response to periods of water quality exceedance. During the second week of February, when it became clear through the use of the forecast models that the salinity at both Vernalis and Crows Landing stations was trending towards potential exceedance of the 30 day running average EC stakeholders were notified directly. The likely date of exceedance was estimated to be March 5 from WARMF and Regression model forecasts made on 23 February 2021. In order to provide stakeholders adequate time to perform remedial actions, we decided to directly engage with stakeholders in the SJRB rather than rely on the USBR's normal weekly posting of flow, EC and 30 day running average EC at the three compliance monitoring stations. Communication with stakeholders was primarily by e-mail to east- and west-side agricultural stakeholder coalitions, directly impacted water district, and representatives of the private, state and federal wetland entities. the San Joaquin Valley Drainage Authority, Grassland Water District, Los Banos Wildlife Management Area, Patterson and West Stanislaus Irrigation Districts, on the east-side Modesto and Turlock irrigation districts and the East SJR Water Quality Coalition. A similar e-mail was sent to the Regional Water Quality Control Board, the basin regulator, that has the power to set fines for water quality objective exceedances.

In retrospect, the timing of the stakeholder outreach was timely and prescient. Although anticipated, programmatic fish migration flows from east-side reservoirs, that started in mid-April, were able to drive down the EC at Vernalis below the 700 μS/cm limit that came into effect on April 30. The Merced River is the only tributary to the SJR upstream of Crows landing and supplemental flows for fish migration were insufficient to prevent the EC at Crows landing from exceeding objectives. During the period of exceedance at the Crows Landing compliance monitoring, there were opportunities to address the excess salt loading to the SJR. During the initial period of exceedance, raising the board elevation at the San Luis Drain outlet (a previously used conveyance facility that carries only subsurface agricultural drainage) and storing drainage return flows in the drain for later release would have reduced salt loading by 100 tons (91 tonnes) per day and eliminated the deficit in SLAC. Most drainage return flows into the drain are from seepage from adjacent agricultural land and wetland and with an average salinity of 2500 μS/cm. However, after the first days of exceedance, the daily EC remained elevated above 1550 μS/cm and the 30 day running average SLAC deficit climbed to a steady state load of approximately negative 1000 tons (907 tonnes) per day.

The exceedance of the Crows Landing EC objective occurred during the wetland drawdown period when the Grassland Water District and adjacent State and Federal refuges are draining ponded surface water to allow germination of swamp timothy, smartweed and water grass food crops that serve overwintering waterfowl. Since the timing of this drawdown is critical for swamp timothy production and the waterfowl that prefer this food source, asking wetlands to curtail drawdown during this period was viewed as unrealistic by wetland resource managers.

Procurement of additional dilution flow from the Merced Irrigation District was also unrealistic given the prevailing drought conditions and anticipated water shortages during the summer of 2021. In addition, some entity would have had to foot the bill for procurement of any additional supply if supply were available.

The Regional Board has taken a "wait and see" approach to this first test of the real-time water quality management system and the newly promulgated upstream EC objectives at Crows Landing and Maze Road compliance monitoring stations. There has been no discussion of fines or allocation of penalties across subareas contributing salt load to the SJR from the Regional Board. There is also the fact to consider that riparian diverters along the northwest-side subarea is the river reach that the upstream objective was promulgated to protect. Fining stakeholders who are being harmed by the elevated EC along this reach of the SJR would be problematic.

At the time of writing, the severe drought conditions in the basin have reduced forecasted flow for 10 June 2021 at the Crows landing compliance monitoring station to under 100 cfs and daily EC is once again over the 1550 threshold EC. The 30 day running average EC is climbing once again and may remain above the objective for the remainder of the irrigation season while drought mitigation actions are in force.

9. Summary and Conclusions

Real-time salinity management is a stakeholder- and water agency-sanctioned program that helps to maximize allowable salt export from the agriculture-dominated SJR Basin. The essential components of the current program that are now in place include the establishment of telemetered sensor networks, a web-based information system for sharing data, a basin-scale salt load assimilative capacity forecasting model and institutional entities tasked with performing weekly forecasts of river SLAC and using these forecasts to improve scheduling of west-side drainage salt load export and the dilution provided by east-side reservoir releases. Two modeling approaches were developed simultaneously, in part to see if a higher level of automation could be introduced in developing SLAC forecasts and if the frequency of these forecasts could be moved from weekly using the WARMF numerical simulation model to a simpler flow-based regression modeling approach run daily. The Regression model relies on a comprehensive statistical analysis of the relationship between flow and salt concentration at three compliance monitoring sites. The WARMF watershed water quality simulation model provided the conventional SLAC forecasting approach. The model is data driven and although model data acquisition is almost fully automated, there is still a need for user involvement for simulation times that may take an hour or more. The results from both models are migrated manually to Excel spreadsheets that are used to produce graphics that are posted to the web daily in the case of the Regression model and weekly for the WARMF model.

The first part of this paper has provided a comprehensive analysis of the model results when used to make 14 day EC forecasts (daily and 30 day running average EC) and an estimate of 14 day river SLAC. Analysis of the results from both model-based forecasting approaches over a period of five years shows that the regression-based forecasting model, run daily Monday to Friday each week, provided marginally better performance. However, the regression-based forecasting model assumes the same general relationship between flow and salinity which breaks down during extreme weather events such as droughts when water allocation cutbacks among stakeholders are not evenly distributed across the basin. A recent test case was used to demonstrate the potential utility of both models in dealing with an exceedance event at the Crows Landing compliance monitoring station. This year is providing an opportunity to test the robustness and reliability of the flow-EC relationship that the regression model relies upon since contract water delivery to USBR contractors is scaled back unequally during times of shortage in association with District water rights. The major lesson learned from the project to date is that a dual modeling approach of using a simple Regression model for daily automated forecasting with weekly simulation model runs using the WARMF model appears to be a good compromise at present that provides sufficient frequency of forecasts to allow stakeholders to make timely decisions (Regression model) while using stakeholder data to eliminate model inconsistencies during periods of unusual or extreme basin hydrology. The use of the WARMF model in this dual modeling approach provides modelers with a tool to more fully understand the current state of the system and to investigate unusual occurrences in basin hydrology and water quality that are only possible with a mechanistic model like the WARMF model.

In the future, it would be desirable that the Regression and WARMF models are both run daily which would eliminate some of the model comparison questions that were addressed in this study. Further automation of WARMF model data pre-processing steps could be combined with similarly automated real-time data quality assurance routines—perhaps enhanced with machine learning procedures to eliminate data gaps, remove sensor drift and data spikes to improve model performance. The lack of a robust and customizable, public domain real-time data quality assurance software tool remains the biggest remaining impediment to water quality forecasting capabilities and if addressed could enhance stakeholder confidence in this instance of model-based environmental decision support.

Author Contributions: Conceptualization, N.W.T.Q.; methodology, N.W.T.Q., M.K.T. and J.L.; software, J.L.; statistical analysis, M.K.T.; data curation, M.K.T., J.L. and N.W.T.Q.; visualization, M.K.T. and N.W.T.Q.; writing—original draft preparation, N.W.T.Q.; writing—review and editing, N.W.T.Q. and M.K.T.; response to reviewer comments, N.W.T.Q. All authors have read and agreed to the published version of the manuscript.

Funding: This research received no external funding.

Institutional Review Board Statement: Not applicable.

Informed Consent Statement: Not applicable.

Data Availability Statement: No new data, models, or code were generated or used during the study—this study has com-piled publicly accessible data and existing model output to create a unique contribution to the literature. Daily model forecasts of flow and EC at the three compliance monitoring sites mentioned in the paper produced using the Regression model are available on the USBR's web portal at http://www.usbr.gov/ptms/ (accessed on 20 September 2021). Weekly WARMF model forecasts may be substituted for the Regression model forecasts when SJR flow data is unavailable or unreliable. These forecasts are posted on the same web portal.

Acknowledgments: The lead author wishes to recognize individual contributions made to the manuscript by the co-authors within the USBR. Much of the statistical analysis in the paper was drawn from an internal agency report provided by Michael Tansey. James Lu developed the EC forecasting algorithm for the Regression model and is responsible for daily web posting of Regression model-based forecasts of EC at the three SJR compliance monitoring sites. WARMF model forecasts are made weekly by Jun Wang and the lead author and are posted by Jun Wang on the same USBR web portal. Recent results from both models were used in the case study presented. The lead author also wishes to acknowledge the support of the San Joaquin Valley Drainage Authority and member water districts including Grasslands Water District, Patterson and West Stanislaus Irrigation Districts that have helped to advance the concept of real-time water quality management. Also, the California Regional Water Quality Control Board, the State regulator, for encouraging this new approach to salt management in the Basin that maximizes the beneficial use of the SJR. Much of the success of the Program is a result of generous State of California Department of Water Resources funding under Proposition 84 which has allowed upgrading and ongoing support of the various flow and EC sensor networks. The USBR has provided essential ongoing support since the inception of the Program.

Conflicts of Interest: The authors declare no conflict of interest.

References

1. United States Environmental Protection Agency (USEPA). Impaired Waters and TMDLs: Overview of Total Maximum Daily Loads (TMDLs). 2018. Available online: http://www.epa.gov/tmdl/overview-total-maximum-daily-loads-tmdls#6 (accessed on 6 March 2020).
2. Ambrose, R.B.; Martin, J.L.; McCutcheon, S.C. *Technical Guidance Manual for Performing Waste Load Allocations, Book III: Estuaries, Part 2: Application of Estuarine Waste Load Allocation Models*; U.S. Environmental Protection Agency Report EPA-823-R-92-003; Office of Water: Washington, DC, USA, 1990. [CrossRef]
3. Martin, J.L.; McCutcheon, S.C. *Hydrodynamics and Transport for Water Quality Modeling*; CRC Press: New York, NY, USA, 1999.
4. McCutcheon, S.C. *Water Quality Modeling: Volume I, River Transport and Surface Exchange*; CRC Press: Boca Raton, FL, USA, 1989.
5. McCutcheon, S.C.; Zhu, D.W.; Bird, S.L. Model calibration, validation and use. In *Technical Guidance Manual for Performing Waste Load Allocations, Book III Estuaries, Part 2 Application of Estuarine Waste Load Allocation Models*; EPA 823 R 92 003; Ambrose, R.B., Martin, J.L., McCutcheon, S.C., Eds.; United States Environmental Protection Office of Water: Washington, DC, USA, 1990; p. 77.
6. Borah, D.K.; Ahmadisharaf, E.; Padmanabhan, G.; Imen, S.; Mohamoud, Y.M. Watershed models for development and implementation of total maximum daily loads. *J. Hydrol. Eng.* **2019**, *24*, 03118001. [CrossRef]
7. Chapra, S.C. Engineering water quality models and TMDLs. *J. Water Resour. Plan. Manag.* **2003**, *129*, 247–256. [CrossRef]
8. Zhang, H.X.; Quinn, N.W.T. Simple models and analytical procedures for total maximum daily load assessment. *J. Hydrol. Eng.* **2019**, *24*, 02518002. [CrossRef]
9. Task Committee on TMDL Modeling and Analysis. *Total Maximum Daily Load Analysis and Modeling: Assessment of the Practice*; Environmental and Water Resources Institute, American Society of Civil Engineers: Reston, VA, USA, 2017.
10. Ahmadisharaf, E.; Camacho, R.A.; Zhang, H.X.; Hantush, M.M.; Mohamoud, Y.M. Calibration and validation of watershed models and advances in uncertainty analysis in TMDL studies. *J. Hydrol. Eng.* **2019**, *24*, 03119001. [CrossRef]
11. Ahmadisharaf, E.; Benham, B. Risk-based decision making to evaluate pollutant reduction scenarios. *Sci. Total Environ.* **2020**, *70*, 135022. [CrossRef] [PubMed]
12. California Environmental Protection Agency. *Total Maximum Daily Load for Salinity and Boron in the Lower San Joaquin River*; Staff Report; Regional Water Quality Control Board: Central Valley Region: Sacramento, CA, USA, 2002.
13. California Regional Water Quality Control Board. *Amendments to the Water Quality Control Plan for the Sacramento River and San Joaquin River Basin*; Draft Final Staff Report and Technical TMDL Report; California Regional Water Quality Control Board: Sacramento, CA, USA, 2004.
14. Quinn, N.W.T. Policy Innovation and Governance for Irrigation Sustainability in the Arid, Saline San Joaquin River Basin. *Sustainability* **2020**, *12*, 4733. [CrossRef]

15. Quinn, N.W.T.; Hanna, W.M. A decision support system for adaptive real-time management of seasonal wetlands in California. *Environ. Modeling Softw.* **2003**, *18*, 503–511. [CrossRef]
16. Quinn, N.W.T.; Karkoski, J. Real-time management of water quality in the San Joaquin River Basin. California. *Am. Water Resour. Assoc.* **1998**, *34*, 1473–1486. [CrossRef]
17. Arax, M.; Wartzman, R. *The King of California: J.G. Boswell and the Making of a Secret American Empire*; Public Affairs: New York City, NY, USA, 2005; ISBN 1-58648-281-5.
18. Oster, J.D.; Wichelns, D.E.W. Hilgard and the history of irrigation in the San Joaquin Valley: Stunning productivity, slowly undone by inadequate drainage. In *Salinity and Drainage in San Joaquin Valley, California: Science, Technology, and Policy*; Chang, A.C., Silva, D.B., Eds.; Springer: Dordrecht, The Netherlands, 2014; pp. 7–46.
19. Quinn, N.W.T. The San Joaquin Valley: Salinity and Drainage Problems and the Framework for a Response. In *Salinity and Drainage in San Joaquin Valley, California: Science, Technology, and Policy, Global Issues in Water Policy 5*; LBNL Topical, Report–LBL-38498; Chang, A.C., Brawer Silva, D., Eds.; Springer Science+Business Media: Dordrecht, The Netherlands, 2014. [CrossRef]
20. Brownell, J. *Potential SJR Daily Salt Discharge Exceedance Fees by Subarea (10-Year Period: 2001–2012)*; Analysis Performed by CRWQCB Staff to Estimate Potential Annual Fee Schedules for Subareas Contributing Salt to the SJR; California Regional Water Quality Control Board: Sacramento, CA, USA, 2012.
21. California Regional Water Quality Control Board. *Amendments to the Water Quality Control Plan (Basin Plan) for the Sacramento River Basin and San Joaquin River Basin (with Amendments) Final Staff Report*; California Regional Water Quality Control Board: Sacramento, CA, USA, 2018.
22. Chen, C.W.; Herr, J.; Gomez, L.E.; Quinn, N.W.T.; Kipps, J.; Landis, P.J.; Cummings, E.W. *Design and Development of a Graphic User Interface for Real-Time Water Quality Management of the San Joaquin River*; California Department of Water Resources: San Joaquin, CA, USA, 1996.
23. Herr, J.; Weintraub, L.H.Z.; Chen, C.W. *User's Guide to WARMF: Documentation of Graphical User Interface*; Topical report; EPRI: Palo Alto, CA, USA, 2001.
24. Herr, J.; Chen, C.W. *San Joaquin River Model: Calibration Report. CALFED Project ERP-02D- P63 Monitoring and Investigations for the San Joaquin River and Tributaries Related to Dissolved Oxygen*; Systech Water Resources Inc.: San Ramon, CA, USA, 2006.
25. Herr, J.W.; Chen, C.W. WARMF: Model use, calibration, and validation. *Trans. ASABE* **2012**, *55*, 1387–1394. [CrossRef]
26. California State Water Resources Control Board. *Regulation of Agricultural Drainage to the San Joaquin River*; Final Report, Order No. WQ 85-1; California State Water Resources Control Board: Sacramento, CA, USA, 1985.
27. Kratzer, C.R.; Pickett, P.J.; Rashimawi, E.A.; Cross, C.L.; Bergerson, K.D. *An Input-Output Model of the San Joaquin River from the Lander Avenue Bridge to the Airport Way Bridge*; Technical Committee Report No. N.Q. 85-1; California Water Resources Control Board: Sacramento, CA, USA, 1987.
28. Quinn, N.W.T.; Amye, O.; Herr, J.; Wang, J.; Raley, E. WARMF-Online—A Web-Based Portal Supporting Real-time Salinity Management in the San Joaquin River Basin. *Open Water J.* **2017**, *4*, 4.
29. Lu, J.; Wang, J.; Raley, E.; Quinn, N.W.T.; Kabir, J. *An Alternative Approach to Salinity Forecasting in the Lower San Joaquin River*; Modern Environmental Science and Engineering; MESE20191018-2; California Water Resources Control Board: Sacramento, CA, USA, 2019.
30. Chen, Y.; Song, L.; Liu, Y.; Yang, L.; Li, D.A. Review of the Artificial Neural Network Models for Water Quality Prediction. *Appl. Sci.* **2020**, *10*, 5776. [CrossRef]

water

MDPI

Article

Supporting Restoration Decisions through Integration of Tree-Ring and Modeling Data: Reconstructing Flow and Salinity in the San Francisco Estuary over the Past Millennium

Paul H. Hutton [1], David M. Meko [2] and Sujoy B. Roy [1,*]

[1] Tetra Tech, Lafayette, CA 94549, USA; paul.hutton@tetratech.com
[2] Laboratory of Tree-Ring Research, University of Arizona, Tucson, AZ 85721, USA; dmeko@arizona.edu
* Correspondence: sujoy.roy@tetratech.com

Abstract: This work presents updated reconstructions of watershed runoff to San Francisco Estuary from tree-ring data to AD 903, coupled with models relating runoff to freshwater flow to the estuary and salinity intrusion. We characterize pre-development freshwater flow and salinity conditions in the estuary over the past millennium and compare this characterization with contemporary conditions to better understand the magnitude and seasonality of changes over this time. This work shows that the instrumented flow record spans the range of runoff patterns over the past millennium (averaged over 5, 10, 20 and 100 years), and thus serves as a reasonable basis for planning-level evaluations of historical hydrologic conditions in the estuary. Over annual timescales we show that, although median freshwater flow to the estuary has not changed significantly, it has been more variable over the past century compared to pre-development flow conditions. We further show that the contemporary period is generally associated with greater spring salinity intrusion and lesser summer–fall salinity intrusion relative to the pre-development period. Thus, salinity intrusion in summer and fall months was a common occurrence under pre-development conditions and has been moderated in the contemporary period due to the operations of upstream reservoirs, which were designed to hold winter and spring runoff for release in summer and fall. This work also confirms a dramatic decadal-scale hydrologic shift in the watershed from very wet to very dry conditions during the late 19th and early 20th centuries; while not unprecedented, these shifts have been seen only a few times in the past millennium. This shift resulted in an increase in salinity intrusion in the first three decades of the 20th century, as documented through early records. Population growth and extensive watershed modification during this period exacerbated this underlying hydrologic shift. Putting this shift in the context of other anthropogenic drivers is important in understanding the historical response of the estuary and in setting salinity targets for estuarine restoration. By characterizing the long-term behavior of San Francisco Estuary, this work supports decision-making in the State of California related to flow and salinity management for restoration of the estuarine ecosystem.

Keywords: tree-ring; flow reconstruction; pre-development; estuarine salinity; salinity intrusion

Citation: Hutton, P.H.; Meko, D.M.; Roy, S.B. Supporting Restoration Decisions through Integration of Tree-Ring and Modeling Data: Reconstructing Flow and Salinity in the San Francisco Estuary over the Past Millennium. *Water* **2021**, *13*, 2139. https://doi.org/10.3390/w13152139

Academic Editor: Nigel W.T. Quinn

Received: 2 June 2021
Accepted: 27 July 2021
Published: 3 August 2021

1. Introduction

Populated estuarine regions worldwide have been subject to a variety of stressors, including the introduction of invasive species, loss of tidal habitat, anthropogenic alterations to the natural hydrologic cycle (including freshwater diversions), impacts to sediment transport resulting from upstream watershed land use modifications, and other water quality impairments [1]. These stressors can adversely affect the estuarine habitat for resident and anadromous aquatic species. Today, there is growing interest in many parts of the world to restore estuaries to more pre-development or natural conditions [2–5]. Although restoration planning must account for multiple interacting stressors, for estuaries subjected to significant hydrologic alterations, restoration of a more natural hydrology and salinity regime is key. To support such restoration planning, pre-development reference conditions may need to be defined, although no formal methodology is proposed in current U.S. regulations. Directly observed data representing reference conditions in a developed estuary are difficult to obtain, especially when the development has occurred over centuries. However, some pre-development characteristics can be inferred from proxy data, notably estimates of precipitation in the estuary watershed through tree-ring measurements of long-lived tree species.

This work seeks to support restoration planning in the San Francisco Estuary, the largest estuary on the Pacific coasts of North and South America, by characterizing the region's pre-development hydrologic and salinity conditions over the past millennium. The estuarine region includes a series of interconnected embayments, rivers, sloughs, marshes as well as the delta formed by the Sacramento and San Joaquin Rivers (hereafter referred to as the "Delta"), which together drain a watershed of 75,000 square miles, more than 40% of the area bounded by the state of California [6,7]. Following European settlement of California in the mid-18th century and the subsequent Gold Rush (circa 1850), the estuary and its watershed have been subject to extensive changes, including land-use conversion to agriculture and urbanization, construction of water storage and diversion facilities on major rivers, channelization and modification of riparian and tidal habitats, and out-of-basin exports of water [7–9]. The estuary is currently the focus of much scientific attention because of its importance to aquatic ecosystems and because large parts of the state's urban and agricultural economies are dependent on water supplies from the Delta [7,10,11].

Freshwater flow to the estuary (termed "Delta outflow") has been identified as a vital planning component for regional sustainability. Delta outflow and salinity have been managed for several decades through the regulation of upstream reservoirs and out-of-basin exports. Maximum salinity levels are prescribed at various locations in the Delta; the broader salinity regime is regulated as the position of the 2 parts per thousand bottom isohaline from Golden Gate (measured in km), commonly referred to as X2 [12–14]; see Figure 1 for isohaline positions). Despite ongoing regulatory efforts, the abundance of many Delta fish species continues to decline from the first formally recorded levels in the 1960s [7,15–17]. In response to these declines, additional freshwater flow and salinity regulations are being considered for future implementation [18]. An improved understanding of the estuary's hydrology and salinity characteristics prior to development, and differences from contemporary conditions, will support decisions related to its future management.

The broader region delimited by the San Francisco Estuary and its upstream Central Valley watershed benefits from the availability of extensive data to reconstruct past flow and salinity conditions. These data include flow and salinity measurements, over a century or more, that represent the intensification of development in the region (e.g., [14]). These data also include tree-ring measurements to characterize watershed precipitation over the past two millennia (e.g., [19–21]). The specific research objectives of this work are to refine and update tree-ring-based reconstructions of Central Valley runoff over the past millennium and reconstruct Delta outflow and salinity over similar millennial timeframes using our runoff estimates within a modeling framework informed by previously published work. This integrated evaluation provides a time-resolved characterization of the estuary's flow–salinity behavior that allows comparison between pre-development and contemporary conditions.

This work builds on previous research that either (i) relies on a contemporary hydrologic sequence to estimate outflow and salinity changes using different modeled representations of the region's level of development [22] or (ii) relies on contemporary salinity data to estimate salinity changes using a tree-ring based hydrologic sequence [23]. By using a tree-ring based hydrologic sequence in conjunction with a modeling approach that estimates pre-development estuarine flow and salinity responses, this work attempts to represent the actual range of flow and salinity conditions over long time horizons and is expected to better support regulatory decision making by providing a baseline to inform future flow regulations and restoration actions in the estuary. Furthermore, this work places the wet and dry flow patterns recorded in the estuary over the past 150 years in the context of flow variations estimated over the past millennium from the tree-ring proxy record.

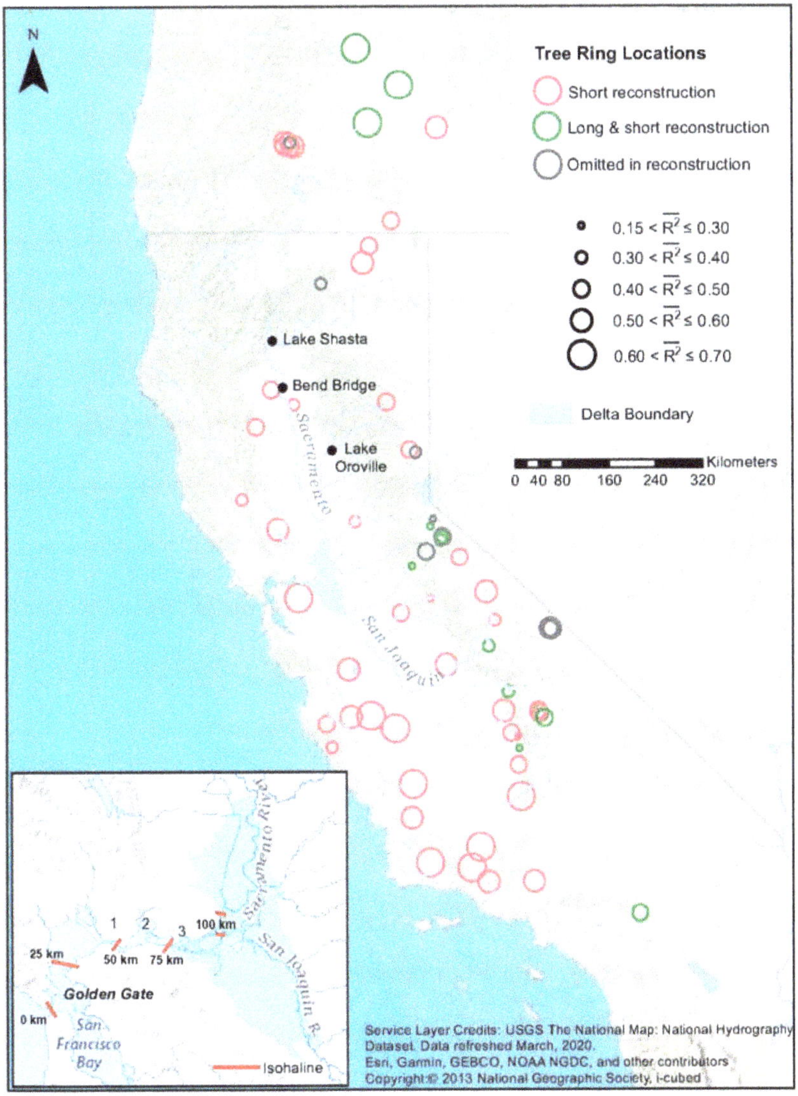

Figure 1. Study location map showing the locations of tree ring sites used in the analysis. Circles mark sites contributing to the short (60 sites) and long (13 sites) reconstructions. Circles sized proportional to percentage of variance explained in regression models for single site reconstructions (SSRs). Sites contributing to long reconstruction marked with green; those as well sites marked with red contribute to the short reconstructions. Nine sites (gray) were screened out and not used in later reconstruction steps.

2. Background

To provide background for this work, we present a brief overview of the study region's geographic setting followed by a review of the region's hydrologic and salinity conditions over the past millennia. This review differentiates between three periods: a "pre-development" period, a "contemporary" period, and an "early development" period that bridges the pre-development and contemporary periods. We define the terminus of the pre-development period as water year (WY) 1850, which roughly aligns with the California Gold Rush and follows previous work [24–26] California water years run from 1 October through 30 September. Furthermore, we define the start of the contemporary period as WY 1912, a date that aligns with availability of Delta outflow estimates [27] but pre-dates the availability of systematic estuarine salinity measurements by about a decade [14]. By default, the intervening early development period spans six decades between WYs 1851 and 1911.

2.1. Geographic Setting

The geographic focus of this paper is the upper portion of the San Francisco Estuary, including Suisun Bay, the Delta, and the Central Valley watershed upstream of the estuary (Figure 1). The Delta is the entry point of over 90% of the freshwater flow to the estuary [28] and drains the Sierra Nevada mountain range and Central Valley—a watershed of approximately 75,000 square miles. The configuration of the estuary formed approximately 5000 years ago when sea level rise stabilized [29]. Sea level rise maintained an average rate of 1.0–1.3 mm/yr [30] through the late Holocene until the late 19th century when it shifted to an average rate of 2 mm/yr [31].

2.2. Pre-Development Conditions

Prior to development of the Central Valley and the San Francisco Estuary, the Sacramento, San Joaquin, and other rivers that drain the region had insufficient capacity to carry peak wet season flows generated by precipitation and snowmelt runoff. Rivers overflowed their natural levees in most years and discharged into adjacent low-lying basins, thus attenuating runoff to the Delta. As these flood flows receded, the low-lying basins would partially drain back to the rivers through smaller channels and sloughs; however, the basins typically remained inundated through late summer [32,33] Seasonal overtopping of the pre-development levees supported inland marshes [24,34–36], while riparian forests existed on natural riverbanks [37] and grasslands interwoven with vernal pools and valley oaks extended from the floodplains to the tree-covered foothills [38–40]. Water use by natural vegetation [41], in combination with the annual cycle of flooding, reduced the amount of precipitation and snowmelt runoff that reached the Delta. As natural levees were raised and wetlands and riparian forests were drained and cleared, water use by agriculture replaced water use by native vegetation in the Central Valley and the Delta. Fox et al. [24] estimated that annual water use from the natural landscape was similar to that of the highly altered contemporary landscape, such that freshwater flow reaching the estuary (i.e., Delta outflow) was minimally changed. In contrast to the Central Valley and Delta, land use changes in the surrounding foothill and mountain watersheds have been relatively minor [24]. The remainder of this section reviews previous efforts to characterize pre-development conditions using tree-ring data and flow–salinity modeling approaches.

2.2.1. Estimates of Pre-Development Central Valley Runoff from Tree-Ring Data

Annually resolved variations in hydroclimate before the start of instrumented weather records can be inferred from tree-ring records. For some tree species and climate regimes, tree growth is limited by drought stress, such that tree-ring chronologies, or standardized indices of ring width closely track the occurrence of wet and dry years [42–44]. A drought atlas from 835 tree-ring chronologies in North America, which covers two millennia, underscores the shortcomings of a relatively short instrumented record for characterizing extremes of hydroclimate [45]. An expanded network of 1285 chronologies identifies unmatched severe, widespread, persistent Southwest droughts in the medieval period [46], and independent tree-ring evidence from exposed stumps in lakes and rivers suggests that two such droughts in the Sierra Nevada may have lasted more than two centuries [47]. Paleo-simulations of Mono Lake from tree-ring data independently corroborate the timing and magnitude of Stine's drought-induced low stands and suggest centennial-average precipitation and river runoff in the central Sierra Nevada as low as 75% of the 20th century values during the medieval period [48].

Most relevant to our characterization of pre-development San Francisco Estuary hydrology are quantitative tree-ring reconstructions of annual discharge or runoff for the Sacramento and San Joaquin Rivers. Streamflow reconstructions from tree-rings are generally done by linear regression, in which a time series of unimpaired runoff is calibrated with time series from a network of indices of annual tree-ring width. Regression approaches, which can vary greatly from one study to another, are reviewed elsewhere [49,50]. Reconstructions for many basins in the western United States are available at https://www.treeflow.info/ (accessed on 9 July 2021). The first such reconstruction, which estimated flow in the Sacramento River at Bend Bridge (see Figure 1), utilized a network of 17 tree-ring chronologies that dated back to 1560. This reconstruction indicated that the wettest (1854–1916) and driest (1917–1950) periods overlapped with the historical period for which gaged flows are available in Earle [51].

The accuracy of Sacramento River runoff reconstructions over the past 500 years was improved by a network of blue oak (*Quercus douglasii*) chronologies whose collection began in the mid-1990s [52]. These blue oak chronologies, along with new collections of western Juniper (*Juniperus occidentalis*), were utilized with other tree-ring chronologies to reconstruct Sacramento River runoff back to 869; this work showed that the instrumented flow record was deficient in representing long duration (e.g., decadal and longer) droughts and wet periods [53]. A more recent effort reconstructed annual runoff for the Sacramento and San Joaquin Rivers and their major tributaries for the interval 900–2012 [20]. In contrast to Earle [51], these reconstructions indicated that, while the instrumented record does not reflect the extreme single-year Central Valley droughts, it does include multi-year droughts of similar magnitude to the most extreme droughts of the long-term record. The reconstructions further indicated exceptionally long multi-decadal swings between wet and dry conditions in the medieval period. More recent work applying the Sacramento River reconstruction [20] underscores the spatial extent of medieval drought: multi-basin coverage of hydrologic drought during the 1100s in the Sierra Nevada as well as the Colorado Rockies [54].

2.2.2. Estimates of Pre-Development San Francisco Estuary Salinity from Tree-Ring Data

As discussed above, tree-ring data have been widely used to extend the instrumented time series of river flow and runoff. Tree-ring data have also been used to extend time series of measured salinity in the estuary, recognizing cause–effect relationships between precipitation, runoff, river flows and estuarine salinity.

Extending an earlier reconstruction [52], Stahle [55] used three blue oak tree-ring chronologies to reconstruct salinity in San Francisco Bay over the 673-year period from 1333 to 2005. The reconstruction was calibrated with near surface salinity (January through July averages) at a stationary location measured near Golden Gate at Fort Point over the period WYs 1922–1952. Based on their salinity reconstruction, the authors concluded that the droughts of 1977 and 1986–1991 were among the most severe in the 673-year record. They observed that their reconstruction systematically underestimated the salinity during most of the verification period WYs 1952–2005, citing anthropogenic changes to Delta outflow through increased water use in the watershed and Delta diversions. Fox et al. [56], in a study of San Francisco Estuary salinity trends, provided an alternative explanation for the fixed location salinity behavior examined by Stahle [52]. Noting that salinity at locations near the ocean are subject to additional drivers besides Delta outflow, Fox et al. [56] concluded that trends at Fort Point (referring to the location as "Presidio") since 1946 were primarily affected by trends in coastal conditions rather than trends in Delta outflow.

Stahle et al. [23] also applied Blue oak tree-ring chronologies to directly reconstruct the longitudinal position of the X2 isohaline in San Francisco for the 625-year period from 1379 to 2003. The reconstruction was calibrated with spring X2 data (February through June averages) that were estimated from instrumented salinity gages over the period 1956–2003. Reporting correlations between reconstructed X2 position, sea surface temperature and atmospheric circulation regimes over the north Pacific, they concluded that X2 minima tended to occur during very strong El Niño events but X2 maxima did not appear to occur during La Niña events. This salinity reconstruction does not represent pre-development X2 conditions; rather, it represents how X2 may have fluctuated under the climatic variability of the past six centuries given a behavior similar to the contemporary estuary.

2.2.3. Models of Pre-Development Central Valley Hydrology and Delta Hydrodynamics

Pre-development Central Valley hydrology and Delta outflow were characterized by the California Department of Water Resources (CDWR) [25] utilizing two models to simulate watershed hydrology. They used the Soil Water Assessment Tool (SWAT) [57] to model precipitation–runoff characteristics of the upper elevation Central Valley watersheds and the (California) Central Valley Simulation Model, or C2VSim [58], an integrated hydrologic model, to simulate groundwater and surface water hydrology on the pre-development Central Valley floor. Land use was based on prior characterizations of natural vegetation [24,59]. Potential evapotranspiration from natural vegetation was estimated using reference evapotranspiration from Orang et al. [60] and vegetation coefficients from Howes et al. [41]. CDWR [25] estimated a long-term annual average pre-development Delta outflow of 23.9 billion cubic meters (BCM) assuming a repeat of a 93-year contemporary climate sequence spanning WYs 1922–2014. Gross et al. [22] utilized these modeled values to compare inter- and intra-annual variability of pre-development and contemporary Delta outflow.

Pre-development salinity conditions in the San Francisco Estuary were investigated and compared to contemporary salinity conditions by Andrews et al. [26] using a three-dimensional hydrodynamic model [61]. Their pre-development model was based on a planform developed by [62] and bathymetry from multiple sources. Their simulation used observed inflow data from February 2006 to October 2008 to represent wet, dry, and critically dry water years. Andrews et al. [26] found the dramatic changes in estuary planform and bathymetry, as well as differences in mean sea level between the pre-development and contemporary conditions, to have limited influence on saltwater intrusion. The pre-development estuary was found to have less saltwater intrusion for the same Delta outflow and a faster response of saltwater intrusion to changes in Delta outflow. Due to the changes in seasonal distribution of Delta outflow, saltwater intrusion was found to be less variable for their contemporary scenario than their pre-development scenario. Changes to the seasonal timing of freshwater flows was reported to have a larger influence on saltwater intrusion than the changes in estuarine planform and bathymetry. Gross et al. [22] utilized this work to compare inter- and intra-annual variability of pre-development and contemporary salinity intrusion in the Delta.

The aforementioned model studies of pre-development conditions used an analysis method termed the "level-of-development" approach [63]. In this approach, landscape, channel geometry, and anthropogenic flow modification through reservoirs or withdrawals are fixed to represent a specific era or scenario (e.g., pre-development conditions, contemporary conditions, planned future conditions) and hydrology is typically represented by a sequence of historically observed precipitation or runoff. Thus, these model studies seek to describe how a pre-development or modern landscape and estuary would respond given contemporary instrumentally derived climatic inputs.

As described later in this paper, our work adopts some level-of-development assumptions to characterize pre-development Central Valley hydrology and Delta hydrodynamics. For example, we assume a stationary pre-development landscape consistent with Fox et al. [24] and CDWR [25] and a stationary pre-development outflow–salinity relationship consistent with Andrews et al. [26]. However, our work deviates from a typical level-of-development analysis in one crucial aspect—the driving hydrology is not simply represented by repeating the sequence of observed runoff over the instrumented period. Rather, it reflects the estimated runoff over a millennial time scale obtained from the tree-ring proxy record from the watershed.

2.3. Early Development Conditions

The pre-development landscape has been radically modified over two centuries, starting in the mid-18th century when Spanish settlers arrived, bringing livestock and range management. The discovery of gold along the American River in 1848 spurred agricultural and urban development in the Central Valley. That same year, the federal government transferred ownership of "swamp and overflowed lands" to California on the condition that they be drained and reclaimed. These permanent wetlands were largely converted to agriculture by 1930.

Regular flooding on major rivers led to the formation of levees and reclamation districts by 1860. Starting in the 1870s, studies were conducted to determine how to reduce flooding and supply irrigation water. The Office of the State Engineer was established in 1878 to further these plans, and in 1880 the legislature approved the Drainage Act, proposing valley-wide flood control. These studies culminated in the Central Valley Project Act in 1933. Water resources were further reconfigured in response to voter approval of the Burns-Porter Act in 1960, financing the State Water Project [64]. Ultimately, the Central Valley was re-plumbed to move water throughout the state in a complex man-made water system with some 1300 miles of aqueduct and 1350 surface reservoirs with 40 million acre-feet (32.4 BCM) of storage [64].

Although this period of early development between WYs 1851 and 1911 is poorly understood hydrologically, limited availability of instrumented data facilitated previous work. Arguably the most significant data set compiled during the latter part of this period was published by the California Department of Public Works [65], the predecessor to CDWR. This document, commonly referred to as Bulletin 5, reports a long-term record of stream flows to the Central Valley beginning in WY 1872.

Moftakhari et al. [66] reconstructed a Delta outflow time series spanning the early development period (beginning in 1858) through correlation with tide gauge data measured at San Francisco. Moftakhari et al. [67] reconstructed a Delta outflow time series beginning in WY 1850 through correlation with Sacramento River stage data measured at Sacramento. River stage data were unavailable over WYs 1863–1881; thus, the authors augmented the reconstructed outflow time series using the work of Moftakhari et al. [66]. MacVean et al. [68] explored the hydrology of the early development period following 1850 by synthesizing reconstructed time series of precipitation, basin inflows, land use, and levee construction in a semi-distributed hydrologic model. They concluded that, in spite of significant anthropogenic modifications to the region's hydrology, by the 1920s Delta outflow remained similar to pre-development conditions, due in part to flow augmentation provided by flood control infrastructure and enhanced channel conveyance. MacVean et al. [68] concluded that levee construction, rather than land use change, had the greatest impact on Delta hydrology during this early development period.

2.4. Contemporary Conditions

Extensive salinity intrusion in the Delta in the early 20th century, caused by a combination of hydrologic variation and upstream land use and hydrologic change, motivated a series of Delta field investigations that led to a better understanding of the relationship between sources of water flows and salinity patterns in the Delta [65]. These findings supported the development of reservoirs in the upstream watershed to store winter and spring flows and supply irrigation water needs in the summer months. Among the various reservoirs built in the Central Valley, the federal government completed construction of the 4.5 million acre-feet (5.6 BCM) Lake Shasta in 1944 as part of the Central Valley Project (CVP) and the state government completed construction of the 3.5 million acre-feet (4.3 BCM) Lake Oroville in 1968 as part of the State Water Project (SWP) (see Figure 1). Over a period of roughly three decades, a complex network of reservoirs, aqueducts, pumps and gates was constructed to facilitate transport of water to other parts of the state for agricultural and municipal use.

Today, regulatory activity related to the management of estuarine flow and salinity is led by the California State Water Resources Control Board (CSWRCB), an agency concerned with both the water quality and water rights adjudication in California. In August 1978, the CSWRCB adopted its Delta Plan and Decision 1485 which set objectives for Delta outflow [69]. CSWRCB updated its Delta Plan in 1995 and adopted Decision 1641 in 2000 [70], which is still in force. The position of the X2 isohaline is a particular focus of salinity regulation in the estuary, and target ranges are defined by season and water year type. The position of the X2 isohaline is managed through control of out-of-basin exports from the Delta and reservoir outflows from major CVP and SWP reservoirs in the watershed.

Based on continued risk to certain endangered aquatic species, additional restrictions were imposed on the system through biological opinions rendered by the U.S. Fish and Wildlife Service in 2008 [71] and the National Marine Fisheries Service [72] under the U.S. Endangered Species Act. Both biological opinions were recently updated [73,74]. Additional flow regulations are being considered as part of the CSWRCB's Delta Plan periodic review [18,75].

A variety of models have been developed to interpret flow and salinity intrusion in the contemporary Delta; these models are used for research, regulatory planning and for CVP and SWP operations support. Jassby et al. [12] is an example of a commonly used empirical X2-outflow model; Rath et al. [76] provides a comprehensive review of this and other published empirical X2-outflow models. Mechanistically based hydrodynamic models of the estuary include the one-dimensional Delta Simulation Model 2 or DSM2 [77] and more complex three-dimensional models such as SCHISM [78], UnTRIM [61,79], and Delft3D [80].

3. Methods

3.1. Data

Observed and synthetic hydrology data associated with the estuary and its contributing watershed were used in this work. Measured tree-ring data from available sites in Northern California and Oregon (Figure 1) were used to develop synthetic annual runoff sequences. These data are described below.

3.1.1. Hydrology Data

The hydrology data used in this work were drawn from California state data sources and include secondary (i.e., processed) sources such as water balances and model simulations. Table 1 presents a summary of the data used in our work, including sources.

A widely used measure of Central Valley hydrology is the Eight River Index (8RI), which constitutes the unimpaired Sierra Nevada runoff to the Sacramento, Feather, Yuba, American, Stanislaus, Tuolumne, Merced, and San Joaquin Rivers. The 8RI represents a theoretical quantity that removes anthropogenic influences such as reservoir impoundments and land use modifications and thus does not reflect actual runoff conditions. The 8RI data for WYs 1906–2018 were obtained from the website of the California Data Exchange Center [81] and served as the predictand (after transformation) in the tree-ring reconstruction model described below. The previously described Bulletin 5 [65] streamflow data, which can be used to compute the 8RI beginning in WY 1872, was used for additional validation of the tree-ring runoff reconstructions.

Table 1. Summary of model calibration and validation data.

Data	Time Period	Source and Description
Eight River Index (8RI) [(1)]	WYs 1872–2018	[65,81]
Tree-Ring Chronologies (long record)	903–2008	Supplemental Materials: A and B
Tree-Ring Chronologies (short record)	1640–2001	Supplemental Materials: A and B
Pre-Development Delta Outflow (simulated)	WYs 1922–2014	[25]
Contemporary Delta Outflow	WYs 1912–2018	[27,82]
Pre-Development X2	WYs 1922–2014	Generated as part of this work using simulated pre-development Delta outflow and an empirical relationship developed by Andrews et al. [26]
Contemporary X2	WYs 1920–2018	Generated as part of this work using contemporary Delta outflow and an empirical relationship developed by Andrews et al. [26]

(1) measure of Central Valley Runoff.

We adopted publicly available simulation output of the valley floor hydrology [25] to calibrate a pre-development relationship between Central Valley unimpaired runoff and Delta outflow on an annual basis. The simulation assumes pre-development land use in the Central Valley and Delta as presented in Fox et al. [24] and associated natural vegetation evapotranspiration as presented in Howes et al. [41]. Furthermore, the simulation uses historical flows from the surrounding upper-elevation watersheds as boundary inputs; these boundary inputs represent unimpaired runoff data corresponding to WYs 1922–2014, a 93-year period inclusive of widely varying hydrologic conditions.

We used historical estimates of freshwater flows to the San Francisco Estuary (i.e., Delta outflow) spanning WYs 1912–2018 to calibrate a contemporary relationship between Central Valley unimpaired runoff and Delta outflow on an annual basis. Due to the complexity of direct observation of Delta outflow, these estimates are not based on tidal flow measurements. Rather, these estimates are computed from a budget of inflows and diversions from the Delta [27,82] As discussed below under Modeling Approach, we used calibrated runoff–outflow relationships (rather than historical data) to reconstruct contemporary Delta outflow conditions. We decided to use reconstructed flows for the contemporary period to provide a homogeneous time series for comparison with pre-development conditions. Otherwise, flow differences between the two periods could be perceived to be due to statistical artifacts, such as the compression of variance of reconstructed flows relative to variance of observed flows over a common period [53,83].

3.1.2. Salinity Data

Several modeling steps were followed to generate salinity data necessary to calibrate pre-development relationships between annual Delta outflow and seasonal average X2. First, the daily outflow time series from the aforementioned 93-year simulation [25] was transformed into a daily "antecedent outflow" time series to represent flow time-history in the estuary [14,84]. This transformed outflow time series was used to generate a daily X2 time series using a pre-development flow–salinity relationship reported by Andrews et al. [26]. Finally, this synthetic daily X2 time series was averaged to develop seasonal (February–June and July–October) average X2 calibration time series.

For internal consistency, rather than using historical data, we followed a similar methodology to generate salinity data necessary to calibrate contemporary relationships between annual Delta outflow and seasonal average X2. Specifically, we used a subset of the daily contemporary Delta outflow time series [27,82] spanning WYs 1920–2018 and a contemporary flow–salinity relationship reported by [26]. The initial years of the contemporary period (WYs 1912–1919) were excluded from model calibration due to lack of historical Delta outflow data at a daily resolution.

3.1.3. Tree-Ring Data

Sixty-nine tree-ring site chronologies of total ring width were assembled as part of our work (Figure 1). Each chronology typically represents many (e.g., 15 or more) trees at a specific location. Initial screening criteria were a minimum time coverage of the period 1636–2003, and geographical location within a box delineated by latitudes 34.5 N to 44.0 N and longitudes 118 W to 125 W.

We started with files of measured ring widths obtained from two studies conducted for CDWR [20,85] and supplemented those with additional files from the International Tree-ring Data Bank [86]. Ring widths were standardized uniformly into site chronologies using Matlab functions following similar protocol to that in the ARSTAN standardization package [87]. This includes fitting ring-width series with a cubic smoothing spline [88], computing core indices as the ratio of ring-width to the smooth spline and averaging the indices over cores to get the site chronology. Trend in variance indistinguishable from age or size effects was removed using the method recommended by [89]. From an assessment of the persistence in the standard chronologies and the annual flows, we decided to use the residual version [90] of the site chronologies in the reconstruction modeling. The residual chronology is an average over core indices whose low-order autocorrelation has been removed fitting the index to an autoregressive (AR) model. We used a modified Akaike information criterion [91] to select the AR order. Site chronologies are averages over fewer and fewer trees toward the early part of the tree-ring record. Adequacy of sample size for each chronology was assessed by the expressed population signal [92]. Secondary screening eliminated any site chronology whose Pearson correlation with annual flows over the available period of data overlap (subset of the WYs 1906–2018 period) was either statistically insignificant or unstable over time at significance level $\alpha = 0.05$. The temporal stability of correlation was tested using a difference-of-correlation test [93] of the null hypothesis that the sample correlations for the first and second halves of the overlap are from the same population.

Additional tree-ring data and processing information are included in the Supplementary Materials section. Sites and metadata are listed in Supplementary Material A. Chronology development is described in more detail in Supplementary Material B. Additional electronic data files are also included Supplementary Materials and identified in Supplementary Material A: files of original tree-ring width measurements; a time series matrix of the residual site chronologies; a time series matrix of observed and reconstructed flows, with confidence intervals on the reconstruction.

3.2. *Modeling Approach*

A composite modeling approach was employed to reconstruct time series of annually varying Central Valley runoff, Delta outflow and salinity spanning more than 1000 years. The modeling approach, as summarized below, consists of three components. The purpose of the first component is to reconstruct a runoff time series from measured tree-ring data. The purpose of the second and third components is to reconstruct time series of Delta outflow and salinity from the modeled runoff time series representing the pre-development and contemporary periods, respectively. The three model components draw from a variety of measured and synthetic (i.e., simulated) calibration data. The model components and time periods are summarized in Table 2.

Table 2. Summary of modeled time periods and key variables.

Time Period	WYs	Central Valley Runoff (Eight River Index)	Delta Outflow	X2 Position (Salinity)
Predevelopment	Prior to 1850	Predicted from tree-ring data using Model 1	Predicted from Equation (2) using runoff estimates and parameters in Table 5 (Model 2)	Predicted from Delta outflow using Equation (3) and parameters in Table 7 (Model 2)
Early development	1851–1911	Predicted from tree-ring data using Model 1	Predicted from Equation (2) using runoff estimates and parameters in Table 5 for pre-development period (Model 2)	Predicted from Delta outflow using Equation (3) and parameters in Table 7 for pre-developmentperiod (Model 2)
Contemporary	Post 1912	Predicted from tree-ring data using Model 1	Predicted from Equations (2) and (4) using runoff estimates and parameters in Tables 5 and 6 (Model 3)	Predicted from Delta outflow using Equation (3) and parameters in Table 7 (Model 3)

3.2.1. Selection of Modeled Time Periods

Our modeling approach differentiates between three eras described earlier: a "pre-development" period, an "instrumented" or "contemporary" period, and an "early development" period that bridges the pre-development and instrumented periods. Tree-ring reconstructions of Central Valley runoff were developed spanning the three time periods. A "long record" reconstruction represents the pre-development period back to WY 903 and a "short record" reconstruction represents the pre-development period back to WY 1640. Both runoff reconstructions cover the early development period spanning WYs 1851–1911; however, a unique Delta outflow and salinity model component was not generated for this period. While some literature is available to characterize the hydrology of this early development period [65–68], associated Delta outflow trends and drivers of change are poorly understood. In light of this uncertainty, we assumed that the early development period was adequately represented by pre-development relationships between runoff, Delta outflow and salinity.

3.2.2. Model 1: Annual Central Valley Runoff Reconstruction from Tree-Ring Data

Separate models were developed to reconstruct 8RI annual flows over the periods 903–2008 (long record) and 1640–2001 (short record). The long record prioritizes reconstruction *length* by making use of a small set of long tree-ring chronologies. The short record prioritizes reconstruction *accuracy* by taking advantage of a larger number of chronologies that, while not of great age, yield improved reconstruction accuracy and several centuries extension of flow beyond the gage record. The procedure described below was repeated for each of the two model periods.

A two-stage reconstruction method, introduced for reconstruction of river basin precipitation [94], and later extended for reconstruction of streamflow [53,95] was modified for this study. The first stage is conversion, by regression, of each of the available N chronologies into a separate single-site reconstruction (SSR) of y_t, the square-root-transformed annual 8RI flows. A square root transform was found adequate to correct problems with violation of assumptions about the regression residuals that occur when using the untransformed flows as the regression predictand in the reconstruction models. The second stage, called multi-site reconstruction (MSR) combines the signals from the individual SSRs to get the final single time series of reconstructed flows. The two stages of reconstruction are outlined below and are described in more detail in Supplementary Material B.

In the first stage, y_t is regressed stepwise on a pool of potential predictors that includes a site chronology, x_t, its square, x_t^2, and lags $t-2$ to $t+2$ on those two variables [96]. Preliminary stepwise modeling using cross-validation [97] is first applied to identify the step, m, beyond which addition of another predictor fails to increase the validation skill as measured by the reduction of error statistic (RE) [98]. The final SSR regression model has only those predictors entered in the first m steps, and substitution of the full available length of tree-ring predictors into the fitted equation gives the SSR reconstruction. Calibration accuracy of the SSR model is summarized by regression R^2. Significance of the regression model is assessed by p_F, the p-value of the overall-F of regression, and calibration uncertainty is summarized by the standard error of the estimate, or root-mean-square error ($RMSE_c$) of calibration [96]. Validation accuracy is summarized by the root-mean-square error of cross-validation ($RMSE_v$) which is computed from the cross-validation errors (see Supplementary Material B).

The product of the first stage of reconstruction in this study is a separate SSR, $\hat{y}_{t,i}$, of transformed flow for each of the $i = 1, \ldots, N$ tree-ring chronologies ($N = 69$) mapped in Figure 1. Those whose SSR calibration signal is not significant ($p_F > 0.05$) or whose SSR has no skill of validation ($RE \leq 0$) were eliminated from the study. Depending on the time coverage (varies over SSRs), the remaining SSRs could contribute in the second stage of reconstruction.

The second stage is a re-calibration of the arithmetic mean of a subset of the N individual SSRs with acceptably strong signal and common time coverage into a final reconstruction—long or short, depending on the particular subset. The regression model for the MSR is

$$y = a + bx_t + e_t, \tag{1}$$

where x_t is the arithmetic average of the SSRs covering either the long or short records, y_t is the square-root-transformed 8RI WY flow, e_t is the error term, and $\{a, b\}$ are regression coefficients. The arithmetic average of the SSRs was preferred as a predictor with the assumption that the flow signal is best represented by the common variation in the SSRs. Since, by definition, the SSRs have variance proportional to the strength of their flow signals, a simple arithmetic average emphasizes chronologies with a strong flow signal. Equation (1) applies to both the long and short reconstructions, but for each x_t is an average over a different set of SSRs. The calibration period is defined by the overlap of 8RI flows (and y_t) with the particular SSR subset: WYs 1906–2008 for the long record and WYs 1906–2001 for the short record. Calibration and validation accuracy for the MSR models were measured by the same statistics already described for the SSR models. Substitution of the full-length SSRs into the fitted equations gave long and short reconstructions covering 903–2008 and 1640–2001, respectively. The $RMSE_v$ of the model was used, along with the assumption that the reconstruction residuals are normally distributed, to place a 50% confidence interval on the annual reconstructed transformed flows, $\hat{y}_t$. As a final step, the MSR reconstructions were back-transformed to original flow units (BCM) before interpretation.

As additional validation, time series plots and the Spearman correlation coefficient [99] were used to check agreement of the long and short reconstructions of 8RI flows spanning WYs 1872–1900 and reported in Bulletin 5 [65]. This step serves as a completely independent verification, as the Bulletin 5 data precede the start of the period used for screening tree-ring data and calibrating and cross-validating the reconstruction models. Supplementary Material C provides the statistics of the SSR models.

Low-frequency features (decadal and longer) are of interest in understanding long-term hydrologic patterns. Severity of droughts and wet periods is also summarized by simple moving averages of reconstructed flows. Consistency of with other work was checked by comparing reconstructions with the sum of separate reconstructions of annual Sacramento River and San Joaquin River flow [20].

3.2.3. Model 2: Pre-Development Outflow and Salinity Reconstruction

The annual 8RI time series tree-ring reconstructions from Model 1 were used as the basis for reconstructing annual Delta outflow volume and salinity in the estuary under pre-development conditions. The modeling logic employed for the long- and short-period reconstructions is presented as a flow chart in Figure 2.

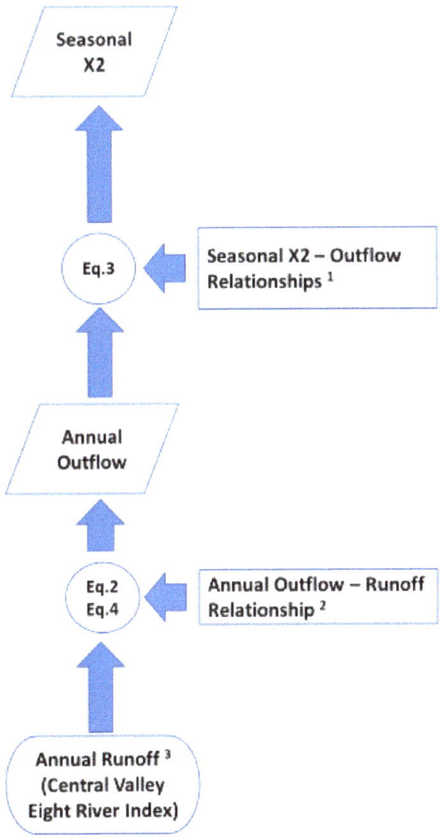

Figure 2. Flow chart for pre-development (Model 2) and contemporary (Model 3) models. [1] Model 2 calibration based on pre-development daily outflow (DWR, 2016) and daily X2-outflow relationship; Model 3 calibration based on historical daily and contemporary daily X2-outflow relationship; [2] Model 2 calibration based on pre-development annual outflow and historical Eight River Index (WYs 1922–2014); Model 3 calibration based on historical annual outflow and historical Eight River Index (WYs 1912–2018); [3] Tree-Ring reconstruction from Model 1 for long record and short record through WY 1850.

Annual pre-development Delta outflow volume was estimated for both reconstruction periods assuming a power law relationship between Delta outflow volume and annual Central Valley runoff volume:

$$Delta\ Outflow = \alpha_1 \times 8RI^{\alpha_2} \quad (2)$$

where outflow and runoff volumes are in units of BCM per year and α_1 and α_2 are fitting parameters determined through least squares analysis [25] simulated pre-development conditions in the Central Valley and Delta assuming a historical runoff pattern measured over the 93-year period spanning WYs 1922–2014. Annual Delta outflow volume from the

CDWR simulation, along with annual Central Valley runoff (as measured by the 8RI over the same 93-year period), were used to calibrate Equation (2). Tree-ring reconstructions of Central Valley runoff (i.e., the 8RI) were used in conjunction with Equation (2) to estimate pre-development annual Delta outflow volumes through WY 1850.

Seasonal (February–June and July–October) average pre-development X2 positions were estimated for the reconstruction periods assuming power law relationships between X2 position and annual Delta outflow volume:

$$X2 = \alpha_3 \times Delta\ Outflow^{\alpha_4} \quad (3)$$

where X2 position is in units of km from Golden Gate and α_3 and α_4 are fitting parameters determined through least squares analysis. The 93-year simulated pre-development Delta outflow time series, as described above, was utilized to calibrate Equation (3). Several modeling steps were followed to generate 93-year X2 calibration data sets. First, daily outflow from the aforementioned CDWR [25] simulation was transformed into a daily "antecedent outflow" time series to represent flow time-history in the estuary [14,84]. This transformed outflow time series was used to generate a daily X2 time series using a pre-development flow–salinity relationship reported by [26]. Finally, this synthetic daily X2 time series was averaged to develop seasonal average X2 calibration time series. Time series of annual pre-development outflow volume, estimated from the tree-ring reconstructions (Model 1), were used in conjunction with Equation (3) to estimate seasonal average X2 positions through WY 1850.

3.2.4. Model 3: Contemporary Outflow and Salinity Reconstruction

Following the methods reported above for pre-development conditions, the annual 8RI time series tree-ring reconstructions from Model 1 were used as the basis for reconstructing annual Delta outflow volume and salinity in the estuary under contemporary conditions. The modeling logic employed for the reconstructions is presented as a flow chart in Figure 2. This logic was applied to both the long- and short-period reconstructions spanning WYs 1912–2008 and WYs 1912–2001, respectively.

Contemporary Delta outflow was estimated for both reconstruction periods assuming the power law relationship provided in (Equation (2)). This relationship, which was calibrated with historical annual runoff and Delta outflow data, does not represent the full contemporary period; rather, it is limited to a relatively stationary period prior to significant increases in water use in the Central Valley and Delta following construction of Shasta Dam in WY 1944. A residual analysis was conducted to address the observed time series trend. Tree-ring reconstructions of Central Valley runoff (from Model 1) were used in conjunction with Equation (2) to estimate contemporary annual Delta outflow volumes for the WYs 1912–1944 period; a modified form of Equation (2), presented later in this text, was used to estimate Delta outflow volumes for the post 1944 period.

Contemporary X2 position was estimated for the long- and short-period reconstruction periods assuming power law relationships between annual Delta outflow volume and seasonal (February–June and July–October) average X2 position. Equation (3) was calibrated for the contemporary period using a data subset spanning WYs 1920–2018. The initial years of the contemporary period (WYs 1912–1919) were excluded from model calibration due to lack of historical Delta outflow data at a daily resolution. Following the methodology used for pre-development model calibration, daily outflow was transformed into a daily antecedent outflow time series and this daily antecedent outflow time series was then transformed into a daily X2 time series using a contemporary flow–salinity relationship reported by Andrews et al. [26]. Finally, this daily X2 time series was averaged to develop seasonal average X2 time series for purposes of model calibration. The time series of annual contemporary outflow volume, estimated from the tree-ring reconstructions, were then used in conjunction with Equation (3) to estimate seasonal average X2 positions for each time series beginning in WY 1912.

4. Results

Following the methods presented in the previous section, we developed multi-century reconstructions of watershed runoff, freshwater flows to the estuary, and the estuary's salinity regime as expressed by intrusion length. Below we summarize the reconstructions, following the logic sequence provided in the flow chart depicted in Figure 2.

4.1. Annual Central Valley Runoff Reconstructions

A total of 60 of the 69 initial chronologies passed screening tests for temporal stability of the runoff signal and significant SSR regression model (Supplementary Material C). Most of these lagged models resulting from stepwise regression have a simple structure. All 60 models include a lag-0 (current year predictor) and 28 models have just one predictor. The median number of predictors is 2 and the maximum is 5. The minimum, median and maximum percentage of calibration-period variance explained by the models are 8%, 29% and 73%, respectively. All models are significant as judged by $p < 0.05$ for the overall F of regression. Blue oak chronologies from the Central Valley or the coastal region tend to have the strongest signal (Figure 1).

Time coverage by SSRs varies according to the coverage of the chronologies themselves. Thirteen of the SSRs have uniform coverage for 903–2008 and comprise the subset for the long reconstruction; all 60 SSRs, with a common period 1640–2001, are available for the short reconstruction. As in previous studies (e.g., Meko et al. [53]), long tree-ring chronologies of western juniper from south-central Oregon are important contributors to the long network (Figure 1).

Regression of y_t on the 13-site-mean and 60-site-mean SSRs yields long and short reconstruction models accounting for 66% and 77% of the calibration-period variance of y_t after adjustment for loss of degrees of freedom (adjusted R^2) (Table 3). Both models have strong validation, as indicated by high positive RE values from cross-validation, and by highly significant correlation of cross-validation predictions with observed flow. Both reconstructions also closely track and have significant correlation with earlier flows (spanning WYs 1872–1900) from gages on the Sacramento, San Joaquin, and other Central Valley rivers that comprise the Bulletin 5 8RI flows (Figure 3) [65]; these earlier data had also been used, with less success, for validation of the first Sacramento River runoff reconstruction [51]. Both models greatly underestimate the flow in WY 1890. Tree-ring reconstruction calibrated by regression with gaged flows tend to be conservative (biased toward the mean) because the variance explained by regression is always less than 100%. This compression of variance theoretically would lead to underestimation of both wet extremes and dry extremes and complicates direct comparison of observed and reconstructed magnitudes of extreme flow events [83]. Moreover, as seen in Figure 3, the magnitude of extreme high flows may be especially difficult to capture because growth of drought-sensitive trees beyond some high level of soil moisture is logically expected to benefit less and less from additional moisture.

Table 3. Statistics for long record and short record tree-ring reconstructions (Model 1). Statistics are for regression models whose predictand is transformed 8RI flow (square root billion cubic meters).

Tree-Ring Record	N [a]	Calibration [b]			Cross-Validation [c]		
		WYs	R^2 adj	RMSE	RMSE	RE	r
Long record 903–2008	13	1906–2008	0.66	0.697	0.712	0.65	0.82
Short record 1640–2001	60	1906–2001	0.77	0.585	0.592	0.77	0.88

[a] Number of contributing tree-ring chronologies. [b] Overall F (not listed) for both models is highly significant ($p <$ 1E-25). [c] RE = reduction-of-error statistic; r = correlation of cross-validation predictions with observed flows. The two correlations listed are both larger than any of the correlations for the 1000 simulated reconstructed flow series ($p < 0.001$)

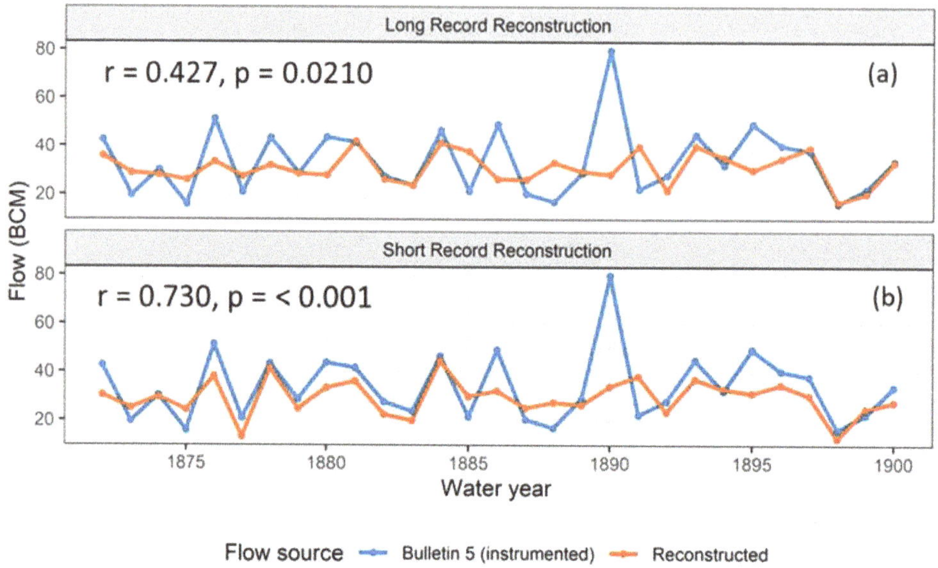

Figure 3. Time series plots (WYs 1872–1900) comparing reconstruction of Central Valley runoff (Model 1) with instrumented flows for (**a**) long record reconstruction and (**b**) short record reconstruction. Spearman correlations and significance annotated. Significance not adjusted for autocorrelation because none of the series are positively autocorrelated. This period, as documented in Bulletin 5 [65], precedes years used for calibrating and validating reconstruction models.

Both the short and long record reconstructions strongly track the sum of individual reconstructions generated previously for the Sacramento and San Joaquin Rivers by Meko et al. [20]. For the 1640–2001 period common to these reconstructions, the Pearson correlation is $r = 0.83$ for the long record reconstruction and $r = 0.95$ for the short record reconstruction. For the earlier 903–1639 period in common with the long reconstruction only, the correlation remains high (r = 0.82). While agreement between reconstructions is limited by differences in tree-ring networks and statistical reconstruction methods, the reconstructions are reasonably consistent in their characterization of droughts and wet periods. The long record reconstruction, for example, includes a period of low runoff values in the mid-1100s that aligns with a period of notable persistent drought in both the Colorado and Sacramento Basins [95,100].

The long and short record 8RI reconstructions are shown in Figure 4, indicating annual and 5-, 10-, 20-, and 100-year center-averaged values. When the late-19th to early 20th century reconstructed flows are compared with reconstructed flows in the preceding centuries, it is apparent that single-year wet and dry extremes are more variable, however, time averaged flows are more consistent over the different periods. This is also summarized in Table 4, which shows that for all of the averaging periods presented, from 5 to 100 years, the range of flows are essentially similar for the full reconstruction and the more recent 1872–2001 period. Importantly, the entirety of the instrumented period, 1872–2018, generally shows a wider range in flows that the reconstructed values. Additionally, low flows over different averaging periods are lower in the instrumental record than in the longer reconstruction period. This comparison suggests that the instrumental flow record is a reasonable representation of the conditions over the past millennium and captures extremes in the low flow periods.

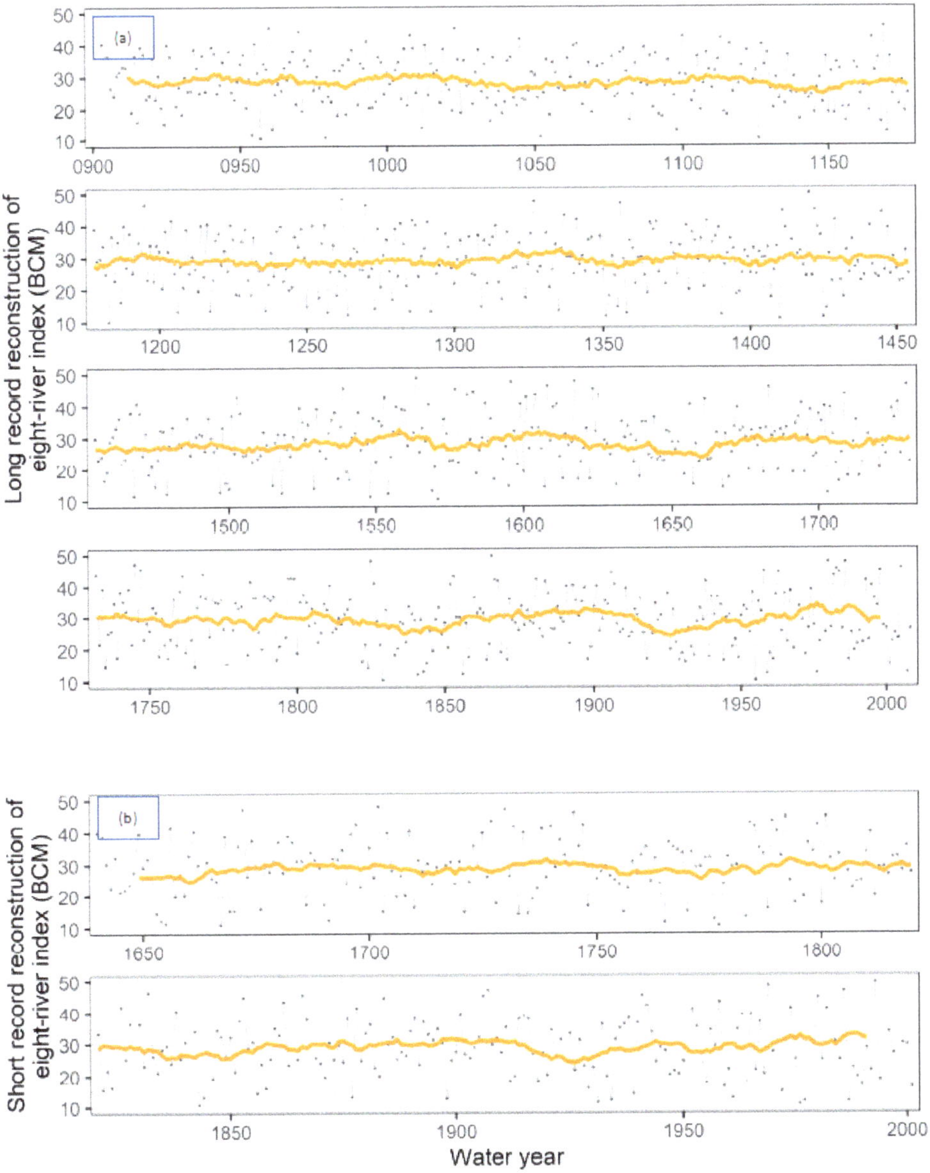

Figure 4. Time series plots of long and short tree-ring reconstructions of Central Valley runoff (Model 1) for (**a**) long record reconstruction, spanning 903–2008 and (**b**) short record reconstruction, spanning 1640–2001. Smooth lines represent 20-year average flows. Other averaging periods, as summarized in Table 4, are excluded from the figure for clarity.

Table 4. Range of reconstructed and instrumented Central Valley Runoff (8RI) for different averaging periods. Units are reported as BCM.

8RI Record	WYs	Center Averaging Period (yr)	Min	Max	Range	5%	10%	25%	50%	75%	90%	95%
Long Record Reconstruction	903–2008	5	16.6	41.4	24.7	23.0	24.3	26.5	28.8	31.0	32.7	33.9
Short Record Reconstruction	1640–2001	5	18.7	39.9	21.2	22.1	23.3	26.0	28.7	31.3	34.0	35.4
Long Record Re-Construction	1872–2001	5	19.5	41.4	21.8	22.0	24.4	26.7	29.7	32.3	34.6	36.3
Short Record Re-Construction	1872–2001	5	19.7	39.9	20.1	22.3	23.4	26.3	29.5	31.9	35.3	37.0
Instrumented Flows	1872–2018	5	16.0	44.5	28.5	19.1	22.3	25.4	31.2	34.5	39.0	41.4
Long Record Re-Construction	903–2008	10	21.3	38.2	16.9	24.8	25.7	27.2	28.7	30.2	31.6	32.4
Short Record Re-Construction	1640–2001	10	21.9	36.3	14.3	24.3	25.3	27.2	28.8	30.4	32.3	33.1
Long Record Re-Construction	1872–2001	10	21.3	38.2	16.8	23.0	25.6	27.4	30.1	31.5	32.7	34.1
Short Record Re-Construction	1872–2001	10	21.9	36.2	14.3	24.1	25.7	27.4	29.4	31.1	32.9	33.4
Instrumented Flows	1872–2018	10	20.2	40.3	20.2	21.7	23.9	27.2	31.1	33.8	36.3	37.5
Long Record Re-Construction	903–2008	20	23.7	33.9	10.2	25.9	26.6	27.6	28.7	29.8	30.7	31.3
Short Record Re-Construction	1640–2001	20	24.1	32.6	8.55	25.9	26.6	27.7	29.0	30.0	30.7	31.2
Long Record Re-Construction	1872–2001	20	23.9	33.9	9.94	25.0	25.7	27.4	30.2	31.4	32.3	32.7
Short Record Re-Construction	1872–2001	20	24.1	32.6	8.55	25.3	26.3	27.9	29.5	30.6	31.3	31.8
Instrumented Flows	1872–2018	20	22.4	38.4	16.0	24.7	25.8	28.1	29.9	33.8	35.4	36.2
Long Record Re-Construction	903–2008	100	27.2	29.8	2.58	27.6	27.8	28.1	28.7	29.1	29.3	29.4
Short Record Re-Construction	1640–2001	100	27.9	29.4	1.50	28.2	28.4	28.6	28.8	29.0	29.2	29.2
Long Record Re-Construction	1872–2001	100	28.8	29.5	0.734	28.9	29.0	29.0	29.1	29.3	29.4	29.4
Short Record Re-Construction	1872–2001	100	28.6	29.4	0.763	28.7	28.7	28.8	28.9	29.1	29.3	29.4
Instrumented Flows	1872–2018	100	28.0	31.2	3.20	28.4	28.6	29.4	29.9	30.7	30.9	31.0

Another way to look at the flow reconstructions is to examine the sequence of wet and dry periods in the record, in comparison to the contemporary period. The long record reconstruction was reviewed to highlight patterns of low and high flows that are of interest for water resources management. Figure 5 shows 20-year center averages associated with four 121-year periods with the greatest variations (970–1090; 1100–1220; 1570–1690; 1850–1970). This figure illustrates that shifts between wet and dry periods have occurred several times in the past millennium, but in most of these instances, the range of flow variation is not greater than the reconstructed flows for late 19th and early 20th century. Actual observed flows during high flow periods are higher that the reconstructed flows, a bias expected from the variance compression inherent in regression, and possibly also to lower sensitivity of tree-growth to soil moisture beyond a threshold level. However, if only reconstructed flows are considered for comparison over different time periods, as done throughout this study, it may be inferred that flow patterns in the instrumental period after the 1870s, especially over decadal time scales, are a reasonable representation of the overall variability seen in the past millennium.

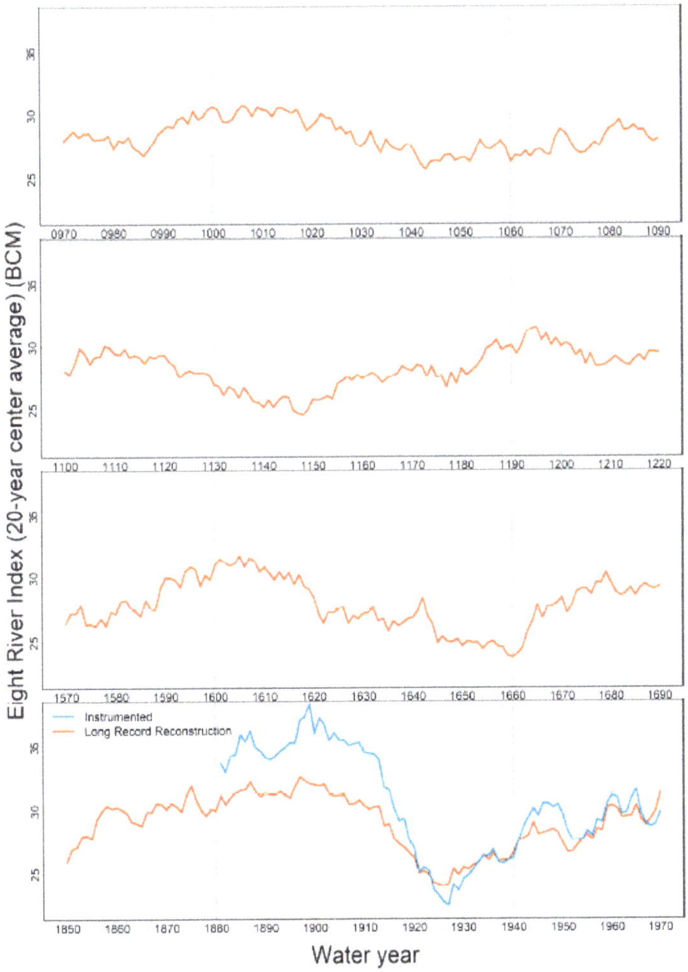

Figure 5. Selected flow periods over the long reconstruction compared with instrumented flow from (20-year centered average for both flow terms).

4.2. *Delta Outflow: Model Calibration and Reconstructions*

4.2.1. Model Calibration

The resulting pre-development relationship between Delta outflow and Central Valley runoff, both expressed in terms of annual flows, is displayed in Figure 6. Equation (2) fitting parameters and regression statistics are summarized in Table 5.

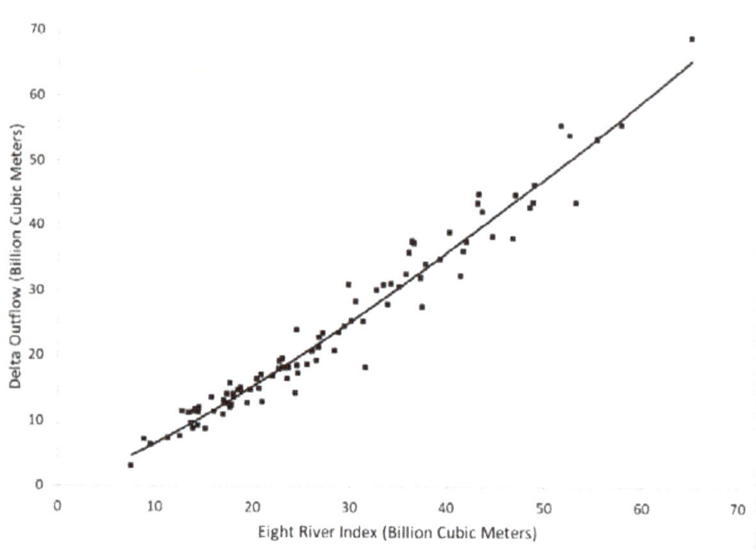

Figure 6. Pre-development annual outflow–runoff relationship (Model 2). See Equation (2) and Table 4 for model fitting parameters.

Table 5. Model fitting parameters and regression statistics for Equation (2): relationships between Delta outflow and Central Valley runoff under pre-development (Model 2) and contemporary (Model 3) conditions. Flow units are in BCM per year.

Model	α_1	α_2	r^2	Std. Error (BCM)
Pre-development (Model 2)	0.380	1.23	0.956	2.7
Contemporary WYs 1912–1944 (Model 3)	0.285	1.38	0.991	1.5
Contemporary WYs 1945–2018 (Model 3)	(1)	(1)	(1)	(1)

(1) The contemporary outflow–runoff relationship for WYs 1945–2018 was adjusted using Equation (4) to reflect significant increases in water use in the Central Valley and Delta following construction of Shasta Dam in WY 1944. See Table 6.

Table 6. Model fitting parameters and regression statistics for Equation (4): adjusted relationships between Delta outflow and Central Valley runoff under WYs 1945–2018 contemporary (Model 3) conditions.

Fitting Parameter/ Regression Statistic	Low Runoff Years 8RI < 24.6 BCM/yr	High Runoff Years 8RI > 24.6 BCM/yr
α1	0.285	0.285
α2	1.38	1.38
α8	−0.000137	0
α9	0.0148	0.00426
α10	0	0.0863
R^2	0.474	0.482

Historical annual Delta outflow was correlated with annual Central Valley runoff (as measured by the 8RI) over a subset of the contemporary period spanning WYs 1912–1944; the resulting model fit is displayed in Supplementary Material D (see Figure D-1). Equation (2) fitting parameters and regression statistics are summarized in Table 5. Model residuals, reported as predicted minus observed, are plotted as a time series in Figure 7a for the full contemporary period spanning WYs 1912–2018. This figure clearly shows that Equation (2) increasingly over-estimates Delta outflow over time following WY 1944, signifying a decreasing trend in Delta outflow relative to the 8RI. Equation (2) residuals were de-trended through the following re-formulation (see Figure D-2 in Supplementary Material D):

$$Delta\ Outflow = \alpha_1 \times 8RI^{\alpha_2} \times \left\{1 - \left[\alpha_8 \times (Water\ Yr - 1944)^2 + \alpha_9 \times (Water\ Yr - 1944) + \alpha_{10}\right]\right\} \quad (4)$$

where α_8, α_9, and α_{10} are dimensionless fitting parameters. Hutton et al. [101] observed that Delta outflow trends, when normalized to the 8RI, were different between low and high runoff years. In high runoff years, they observed a decreasing trend in normalized outflow. However, they reported that:

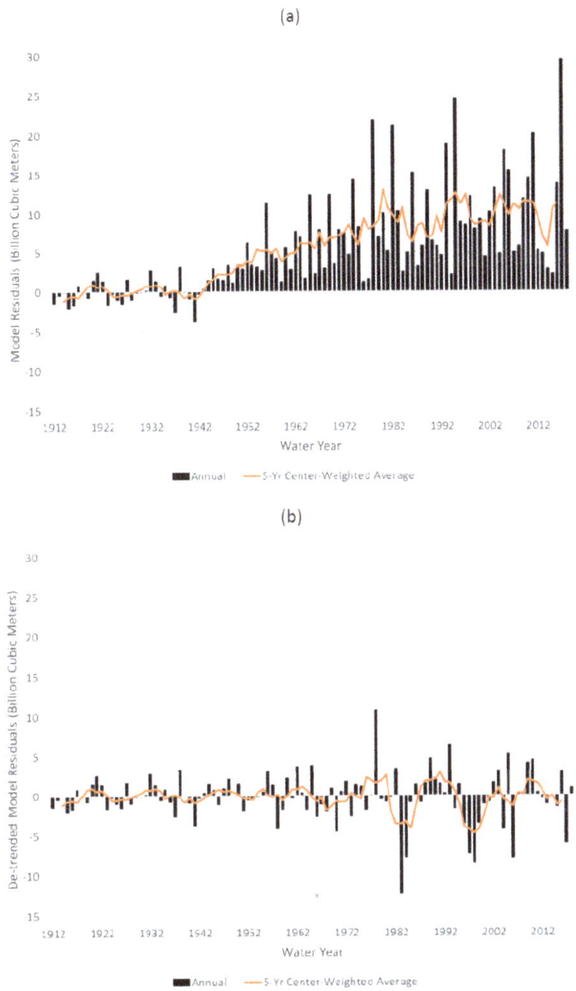

Figure 7. Time series of model residuals associated with contemporary annual runoff–outflow relationship (Model 3). Residuals from Equation (2) applied to the entire WYs 1912–2018 contemporary period are shown in (**a**). De-trended model residuals from Equation (4) applied to WYs 1945–2018 are shown in (**b**).

In drier years, the downward trend in normalized Delta outflow appears to have been curbed (and possibly reversed) over the last few decades due to more restrictive water management (i.e., lower normalized Delta exports) in the estuary and a leveling of water use in the upstream watershed.

Following the observations of [101], Equation (4) was independently calibrated for low runoff years (8RI < 24.6 BCM/yr) and high runoff years (8RI > 24.6 BCM/yr) with a combined standard error of 3.0 BCM. Fitting parameters and regression statistics are provided in Table 6. Model residuals, reported as predicted minus observed, are plotted as a time series in Figure 7b. This figure shows no apparent time trend in the de-trended model residuals. The tree-ring reconstructions of Central Valley runoff (i.e., the 8RI) were used in conjunction with Equations (2) and (4) to estimate contemporary annual Delta outflow volumes for each time series beginning in WY 1912.

4.2.2. Delta Outflow Reconstructions

The annual Central Valley runoff reconstructions (long record and short record) were used as the basis for reconstructing Delta outflow volume under pre-development and contemporary conditions using Models 2 and 3 described above. The results are shown in the form of exceedance frequencies in Figure 8. Through WY 1850, the plot shows little difference between the long and short record reconstructions to estimate pre-development annual Delta outflow volumes. Annual outflow volumes for this period are approximately 35–37 BCM, 24 BCM and 11–13 BCM at the 10th, 50th and 90th percentiles, respectively. For the contemporary period beginning in WY 1912, the plot shows small differences between the long and short record reconstructions. Annual outflow volumes for this period are approximately 40–44 BCM, 24 BCM and 9–11 BCM at the 10th, 50th and 90th percentiles, respectively.

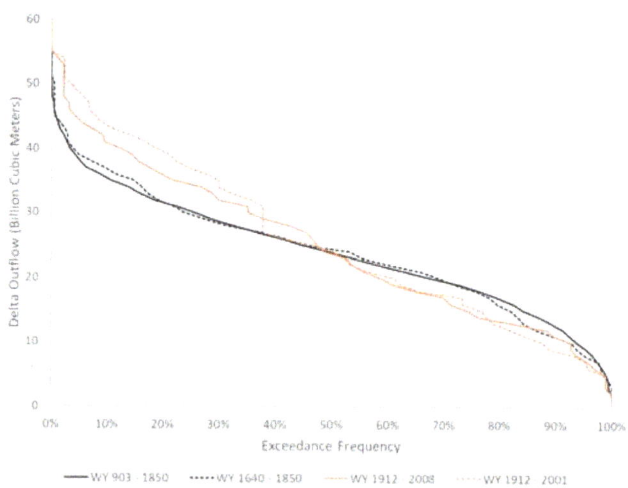

Figure 8. Reconstructed delta outflow exceedance frequency under pre-development (Model 2) and contemporary (Model 3) conditions, for the long and short record reconstructions.

Differences between pre-development and contemporary Delta outflow conditions reflect differences observed in the Central Valley runoff reconstructions as well as differences in water use on the valley floor and in the Delta. For example, assuming a common historical runoff sequence from WYs 1922–2003, Gross et al. [22] reported mean annual Delta outflows of 24.5 BCM and 19.4 BCM under pre-development and contemporary conditions, with the difference approximately equal to CVP and SWP exports from the south Delta, which together average approximately 6.1 BCM [22]. Our work shows similar mean annual Delta outflow conditions (24 BCM) for the pre-development and contemporary periods. However, the contemporary period is associated with a more variable outflow regime relative to the pre-development period, with higher outflows in the low end of the exceedance frequency domain and lower outflows in the high end of the exceedance frequency domain.

Contemporary outflow model residuals associated with Equations (2) and (4) are highly correlated (R^2 = 0.94) with reconstructed Central Valley runoff residuals. Residuals are computed as the difference between reconstructed values and historical values. Figure 9 shows scatter plots and regression lines for long record (WYs 1912–2008) and short record (WYs 1912–2001) relationships, with residuals shown as 5-year center weighted averages.

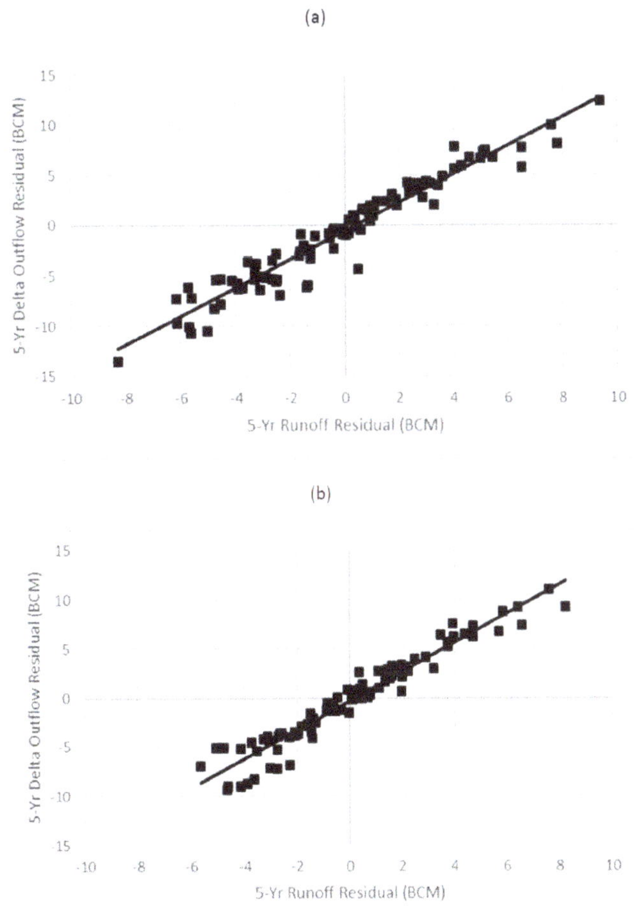

Figure 9. Relationship between reconstructed annual delta outflow residuals and reconstructed central valley runoff residuals for the contemporary period. Residuals are presented as 5-yr center weighted averages. The long record (WYs 1912–2008) and short record (WYs 1912–2001) relationships are shown in (**a**) and (**b**), respectively.

4.3. *X2: Model Calibration and Reconstructions*

4.3.1. Model Calibration

The resulting pre-development relationships between seasonal average X2 and annual average Delta outflow are displayed in Figure 10. Equation (3) fitting parameters and regression statistics are summarized in Table 7. Step changes in contemporary relationships between annual Delta outflow volume and seasonal average X2 position were observed following construction of Shasta Dam in 1944. Hutton et al. [101] observed statistically significant trends in seasonal outflows, with decreasing trends observed in four months (February, April, May and November) and increasing trends observed in two months (July and August). The authors discussed linkages between outflow trends and changes in upstream flows and coincident developments such as reservoir construction and operation, out-of-basin imports and exports, and expansion of irrigated agriculture. The resulting contemporary pre- and post-WY 1945 relationships between seasonal average X2 and annual Delta outflow volume are shown in Supplementary Material D (Figures D-3 and D-4) and regression statistics are summarized in Table 7. Although a physical basis exists for developing independent correlations for the pre- and post-1945 February–June relationships, we note that the derived fitting parameters are not statistically different from one another.

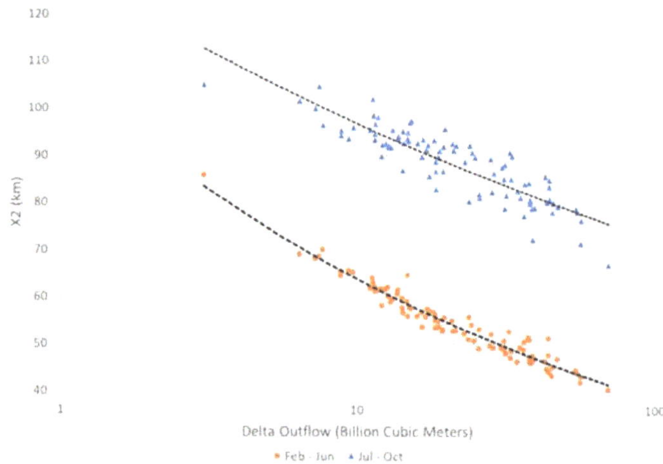

Figure 10. Pre-development seasonal X2–outflow relationships (Model 2). See Equation (3) and Table 5 for model fitting parameters.

Table 7. Model fitting parameters and regression statistics for Equation (3): relationships between X2 position and Delta outflow under pre-development (Model 2) and contemporary (Model 3) conditions X2 is in units of km from Golden Gate and outflow units are in BCM per year.

Model	Season	$\alpha 3$	$\alpha 4$	r^2	Std. Error (km)
Pre-development (Model 2)	February–June	107	−0.223	0.943	1.7
Pre-development (Model 2)	July–October	130	−0.128	0.763	3.7
Contemporary WYs 1912–1944 (Model 3)	February–June	120	−0.242	0.974	1.5
Contemporary WYs 1912–1944 (Model 3)	July–October	142	−0.137	0.760	4.8
Contemporary WYs 1945–2018 (Model 3)	February–June	122	−0.228	0.886	3.6
Contemporary WYs 1945–2018 (Model 3)	July–October	113	−0.106	0.709	4.0

4.3.2. X2 Reconstructions

The annual Delta outflow reconstructions (long record and short record) were used as the basis for reconstructing seasonal (February–June and July–October) average X2 position under pre-development and contemporary conditions, using Models 2 and 3 described above. Figure 11 shows pre-development seasonal average X2 exceedance frequencies for each reconstruction. Little difference is observed between the pre-development long and short period reconstructions except in the 10th–20th percentile range. As shown in the top panel, February–June average X2 positions are approximately 60–63, 53 and 48 km at the 10th, 50th and 90th percentiles, respectively. X2 is further downstream (smaller values) in the spring, indicative of relatively higher flow conditions. As shown in the bottom panel, July–October average X2 positions are approximately 94–96, 86–87 and 82–83 km at the 10th, 50th and 90th percentiles, respectively. X2 is further upstream (larger values) in the summer and fall, indicative of relatively lower flow conditions.

(**a**)

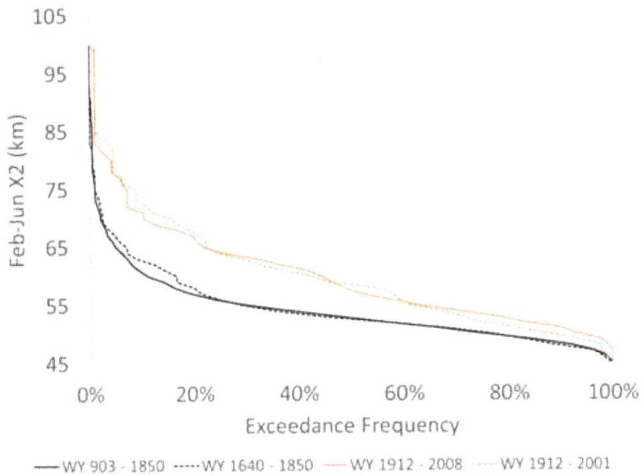

(**b**)

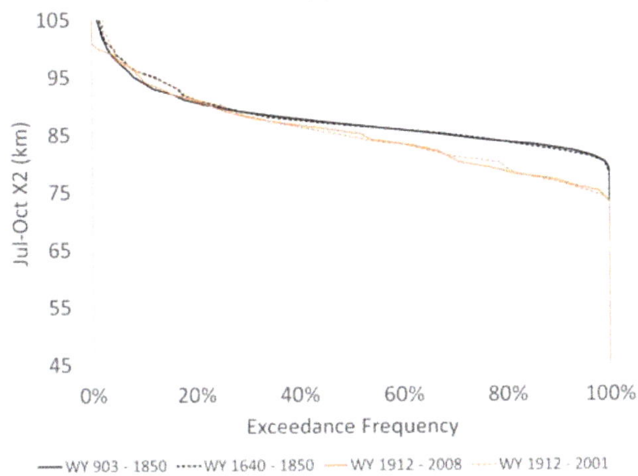

Figure 11. Reconstructed seasonal X2 exceedance frequency under pre-development (Model 2) and contemporary (Model 3) conditions. February–June and July–October X2 exceedance curves are shown in (**a**) and (**b**), respectively. Exceedance curves are provided for long and short record reconstructions.

This same figure shows small differences between the reconstructions throughout the exceedance frequency domain for the contemporary period beginning in WY 1912. February–June average X2 positions are approximately 70–73, 58–59 and 50–51 km at the 10th, 50th and 90th percentiles, respectively. July–October average X2 positions are approximately 94–96, 84–86 and 76–77 km at the 10th, 50th and 90th percentiles, respectively. As under pre-development conditions, contemporary X2 is further downstream (smaller values) in the spring and further upstream (larger values) in the summer and fall.

Differences between pre-development and contemporary X2 conditions reflect differences observed in the Delta outflow reconstructions and differences in interannual pattern of water use. Differences in X2 also reflect natural and anthropogenic drivers that modified the estuary's flow–salinity regime, resulting in greater outflow requirements under contemporary conditions to repel salinity intrusion [22,26] The contemporary period is generally associated with greater spring (February–June) salinity intrusion and lesser summer–fall (July–October) salinity intrusion relative to the pre-development period. Seasonal differences between pre-development and contemporary X2 conditions are indicative of upstream reservoir operations that store water in winter and spring months and release water in summer and fall months.

5. Discussion

Flow and salinity have been the subject of scientific observation in San Francisco Estuary over more than a century [14,65]. These observations, which have provided a reasonable understanding of contemporary conditions and associated trends in the estuary over the past century, have also supported decision-making related to freshwater flow management in the estuary. However, important knowledge gaps exist that cannot be addressed solely by the available observations. These gaps relate to the fact that large-scale changes in the estuary and its watershed pre-date the observational record by several decades; these gaps also relate to the fact that California is known to have been subject to highly variable climatic conditions over the past millennium. Additional data collection is thus insufficient to provide an understanding of how the system behaved prior to the initiation of large-scale disturbances after the mid-19th century. This work is an attempt to fill these gaps by (i) using an updated set of tree-ring chronologies to represent annual runoff into the Central Valley from the surrounding higher-elevation watersheds over the past millennium and (ii) utilizing a modeling approach to relate runoff to freshwater flow to the estuary and to salinity intrusion in the Delta. This integration of tree-ring based estimates of runoff with models of flow and salinity representing different configurations of the system (pre-development and contemporary) allows for a more nuanced exploration of flow and salinity changes over periods much longer than covered by the instrumented record. This information is of scientific and practical importance because it can help guide decisions related to the restoration of the estuarine ecosystem. These decisions have recently focused on potential changes to flow and salinity management in the estuary [18], decisions with major environmental and economic consequences for California.

The updated tree-ring-based reconstruction shows a mean annual Central Valley runoff (8RI) of approximately 29 BCM, a quantity that is similar to that observed in the contemporary system. The reconstruction also shows large single-year anomalies from the mean, although multi-year anomalies over averaging periods of 5–100 years are minimal. An important observation is that, while high runoff extremes overlap with the instrumented record, their magnitudes are lower than observed. The reconstruction indicates the occurrence of individual years with runoff significantly lower than seen in the gauged record; however, over longer averaging periods, pre-development runoff variations are of similar magnitude to those in the instrumented period and represent broadly similar patterns of wet and dry periods. Our findings are contrary to some prior reconstructions—notably Stine [47] based on the position of tree stumps at Mono Lake and Graham and Hughes [48] focused on Merced River runoff and the Mono Lake Basin inflow—that indicate evidence of more severe droughts over the past millennium than any seen in the instrumented period. However, our findings are broadly consistent with the work of Meko et al. [20] that focused on a smaller set of tree-ring data from the Sacramento and San Joaquin River basins. Our work confirms that major long-term deviations from the mean runoff in California's Central Valley are often seen in the tree ring record, but the extended drought of the late 1920s and early 1930s compares with the most extreme in the past millennium.

Consistent with the historical pattern of flow variability, California at the end of WY 2020 appears to be in the midst of another severe long-term drought. The Eight River Index, averaged over the preceding 20-years from WY 2020, stands at 25.6 BCM; long-term average runoff of lower magnitude was last seen in the severe drought of the 1920s and 1930s, when 20-year average flows from WY 1931 to 1939 ranged from 22.4 to 25.1 BCM.

Integration of the tree-ring based runoff with downstream flow models shows that, although median freshwater flows into the estuary have not changed significantly, Delta outflow has been more variable in the contemporary period compared to pre-development conditions. Extending to salinity, the contemporary period is associated with greater spring (February–June) salinity intrusion and lesser summer–fall (July–October) salinity intrusion relative to the pre-development period. Both outflow and salinity intrusion are directly affected by contemporary reservoir storage and release patterns, which tends to smoothen the intra-annual extremes that are seen in the pre-development system.

Our work can also be compared with similar research by Stahle et al. [23] who used tree-ring chronologies to relate to salinity intrusion in the Delta. Using a 625-year (1379–2003) tree ring chronology to reconstruct the February–June average X2 position in San Francisco Estuary, the authors were able to explain 73% of the variance in the observed X2 data over a 1956–2003 calibration period. The authors used their reconstructed salinity record to examine return intervals between single-year X2 extremes and to quantify the frequency of consecutive seasonal maxima and minima over the period of record. [23] recognized that their reconstruction does not mimic pre-development salinity conditions in the estuary. Rather, the authors concede that their X2 time series:

> ... *provides an estimate for variability in the salinity gradient on interannual-to-decadal time scales, given the present land cover, stream morphology, and regulated flow environment, in response to the range of modern and prehistoric seasonal precipitation totals registered in the tree-ring record over the past 625 years.*

In other words, the authors employed a "level-of-development" methodological approach [22,24,63] and their 625-year X2 reconstruction represents salinity variability under a contemporary level of development. Stahle et al. [23] found the hydroclimatic signal from tree growth to be approximately stationary over the past six centuries; thus, we expect that their estimate would be consistent with the contemporary level X2 time series presented by Gross et al. [22] that utilized a shorter, more recent climate sequence spanning WYs 1922–2003.

In contrast with [23], our work explicitly attempts to reconstruct an extended record of historic salinity conditions in the estuary. Our work relies on modeled relationships between Central Valley runoff, Delta outflow and estuarine salinity under natural and anthropogenically altered hydrologic conditions as they occurred over the period spanning 903–2008 (long record) and 1640–2001 (short record). We expect that our pre-1850 estimates would be consistent with the pre-development level X2 time series presented by Gross et al. [22] similarly, we expect that our post-1944 estimates (following construction of Shasta Dam) would be consistent with the contemporary level X2 time series presented by [22]. Consistent with Gross et al. [22], our work shows that, in spite of a relatively stationary hydroclimatic signal from tree growth (a proxy for Central Valley runoff), the San Francisco Estuary's seasonal salinity pattern has changed due to anthropogenic alterations. Specifically, both studies suggest that the estuary is now more saline in the late winter and spring than it was under pre-development conditions because of contemporary water management. Both studies also show that the contemporary estuary is less saline in the late summer and fall than it was under pre-development conditions because of water management.

Flows in the observed record from WY 1872 onwards display significant change, with a wet period from the late 19th century to the first decade of the 20th century, followed by a severe drought in the late 1920s and early 1930s. As noted earlier, these were drivers of the early water resources engineering activities in California, with a focus on flood control in the late 19th century to be followed by a focus on regulation and storage by the 1920s and beyond. The longer flow reconstruction in this work highlights that this period in the observed record captures the types of extreme shifts that have occurred over the past millennium. These reconstructed data therefore show that widely used level-of-development modeling approaches, that repeat the instrumental sequence of flows for different estuary and watershed configurations, appear to be an appropriate methodology for representing a range of pre-development conditions.

The agreement between the tree-ring record with the earliest hydrologic observations confirms and draws attention to a period of dramatic hydrologic change during the late 19th and early 20th centuries, a shift driven by natural factors coupled with rapid regional development. Thus, population growth and extensive watershed modification was overlaid on the underlying hydrologic shift from very wet to very dry conditions, which complicates the task of inferring estuarine changes in the early development period (WYs 1850–1911). Putting this hydrologic shift in the context of other anthropogenic drivers is important in understanding how the estuary responded during this early period and in setting salinity targets for estuarine restoration.

Although tree-ring proxies from long-lived species with high sensitivity to drought are a powerful tool for water resources planning in California, some important caveats are noteworthy. As indicated by the reconstruction statistics, tree-ring width is an imperfect proxy for Central Valley precipitation. In our work, uncertainty is greatest in the early part of the tree-ring record because the data network thins. In particular, the tree-ring record suffers from limited blue oak data, a tree which is shorter-lived than several other less moisture-sensitive species. Our Central Valley runoff reconstructions are especially uncertain in extremely wet years, likely due to a weakening in response of tree growth to changes in precipitation in very wet years [19]. The gaged flows themselves may also be more uncertain under high-flow conditions. The phenomenon can lead to problems with non-normality and non-constant variance (as a function of predicted values) or regression residuals. Transformation of flow before regression (employed in this work) can partly help fix such problems with residuals. Despite these efforts, however, error bars are generally wider in wet years than in dry years [53]).

Another aspect of uncertainty in our Central Valley runoff reconstructions that is not necessarily summarized by calibration and validation statistics is the possible lack of detection of runoff variations at very low frequencies (e.g., wavelengths > 200 years); this results from detrending techniques used to standardize ring widths into annual indices of growth. Specifically, climate trends that span periods longer than the time series of ring width from longest-lived individual trees are removed by detrending. Uncertainty associated with reconstructed flows at very low frequencies could be addressed with alternative detrending methods, such as regional curve standardization or age-banding, which rely on estimation of the curve of expected ring width as a function of tree-ring age [102,103]. Such methods require intensive sampling, demand representative age classes over the entire tree-ring records, and may be possible for a small number of tree species in the study area. The tree-ring data set could be extended in the medieval period (and possibly earlier). Such an extension may be possible if remnant preserved wood can be found for species with strong moisture signals—e.g., *Quercus douglasii*, *Psuedotsuga macrocarpa*, and *Pinus balfouriana*.

The flow–salinity models used in this reconstruction generally assume stationary sea level conditions, although sea level and thus salinity intrusion is expected to have changed over the reconstruction periods. An exception is Andrews et al. [26] in which the pre-development model used a (single) lower sea level value than the contemporary model. Additional detailed mechanistic salinity intrusion modeling may help resolve the impact of continuous sea level changes in the past millennium and allow refinement of the empirical models employed in this work. Although technically feasible, this is generally limited by the cost and computational complexity of running the mechanistic models under a range of observed sea level values.

In conclusion, this work is an important step in integrating tree-ring data with models of flow and salinity in the San Francisco Estuary and its watershed to develop millennial-scale ranges of pre-development conditions. Even though refinements are possible, we believe that our findings support regulatory decision making by providing a baseline to inform future flow regulations and restoration activity in the estuary. While it is helpful to have a target reference range, attainment in future years will continue to be a challenge: much in the system remains highly dynamic and will continue to evolve over time, including sea level, precipitation, snowmelt, and runoff patterns, all of which are expected to be affected by climate change, with limited predictability at present.

Supplementary Materials: The following are available online at https://www.mdpi.com/article/10.3390/w13152139/s1, Supplementary Material A. Metadata of tree-ring chronologies (xlsx file) used in development of Model 1. Two sheets—a table followed by key to columns. Supplementary Material B. Tree-ring Standardization and Flow Reconstruction (pdf file). Descriptions of how tree-ring widths were standardized into site chronologies and how those chronologies were converted to estimated WY flow related to Model 1. Supplementary Material C. Statistics of Single Site Reconstructions (SSRs) (pdf file). Statistics apply to the regression models as calibrated by regression of square-root-transformed annual flow on lagged residual tree-ring chronologies and their squares. Identifies which chronologies are used in Model 1 long record and short record runoff reconstructions. Supplementary Material D. Supplemental figures on outflow and salinity modeling related to Model 3. Supplementary Material E (Referenced in Supplementary Material A). Listing of Annual Reconstructed Flow (tab-separated ASCII text file). Listing of annual observed and reconstructed Central Valley runoff (8RI), along with 50% confidence interval for Model 1 long record and short record runoff reconstructions. Supplementary Material F (Referenced in Supplementary Material A). Compressed tree-ring measurement files. These are compressed ASCII text files of the measured ring widths used to develop the tree-ring chronologies considered for use in the Model 1 annual runoff reconstructions. Individual file names are linked to the site information (metadata in Supplementary Material A) through column "FilePrefix." Supplementary Material G (Referenced in Supplementary Material A). Listing of annual values of residual site tree-ring chronologies. Tab-separated ascii time series matrix of tree-ring chronologies listed in Supplementary Material A. Data columns in same order as chronologies are listed in Supplementary Material A. Row 1 is header that is the column number followed by the SiteCode field from Supplementary Material A. First column is the year. Following rows correspond to tree-ring residual chronologies data for years 899–2016 at each of the 69 sites.

Author Contributions: P.H.H. conceptualized the study; D.M.M. and P.H.H. performed the tree ring and salinity modeling, respectively. P.H.H., D.M.M., and S.B.R. synthesized the findings wrote the manuscript. All authors have read and agreed to the published version of the manuscript.

Funding: Funding for this work was provided by the Metropolitan Water District of Southern California through Agreement Number 143878 with Tetra Tech Inc. (Pasadena, CA, USA).

Institutional Review Board Statement: Not applicable.

Informed Consent Statement: Not applicable.

Data Availability Statement: The data presented in this study are available in Supplementary Materials A–C and E–G.

Acknowledgments: The authors would like to acknowledge David Stahle for sharing his historical salinity data and Michelle Goman for her insights on paleosalinity conditions in the estuary.

Conflicts of Interest: The authors declare no conflict of interest. The funders reviewed the overall research plan and reviewed a draft of the manuscript; the study design, analyses and interpretation of the results were performed by the authors.

References

1. Kennish, M.J. Environmental threats and environmental future of estuaries. *Environ. Conserv.* **2002**, *29*, 78–107. [CrossRef]
2. Zedler, J.B. What's new in adaptive management and restoration of coasts and estuaries? *Estuaries Coasts* **2017**, *40*, 1–21. [CrossRef]
3. European Commission. *Directive 2000/60/EC of the European Parliament and of the Council of 23 October 2000 Establishing a Framework for Community Action in the Field of Water Policy*; Cambridge University Press: Cambridge, UK, 2000.
4. European Commission. Common Implementation Strategy for The Water Framework Directive (2000/60/Ec). 2003. Available online: https://ec.europa.eu/environment/water/water-framework/objectives/pdf/strategy2.pdf (accessed on 1 July 2019).
5. Carvalho, L.; Mackay, E.B.; Cardoso, A.C.; Baattrup-Pedersen, A.; Birk, S.; Blackstock, K.L.; Borics, G.; Borja, A.; Feld, C.K.; Ferreira, M.T.; et al. Protecting and restoring Europe's waters: An analysis of the future development needs of the Water Framework Directive. *Sci. Total Environ.* **2019**, *658*, 1228–1238. [CrossRef] [PubMed]
6. U.S. Environmental Protection Agency (USEPA). San Francisco Bay Delta: About the Watershed. 2018. Available online: https://www.epa.gov/sfbay-delta/about-watershed (accessed on 1 May 2020).
7. Lund, J.R.; Hanak, E.; Fleenor, W.E.; Bennett, W.A.; Howitt, R.E.; Mount, J.F.; Moyle, P.B. *Comparing Futures for the Sacramento-San Joaquin Delta*; University of California Press: Berkeley, CA, USA, 2010.
8. Hundley, N., Jr. *The Great Thirst: Californians and Water–A History*; University of California Press: Berkeley, CA, USA, 2001.
9. Kelley, R. *Battling the Inland Sea: Floods, Public Policy, and the Sacramento Valley*; University of California Press: Berkeley, CA, USA, 1989.
10. Luoma, S.N.; Dahm, C.N.; Healey, M.; Moore, J.N. Challenges facing the Sacramento–San Joaquin Delta: Complex, chaotic, or simply cantankerous? *San Franc. Estuary Watershed Sci.* **2015**, *13*. [CrossRef]
11. Delta Stewardship Council. The Delta plan: Ensuring a Reliable Water Supply for California, a Healthy Delta Ecosystem, and a Place of Enduring Value. 2013. Available online: http://deltacouncil.ca.gov/delta-plan-0 (accessed on 1 July 2019).
12. Jassby, A.D.; Kimmerer, W.J.; Monismith, S.G.; Armor, C.; Cloern, J.E.; Powell, T.M.; Schubel, J.R.; Vendlinski, T.J. Isohaline Position as a Habitat Indicator for Estuarine Populations. *Ecol. Appl.* **1995**, *5*, 272–289. [CrossRef]
13. California State Water Resources Control Board (CSWRCB). Water Quality Control Plan for the San Francisco Bay/Sacramento-San Joaquin Delta Estuary. Division of Water Rights, December 13. 2006. Available online: http://www.waterboards.ca.gov/waterrights/water_issues/programs/bay_delta/wq_control_plans/2006wqcp/docs/2006_plan_final.pdf (accessed on 1 July 2019).
14. Hutton, P.H.; Rath, J.S.; Chen, L.; Ungs, M.L.; Roy, S.B. Nine decades of salinity observations in the San Francisco Bay and Delta: Modeling and trend evaluation. *J. Water Res. Plan. Manag.* **2015**, *142*, 04015069. [CrossRef]
15. Feyrer, F.; Nobriga, M.L.; Sommer, T.R. Multidecadal trends for three declining fish species: Habitat patterns and mechanisms in the San Francisco Estuary, California, USA. *Can. J. Fish. Aquat. Sci.* **2007**, *64*, 723–734. [CrossRef]
16. Moyle, P.B. The future of fish in response to large-scale change in the San Francisco Estuary, California. *Am. Fish. Soc. Symp.* **2008**, *64*, 357–374.
17. Mahardja, B.; Farruggia, M.J.; Schreier, B.; Sommer, T. Evidence of a shift in the littoral fish community of the Sacramento-San Joaquin Delta. *PLoS ONE* **2017**, *12*, e0170683. [CrossRef]
18. California State Water Resources Control Board (CSWRCB). San Francisco Bay/Sacramento–San Joaquin Delta Estuary (Bay-Delta) Program, Phase II Update of the Bay-Delta Plan: Delta Outflows, Sacramento River and Delta Tributary Inflows, Cold Water Habitat and Interior Delta Flows. 2018. Available online: https://www.waterboards.ca.gov/waterrights/water_issues/programs/bay_delta/comp_review.shtml (accessed on 16 May 2018).
19. Meko, D.M.; Stahle, D.W.; Griffin, D.; Knight, T.A. Inferring precipitation anomaly gradients from tree-rings. *Quat. Int.* **2011**, *235*, 89–100. [CrossRef]
20. Meko, D.M.; Woodhouse, C.A.; Touchan, R. *Klamath/San Joaquin/Sacramento Hydroclimatic Reconstructions from Tree-rings, Final Report to California Department of Water Resources*; Agreement 4600008850; California Department of Water Resources: Sacramento, CA, USA, 2014; Available online: https://cwoodhouse.faculty.arizona.edu/sites/cwoodhouse.faculty.arizona.edu/files/FinalCAWDRreport.pdf (accessed on 1 July 2019).
21. Griffin, D.; Anchukaitis, K.J. How unusual is the 2012–2014 California drought? *Geophys. Res. Lett.* **2014**, *41*, 9017–9023. [CrossRef]
22. Gross, E.S.; Hutton, P.H.; Draper, A.J. A Comparison of Outflow and Salt Intrusion in the Pre-Development and Contemporary San Francisco Estuary. *San Franc. Estuary Watershed Sci.* **2018**, *16*. [CrossRef]
23. Stahle, D.W.; Griffin, R.D.; Cleaveland, M.K.; Edmondson, J.R.; Fye, F.K.; Burnette, D.J.; Abatzoglou, J.T.; Redmond, K.T.; Meko, D.M.; Dettinger, M.D.; et al. A tree-ring reconstruction of the salinity gradient in the northern estuary of San Francisco Bay. *San Franc. Estuary Watershed Sci.* **2011**, *9*. [CrossRef]
24. Fox, P.; Hutton, P.H.; Howes, D.J.; Draper, A.J.; Sears, L. Reconstructing the Natural Hydrology of the San Francisco Bay-Delta Watershed. *Hydrol. Earth Syst. Sci.* **2015**, *19*, 4257–4274. [CrossRef]
25. California Department of Water Resources (CDWR). Estimates of Natural and Unimpaired Flows for the Central Valley of California: Water Years 1922–2014. 2016. Available online: https://msb.water.ca.gov/documents/86728/a702a57f-ae7a-41a3-8bff-722e144059d6 (accessed on 1 July 2019).
26. Andrews, S.; Gross, E.; Hutton, P.H. Modeling salt intrusion in the San Francisco Estuary prior to anthropogenic influence. *Cont. Shelf Res.* **2017**, *146*, 58–81. [CrossRef]

27. Hutton, P.H.; Roy, S.B. Characterizing Early 20th Century Delta Outflow and Salinity Intrusion in the San Francisco Estuary. *San Franc. Estuary Watershed Sci. J.* **2019**, *17*. [CrossRef]
28. Cheng, R.T.; Casulli, V.; Gartner, J.W. Tidal, Residual, Intertidal Mudflat (TRIM) Model and its Application to San Francisco Bay, California. *Estuar. Coast. Shelf Sci.* **1993**, *36*, 235–280. [CrossRef]
29. Atwater, B.F.; Hedel, C.W.; Helley, E.J. *Late Quaternary depositional history, Holocene Sea-Level Changes, and Vertical Crustal Movement, Southern San Francisco Bay*; United States Geological Survey: Reston, VA, USA, 1977; pp. 1014–1015.
30. Byrne, R.; Ingram, B.; Starratt, S.; Malamud-Roam, F.; Collins, J.; Conrad, M. Carbon-Isotope, Diatom, and Pollen Evidence for Late Holocene Salinity Change in a Brackish Marsh in the San Francisco Estuary. *Quat. Res.* **2001**, *55*, 66–76. [CrossRef]
31. Sweet, W.; Park, J.; Marra, J.; Zervas, C.; Gill, S.; Sea Level Rise and Nuisance Flood Frequency Changes around the United States. National Oceanic and Atmospheric Administration (NOAA); Technical Report NOS CO-OPS 073. 2014. Available online: NOAA_Technical_Report_NOS_COOPS_073.pdf (accessed on 1 July 2019).
32. Hall, W.H. *Report of the State Engineer to the Legislature of California*; Session of 1880, Part 2; California State Printer: Sacramento, CA, USA, 1880.
33. Grunsky, C.E. The relief outlets and by-passes of the Sacramento Valley flood-control project. *Trans. ASCE* **1929**, *93*, 791–811. [CrossRef]
34. Alexander, B.S.; Mendell, G.H.; Davidson, G. *Report of the Board of Commissioners on the Irrigation of the San Joaquin, Tulare, and Sacramento Valleys of the State of California*; 43rd Congress, 1st Session, House of Representation. Ex. Doc. No. 290; Government Printing Office: Washington, DC, USA, 1874.
35. Hall, W.H. *Topographical and Irrigation Map of the Great Central Valley of California Embracing the Sacramento, San Joaquin, Tulare and Kern Valleys and the Bordering Foothills*; California State Engineering Department: Sacramento, CA, USA, 1887.
36. Garone, P. *The Fall and Rise of the Wetlands of California's Great Central Valley*; University of California Press: Berkeley, CA, USA, 2011.
37. Katibah, E.F. A brief history of riparian forests in the Central Valley of California. In *California Riparian Forests*; Warner, R.E., Hendrix, K.M., Eds.; University of California Press: Berkeley, CA, USA, 1984; pp. 23–29.
38. Holland, R.F. *The Geographic and Edaphic Distribution of Vernal Pools in the Great Central Valley, California*; California Native Plant Society Special Publications No. 4; California Native Plant Society: Fair Oaks, CA, USA, 1978.
39. Burcham, L.T. *California Range Land: An Historical–Ecological Study of the Range Resource of California*; Department of Natural Resources, Division of Forestry: Sacramento, CA, USA, 1957.
40. Dutzi, E.J. Valley Oaks in the Sacramento Valley: Past and Present Distribution. Master's Thesis, University of California, Los Angeles, CA, USA, 1978.
41. Howes, D.J.; Fox, P.; Hutton, P.H. Evapotranspiration from Natural Vegetation in the Central Valley of California: Grass Reference-Based Vegetation Coefficients and the Dual Crop Coefficient Approach. *J. Hydrol. Eng.* **2015**, *20*, 04015004. [CrossRef]
42. Schulman, E. *Tree-Ring Hydrology of the Colorado River Basin, University of Arizona Bulletin XVI (4)*; University of Arizona: Tucson, AR, USA, 1945; p. 51.
43. Fritts, H.C. *Tree-Rings and Climate*; Academic Press: London, UK, 1976; p. 567.
44. Hughes, M.K.; Kelly, P.M.; Pilcher, J.R.; LaMarche, J.V.C. (Eds.) *Clim. From Tree-Rings*; Cambridge University Press: Cambridge, UK, 1982; p. 223.
45. Cook, E.R.; Seager, R.; Cane, M.A.; Stahle, D.W. North American drought: Reconstructions, causes, and consequences. *Earth Sci. Rev.* **2007**, *81*, 93–134. [CrossRef]
46. Cook, E.R.; Seager, R.; Heim, R.R., Jr.; Vose, R.S.; Herweijer, C.; Woodhouse, C. Megadroughts in North America: Placing IPCC projections of hydroclimatic change in a long-term paleoclimate context. *J. Quat. Sci.* **2010**, *25*, 48–61. [CrossRef]
47. Stine, S. Extreme and persistent drought in California and Patagonia during mediaeval time. *Nature* **1994**, *369*, 546–549. [CrossRef]
48. Graham, N.E.; Hughes, M.K. Reconstructing the mediaeval low stands of Mono Lake, Sierra Nevada, CA, USA. *Holocene* **2007**, *17*, 1197–1210. [CrossRef]
49. Loaiciga, H.A.; Haston, L.; Michaelsen, J. Dendrohydrology and long-term hydrologic phenomena. *Rev. Geophys.* **1993**, *31*, 151–171. [CrossRef]
50. Meko, D.M.; Woodhouse, C.A. Application of streamflow reconstruction to water resources management. In *Dendroclimatology, Progress and Prospects, Developments in Paleoenvironmental Research*; Hughes, M.K., Swetnam, T.W., Diaz, H.F., Eds.; Springer: Amsterdam, The Netherlands, 2011; Volume 11, pp. 231–261.
51. Earle, C.J. Asynchronous droughts in California streamflow as reconstructed from tree-rings. *Quat. Res.* **1993**, *39*, 290–299. [CrossRef]
52. Stahle, D.W.; Griffin, R.D.; Meko, D.M.; Therrell, M.D.; Edmondson, J.R.; Cleaveland, M.K.; Stahle, L.N.; Burnette, D.J.; Abatzoglou, J.T.; Redmond, K.T.; et al. The ancient blue oak woodlands of California: Longevity and hydroclimatic history. *Earth Interact.* **2013**, *17*, 1–23. [CrossRef]
53. Meko, D.M.; Therrell, M.D.; Baisan, C.H.; Hughes, M.K. Sacramento River flow reconstructed to A.D. 869 from tree-rings. *J. Am. Water Resour. Assoc.* **2001**, *37*, 1029–1040. [CrossRef]
54. Woodhouse, C.A.; Meko, D.M.; Bigio, E.R. A long view of Southern California water supply: Perfect droughts revisited. *J. Am. Water Resour. Assoc.* **2020**, *56*, 212–229. [CrossRef]

55. Stahle, D.W.; Therrell, M.D.; Cleaveland, M.K.; Cayan, D.R.; Dettinger, M.D.; Knowles, N. Ancient blue oaks reveal human impact on San Francisco Bay salinity. *Eos Trans. Am. Geophys. Union* **2001**, *82*, 141–145.
56. Fox, J.P.; Mongan, T.R.; Miller, W.J. Long-term annual and seasonal trends in surface salinity of San Francisco Bay. *J. Hydrol.* **1991**, *122*, 93–117. [CrossRef]
57. Arnold, J.G.; Moriasi, D.N.; Gassman, P.W.; Abbaspour, K.C.; White, M.J.; Srinivasan, R.; Santhi, C.; Harmel, R.D.; Van Griensven, A.; Van Liew, M.W.; et al. SWAT: Model use, calibration, and validation. *Trans. ASABE* **2012**, *55*, 1491–1508. [CrossRef]
58. Brush, C.F.; Dogrul, E.C.; Kadir, T.N. *Development and Calibration of the California Central Valley Groundwater-Surface Water Simulation Model (C2VSim)*; Version 3.02-CG; Bay-Delta Office, California Department of Water Resources: Sacramento, CA, USA, 2013.
59. Küchler, A.W. Natural vegetation of California, pocket map. In *Terrestrial Vegetation of California*; Barbour, M.G., Major, J., Eds.; John Wiley & Sons, Inc.: New York, NY, USA, 1977; pp. 909–938.
60. Orang, M.N.; Snyder, R.L.; Geng, S.; Hart, Q.J.; Sarreshteh, S.; Falk, M.; Beaudette, D.; Hayes, S.; Eching, S. California simulation of evapotranspiration of applied water and agricultural energy use in California. *J. Integr. Agric.* **2013**, *12*, 1371–1388. [CrossRef]
61. Casulli, V.; Walters, R.A. An unstructured grid, three-dimensional model based on the shallow water equations. *Int. J. Numer. Methods Fluids* **2000**, *32*, 331–348. [CrossRef]
62. Whipple, A.A.; Grossinger, R.M.; Rankin, D.; Stanford, B.; Askevold, R.A. Sacramento-San Joaquin Delta Historical Ecology Investigation: Exploring Pattern and Process. In *Prepared for the California Department of Fish and Game and Ecosystem Restoration Program*; A Report of SFEI-ASC's Historical Ecology Program, Publication #672; San Francisco Estuary Institute-Aquatic Science Center: Richmond, CA, USA, 2012.
63. Draper, A.J.; Munevar, A.; Arora, S.K.; Reyes, E.; Parker, N.L.; Chung, F.I.; Peterson, L.E. CalSim: Generalized model for reservoir system analysis. *J. Water Resour. Plan. Manag.* **2004**, *130*, 480–489. [CrossRef]
64. Kahrl, W.L. *The California Water Atlas, State of California*; Governor's Office of Planning and Research; CA DWR: Sacramento, CA, USA, 1979.
65. California Department of Public Works (CDPW). Flow in California Streams–Report to the Legislature of 1923 on the Water Resources of California, Bulletin 5, Appendix A, Division of Engineering and Irrigation. 1923. Available online: http://wdl.water.ca.gov/waterdatalibrary/docs/historic/bulletins.cfm (accessed on 1 July 2019).
66. Moftakhari, H.R.; Jay, D.A.; Talke, S.A.; Kukulka, T.; Bromirski, P.D. A novel approach to flow estimation in tidal rivers. *Water Resour. Res.* **2013**, *49*, 4817–4832. [CrossRef]
67. Moftakhari, H.; Jay, D.; Talke, S.; Schoellhamer, D. Estimation of historic flows and sediment loads to San Francisco Bay, 1849–2011. *J. Hydrol.* **2015**, *529*, 1247–1261. [CrossRef]
68. MacVean, L.; Thompson, S.; Hutton, P.; Sivapalan, M. Reconstructing Early Hydrologic Change in the California Delta and its Watersheds. *Water Resour. Res.* **2018**, *54*, 7767–7790. [CrossRef]
69. California State Water Resources Control Board (CSWRCB). Water Right Decision 1485, August. 1978. Available online: http://www.waterboards.ca.gov/waterrights/board_decisions/adopted_orders/decisions/d1450_d1499/wrd1485.pdf (accessed on 1 July 2019).
70. California State Water Resources Control Board (CSWRCB). Revised Water Right Decision 1641, March 15. 2000. Available online: https://www.waterboards.ca.gov/waterrights/board_decisions/adopted_orders/decisions/d1600_d1649/wrd1641_1999dec29.pdf (accessed on 1 July 2019).
71. U.S. Fish and Wildlife Service (USFWS). *Formal Endangered Species Act Consultation on the Proposed Coordinated Operations of the Central Valley Project (CVP) and State Water Project (SWP). Technical Memorandum from USFWS to US Bureau of Reclamation.*; 2008. Available online: http://www.fws.gov/sfbaydelta/documents/swp-cvp_ops_BO_12-15_final_OCR.pdf (accessed on 1 July 2019).
72. National Marine Fisheries Service (NMFS). Biological Opinion and Conference Opinion on the Long-Term Operations of the Central Valley Project and State Water Project. In *Technical memorandum from NMFS to U.S. Bureau of Reclamation*; 2009. Available online: https://nrm.dfg.ca.gov/FileHandler.ashx?DocumentID=21473 (accessed on 19 October 2017).
73. U.S. Fish and Wildlife Service (USFWS). Biological Opinion on the Reinitiation of Consultation on the Coordinated Operations of the Central Valley Project and State Water Project. 2019. Available online: https://www.fws.gov/sfbaydelta/cvp-swp/documents/10182019_ROC_BO_final.pdf (accessed on 1 July 2019).
74. National Marine Fisheries Service (NMFS). Biological Opinion on Long-Term Operation of the Central Valley Project and the State Water Project. 2019. Available online: https://www.fisheries.noaa.gov/resource/document/biological-opinion-reinitiation-consultation-long-term-operation-central-valley (accessed on 1 July 2019).
75. Fleenor, W.; Goodwin, P.; Monsen, N.; Ruhl, C. *An Independent Peer Review Report for the Delta Science Program on the DWR Report on Estimating Net Delta Outflow (NDO): Approaches to Estimating NDO in the Sacramento-San Joaquin Delta*; Delta Stewardship Council: Sacramento, CA, USA, 2016.
76. Rath, J.S.; Hutton, P.H.; Ateljevich, E.A.; Roy, S.B. A Survey of X2 Isohaline Empirical Models for the San Francisco Estuary. *San Franc. Estuary Watershed Sci* **2021**. in review.
77. California Department of Water Resources (CDWR), 2021. DSM2: Delta Simulation Model II. Available online: https://water.ca.gov/Library/Modeling-and-Analysis/Bay-Delta-Region-models-and-tools/Delta-Simulation-Model-II (accessed on 1 July 2019).

78. Chao, Y.; Farrara, J.D.; Zhang, H.; Zhang, Y.J.; Ateljevich, E.; Chai, F.; Davis, C.O.; Dugdale, R.; Wilkerson, F. Development, implementation, and validation of a modeling system for the San Francisco Bay and Estuary. *Estuar. Coast. Shelf Sci.* **2017**, *194*, 40–56. [CrossRef]
79. Casulli, V.; Zanolli, P. High resolution methods for multidimensional advection-diffusion problems in free-surface hydrodynamics. *Ocean Model.* **2005**, *10*, 137–151. [CrossRef]
80. Martyr-Koller, R.C.; Kernkamp, H.W.J.; Van Dam, A.; van der Wegen, M.; Lucas, L.V.; Knowles, N.; Jaffe, B.; Fregoso, T.A. Application of an unstructured 3D finite volume numerical model to flows and salinity dynamics in the San Francisco Bay-Delta. *Estuar. Coast. Shelf Sci.* **2017**, *192*, 86–107. [CrossRef]
81. California Data Exchange Center (CDEC). 2019. Available online: http://cdec.water.ca.gov/ (accessed on 1 July 2019).
82. DAYFLOW Program. 2019. CDWR. Available online: http://www.water.ca.gov/dayflow/ (accessed on 1 July 2019).
83. Robeson, S.M.; Maxwell, J.T.; Ficklin, D.L. Bias correction of paleoclimatic reconstructions: A new look at 1200+ years of Upper Colorado River flow. *Geophys. Res. Lett.* **2020**, *47*, e2019GL086689. [CrossRef]
84. Denton, R.A. Accounting for antecedent conditions in seawater intrusion modeling—Applications for the San Francisco Bay–Delta. In Proceedings of the 1993 Hydraulic Division National Conference, San Francisco, CA, USA, 25–30 July 1993; Shen, H.W., Ed.; ASCE: Reston, VA, USA, 1993; pp. 821–826.
85. Meko, D.M.; Woodhouse, C.A.; Bigio, E.R. *Southern California Tree-ring Study, Final Report to California Department of Water Resources*; Agreement 4600011071; California Department of Water Resources: Sacramento, CA, USA, 2018; Available online: https://cwoodhouse.faculty.arizona.edu/sites/cwoodhouse.%20faculty.arizona.edu/files/FinalReport2018NoAppendices.pdf. (accessed on 1 July 2019).
86. Zhao, S.; Pederson, N.; D'Orangeville, L.; HilleRisLambers, J.; Boose, E.; Penone, C.; Bauer, B.; Jiang, Y.; Manzanedo, R.D. The International Tree–Ring Data Bank (ITRDB) revisited: Data availability and global ecological representativity. *J. Biogeogr.* **2018**, *46*, 355–368. [CrossRef]
87. Cook, E.R.; Krusic, P.J.; Holmes, R.H.; Peters, K. Program ARSTAN, Version 41d. 2007. Available online: www.ldeo.columbia.edu/tree-ring-laboratory (accessed on 1 July 2019).
88. Cook, E.R.; Peters, K. The smoothing spline: A new approach to standardizing forest interior tree-ring width series for dendroclimatic studies. *Tree-Ring Bull.* **1981**, *41*, 45–53.
89. Osborn, T.J.; Briffa, K.R.; Jones, P.D. Adjusting variance for sample-size in tree-ring chronologies and other regional mean times series. *Dendrochronologia* **1997**, *15*, 89–99.
90. Cook, E.R.; Kairiukstis, L.A. (Eds.) *Methods of Dendrochronology: Applications in the Environmental Sciences*; Kluwer Academic Publishers: Dordrecht, The Netherlands, 1990.
91. Hurvich, C.M.; Tsai, C.L. Regression and time series model selection in small samples. *Biometrika* **1989**, *76*, 297–307. [CrossRef]
92. Wigley, T.M.L.; Briffa, K.R.; Jones, P.D. On the average value of correlated time series, with applications in dendroclimatology and hydrometeorology. *J. Appl. Meteorol. Climatol.* **1984**, *23*, 201–213. [CrossRef]
93. Snedecor, G.W.; Cochran, W.G. *Statistical Methods*, 8th ed.; Iowa State University Press: Ames, IA, USA, 1989; p. 503.
94. Meko, D.M. Dendroclimatic reconstruction with time varying predictor subsets of tree indices. *J. Clim.* **1997**, *10*, 687–696. [CrossRef]
95. Meko, D.M.; Woodhouse, C.A.; Baisan, C.H.; Knight, T.; Lukas, J.J.; Hughes, M.K.; Salzer, M.W. Medieval drought in the Upper Colorado River Basin. *Geophys. Res. Lett.* **2007**, *34*. [CrossRef]
96. Weisberg, S. *Applied Linear Regression*, 2nd ed.; John Wiley: New York, NY, USA, 1985; p. 324.
97. Michaelsen, J. Cross-validation in statistical climate forecast models. *J. Appl. Meteorol.*. **1987**, *26*, 1589–1600. [CrossRef]
98. Fritts, H.C.; Guiot, J.; Gordon, G.A. Verification. In *Methods of Dendrochronology, Applications in the Environmental Sciences*; Cook, E.R., Kairiukstis, L.A., Eds.; Kluwer Academic Publishers: Dordrecht, The Netherlands, 1990; pp. 178–185.
99. Panofsky, H.A.; Brier, G.W. *Some Applications of Statistics to Meteorology*; The Pennsylvania State University: State College, PA, USA, 1968; p. 224.
100. Meko, D.M.; Woodhouse, C.A.; Morino, K. Dendrochronology and links to streamflow. *J. Hydrol.* **2012**, *412–413*, 200–209. [CrossRef]
101. Hutton, P.H.; Rath, J.S.; Roy, S.B. Freshwater flow to the San Francisco Bay-Delta Estuary over nine decades. Part 1: Trend evaluation, hydrological processes. *Hydrol. Process.* **2017**, *31*, 2500–2515. [CrossRef]
102. Briffa, K.R.; Jones, P.D.; Bartholin, T.S. Fennoscandian summers from AD 500: Temperature changes on short and long timescales. *Clim. Dyn.* **1992**, *7*, 111–119. [CrossRef]
103. Briffa, K.R.; Osborn, T.J.; Schweingruber, F.H.; Harris, I.C.; Jones, P.D.; Shiyatov, S.G.; Vaganov, E.A. Low-frequency temperature variations from a northern tree ring density network. *J. Geophys. Res.* **2001**, *106*, 2229–2941. [CrossRef]

water MDPI

Review

The Municipal Water Quality Investigations Program: A Retrospective Overview of the Program's First Three Decades

Paul H. Hutton [1], Sujoy B. Roy [1,*], Stuart W. Krasner [2] and Leslie Palencia [3]

1 Tetra Tech, Lafayette, CA 94549, USA
2 Independent Researcher, La Verne, CA 91750, USA
3 Palencia Consulting Engineers, Fullerton, CA 92835, USA
* Correspondence: sujoy.roy@tetratech.com

Abstract: This paper presents the history and evolution of the California Department of Water Resources' Municipal Water Quality Investigations (MWQI) program. This program tracks source water quality in the Sacramento–San Joaquin Delta (Delta) for drinking water supply for nearly two-thirds of California. The program provides early warning of changing conditions in source water quality, provides data and knowledge-based support for operational decision making, and provides scientific support to a variety of urban water users. This retrospective (i) documents program formation, (ii) describes its evolution in response to regulations and technological advances in water treatment and field monitoring, and (iii) notes how the development of federal drinking water quality regulations such as the Disinfection By-Products Rule impacted the program. The MWQI program is believed to be the first drinking water supply program in the United States to conduct continuous, real-time monitoring of organic carbon, bromide, and other anions and to report these data on the internet. In addition to its regular use for operational decision making, the data may be used for evaluating long-term trends and responses to specific changes in the Delta and its watershed. Future program directions will likely be guided by factors that may trigger changes in treatment plant processes and operations, such as emerging contaminants, changes in land and water management practices, permanent Delta island flooding, sea level rise, and climate change. While this retrospective focuses on one region, its multi-decade interplay of science, treatment and monitoring technology, and regulations (as well as practical aspects of managing such a large-scale program) are broadly relevant to professionals engaged in drinking water quality management in other urbanized and developed regions of the world.

Keywords: drinking water quality; operational decision-making; real-time monitoring; Sacramento-San Joaquin Delta

Citation: Hutton, P.H.; Roy, S.B.; Krasner, S.W.; Palencia, L. The Municipal Water Quality Investigations Program: A Retrospective Overview of the Program's First Three Decades. *Water* **2022**, *14*, 3426. https://doi.org/10.3390/w14213426

Academic Editors: Nigel W.T. Quinn, Ariel Dinar, Iddo Kan and Vamsi Krishna Sridharan

Received: 10 September 2022
Accepted: 22 October 2022
Published: 28 October 2022

1. Introduction

Estuaries, which provide freshwater for drinking water consumption and agricultural production, have historically served as outstanding locations for human communities. Estuaries provide access to both rivers and oceans, thereby enhancing opportunities for trade and communication [1]. Because they are highly productive, estuaries have also been an important food source for human habitation [2]. In fact, the earliest civilizations in the world developed around estuaries. Many modern cities have grown near estuaries, including Jakarta, New York City, and Tokyo. Of the 32 largest cities in the world in the early 1990s, 22 were located on estuaries [3].

Estuaries, by definition, exhibit a water quality spectrum between seawater and riverine that varies with freshwater inflows and geometry [1]. The riverine ends of estuaries are often used as drinking water resources for the communities that have grown around them. However, because of the proximity to population centers, these waters tend to exhibit high concentrations of nutrients, pathogens, and other contaminants (in addition to ocean-derived salts and organic matter from contributing watersheds), which create a challenge to drinking water suppliers that treat such waters for human consumption.

The San Francisco Estuary, including the delta formed by the Sacramento and San Joaquin Rivers (hereafter Delta) (Figure 1), plays a major role in California's prosperity and is shaped by the large population centers that have developed around it. In addition to serving as an important habitat to more than 750 animal and plant species [4], the Delta is the largest single water supply source in California and its waters are transported across river basin boundaries to support major urban and agricultural centers in the state. To accomplish these inter-basin transfers, large water projects were constructed in the 1940s by the federal government (i.e., the Central Valley Project or CVP) and in the 1960s by the State of California (i.e., the State Water Project or SWP) [5]. These projects consist of a network of dams in upper elevations of the Delta watershed, combined with aqueducts and pump stations for long-range conveyance. The CVP generally serves agricultural water users, whereas the SWP is primarily devoted to municipal supply (approximately 70%). The largest source of SWP water is the Feather River, which is impounded by Lake Oroville [5]. SWP water, exported from the Delta via the California, North Bay, and South Bay Aqueducts for irrigation and municipal use [6], is allocated through long-term contracts to 29 water agencies (termed "contractors") who are responsible for water delivery to communities and irrigation districts within their jurisdiction [5]. Allocations are developed by the California Department of Water Resources (CDWR) early in each water year (which begins on 1 October) and updated as additional hydrologic information becomes available. Releases from Lake Oroville are managed by CDWR for water exports and to meet various water quality standards in the Delta, notably standards mandated to limit the intrusion of saline water [7]. The SWP is one of the largest water conveyance systems in the world, with an average of 2.9 million acre-feet of water delivered annually in the decade ending in 2016 [5].

Collectively, Delta exports from the SWP and CVP support a $32 billion agricultural industry and serve as an important source of drinking water to almost 27 million residents [4]. Saltwater intrusion was one of the earliest water quality concerns for human uses of Delta water. In 1920, the City of Antioch sued upstream irrigators to protect the city's intake from salinity intrusion [8,9]. In response to this lawsuit, the State of California implemented a monitoring program and published the first authoritative review of Delta salinity and its control in 1931 [10]. Early plans envisaged control of Delta salinity by means of storage regulation on the Sacramento River [10] and a saltwater barrier in the estuary near Carquinez Strait (see Figure 1) [8,11]. Later, as part of the SWP planning process, the California Water Plan [12] and subsequent investigations considered re-routing low salinity Sacramento River flows through and around the Delta (see additional discussion in [8]).

Water quality concerns broadened over time to include a suite of chemical constituents related to natural and anthropogenic sources in the Delta and its watershed. Water quality management in the Delta occurs through a complex framework of federal and state laws, such as the Clean Water Act (1973), the state Porter-Cologne Water Quality Control Act (1969), and the Safe Drinking Water Act (1974), with the goals of supporting beneficial uses for ecosystems, municipal use, and agricultural use. Here, we focus on a program that has evolved to track a subset of constituents in Delta waters that are of concern from the standpoint of drinking water supply.

Since the late 1970s and early 1980s, Delta water quality concerns have included natural organic matter (NOM) [13] and bromide [14]. NOM and bromide are persistent in Delta waters due to agricultural return flows from the region's organic peat soils and seawater intrusion, respectively. NOM promotes the formation of trihalomethanes (THMs) and other carcinogenic disinfection by-products (DBPs) when Delta waters are chlorinated during drinking water treatment [15,16]. In the presence of bromide, brominated DBPs are also formed [17]. Bromine-containing DBPs are of greater health concern than their chlorine-containing analogs [18]. In addition, the estuary receives wastewater discharges from more than 9 million people along its periphery and other pollutant loads (including pesticides, herbicides, and nutrients) from the developed watersheds upstream [19,20].

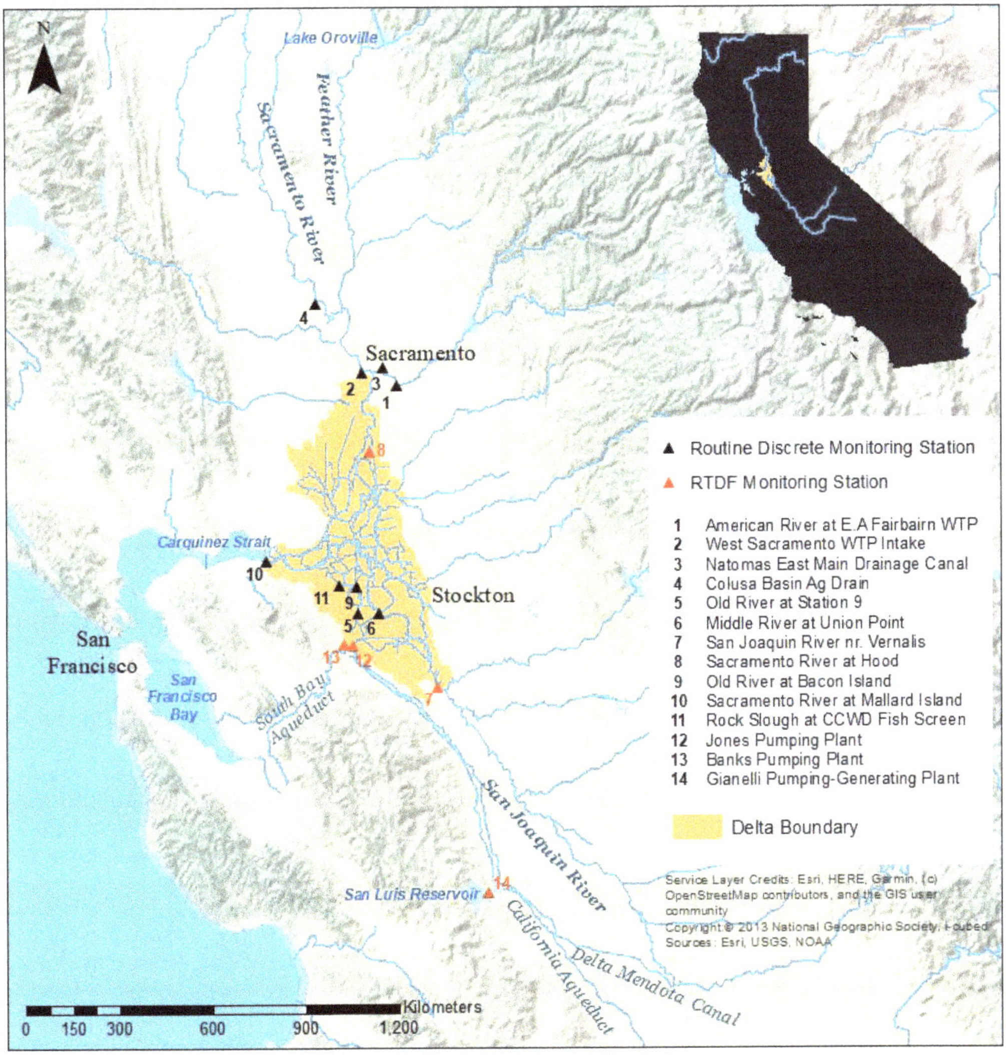

Figure 1. Map of study area identifying MWQI routine discrete and real time monitoring locations (noted as RTDF stations, for real time data forecasting).

Given this confluence of factors, water quality in the estuary related to constituents of drinking water interest has been extensively studied (e.g., [13,15,21–24]). Among all state and local agencies monitoring water quality in the Delta and its tributaries, CDWR's Municipal Water Quality Investigations (MWQI) program is the most extensive and cohesive program established to investigate the quality of Delta source water with respect to its suitability for production of drinking water. MWQI program elements have evolved over three decades in response to advances in science, water treatment technology and regulations, and emerging contaminants, paralleling major investments made by local water supply agencies (Figure 2).

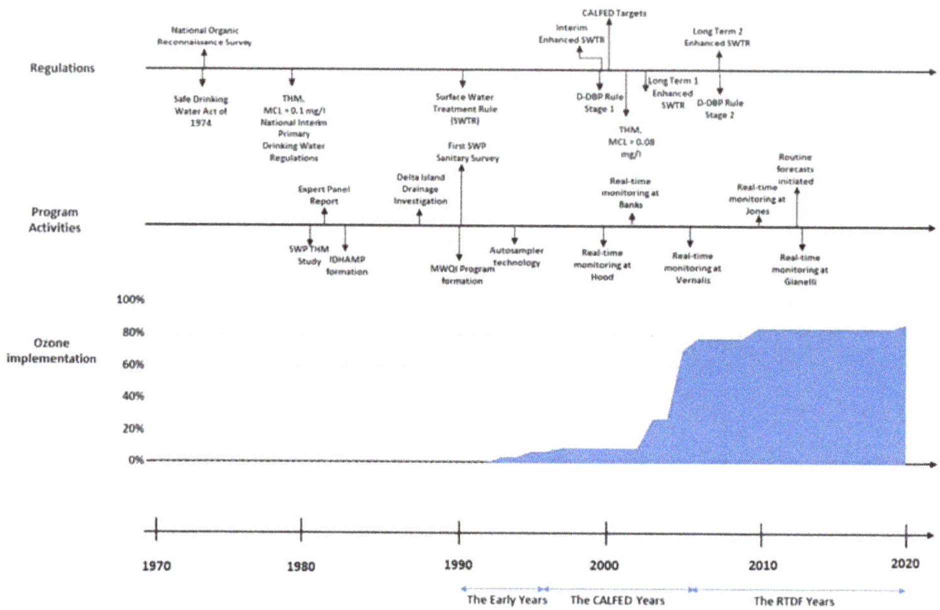

Figure 2. Timeline showing the evolution of drinking water regulations, MWQI program activities, and ozone implementation by participating MWQI water contractors.

The primary objective of this paper is to provide a retrospective overview of CDWR's MWQI program, highlighting its evolution in response to changing understanding of Delta water quality water constituents, regulatory drivers, and new technologies. It is not our intention to provide a subjective critique of the program's perceived utility, benefits, successes, and/or failures. This retrospective builds on a prior summary of the program's history [25] with a focus on program evolution and its utility to SWP operations, drinking water quality regulators, and organizations responsible for municipal water supply. Here, we describe the program, including purpose, organization and funding, and key elements. This description is followed by a historical account of the program origin, which can be traced back to federal and state regulatory activities in the 1970s and early studies by CDWR and other state and local agencies in the 1980s. We then chronicle how the program evolved from its formation in 1990 to its current configuration. We conclude this paper with a discussion of key program accomplishments and future directions. This paper is focused on program-level activities, rather than specific study results or interpretation of data collected through the program. Although this paper focuses on one region, its multi-decade interplay of science, treatment and monitoring technology, and regulations (as well as practical aspects of managing such a large-scale program) are broadly relevant to professionals engaged in drinking water quality management in other urbanized and developed regions of the world.

2. Program Description

The MWQI program supports the use of SWP water for municipal supply through monitoring, forecasting, and reporting of Delta and SWP water quality data. The program provides early warning of changing conditions in source water quality, provides data and knowledge-based support for SWP operational decision-making, and provides scientific support to CDWR, water contractors, and other governmental entities.

2.1. Program Organization & Funding

The MWQI program activities include water quality monitoring, water quality forecasts, and data management. Data generated by the MWQI program, along with related data generated by other programs, are distributed via daily and weekly electronic reports at: http://rtdf.info/ (accessed on 8 October 2022). Program staff administer tasks such as data quality assurance and quality control and participate in other multi-agency water quality management activities across the Delta. MWQI program organization is shown schematically in Figure A1 (see Appendix A) and includes support from groups within CDWR that perform monitoring, laboratory analysis, and modeling. The MWQI program budget was approximately $3.1 million in 2020 [26]. CDWR manages the MWQI program with input and funding from participating urban water agencies.

2.2. Program Elements

In its early years, the focus of MWQI was to establish and maintain a discrete or grab-sample monitoring program that could identify sources of contaminants to the Delta, elucidate how contaminants from each source are transported through the system, and evaluate how they affect concentrations at drinking water intakes. Program data were used to build scientific understanding and to inform a variety of regulatory efforts such as SWP watershed sanitary surveys. As the program evolved, its scope has expanded to include real time data collection, water quality forecasting, and targeted scientific studies.

2.3. Watershed Sanitary Surveys

Watershed sanitary surveys are required under the California Surface Water Treatment Rule (SWTR) and are submitted to the CSWRCB's Division of Drinking Water, formerly the California Department of Public Health and the California Department of Health Services. In February 1988, the California Department of Health Services requested that a sanitary survey of the SWP be conducted to enable Department of Health Services and water agencies treating SWP water to appraise the effectiveness of the operation of existing water treatment plants and to adequately evaluate new treatment plant design requirements. The SWC decided to conduct a single survey of the entire SWP watershed rather than conducting independent surveys when they applied for new water supply permits or amended existing permits. As a result, SWP watershed sanitary surveys cover almost two thirds of the State of California, starting with the upper reaches of the Sacramento and San Joaquin River watersheds and extending to the terminal reservoirs of the SWP in southern California. Since 1990, SWP watershed sanitary surveys have nominally been conducted every five years; this 1990 survey and its updates are summarized in Table A1 (see Appendix A).

3. Program Genesis

As far back as the early 20th century, salinity and inorganic constituents have been the primary constituents of concern in the municipal and agricultural beneficial uses of Delta source waters. Federal legislation in the 1970s, including the Clean Water Act of 1972 [27] and the Safe Drinking Water Act of 1974 [28] led to the regulation of a broad range of water quality constituents of concern in drinking water supplies and in treated drinking water. Coincident with this period, researchers demonstrated the formation of THMs during chlorination [29,30], a conventional treatment process for disinfecting drinking water. Later investigations suggested a potential relationship between THM occurrence and increased incidence of cancer among exposed populations [31]. In 1979, the USEPA promulgated a maximum contaminant level (MCL) of 100 μg/L for total THMs [32].

3.1. SWP THM Study

In anticipation of this new federal THM regulation, the California Department of Health Services (now CSWRCB's Division of Drinking Water), Contra Costa Water District, and the Metropolitan Water District of Southern California conducted independent studies of Delta source waters [14,33]. These studies found that Delta source waters had significant potential to produce THMs when chlorinated as part of conventional water treatment processes. Furthermore, these studies found that treated Delta waters, especially during drought conditions when bromine-containing THMs sharply increased, could exceed the MCL established under the 1979 regulation. These findings led to questions concerning the sources and magnitudes of THM precursors in the SWP and the possible benefits of a peripheral canal or other through-Delta conveyance facility in reducing precursor concentrations in the SWP. To answer these and other questions, CDWR conducted a study from September 1981 to January 1982 to determine the sources of THM precursors in the Sacramento River, Delta, and the SWP [33]. This study, which confirmed high concentrations of THM precursors in Delta source waters, concluded that waters exported from the southern Delta are higher in THM-producing substances than are waters tributary to the northern Delta. Specifically, the study identified Delta island agricultural drainage and seawater intrusion as key contributors of NOM and bromide, respectively, in southern Delta channels and drinking water intakes. Because Delta island peat soils are rich in NOM, they are high in THM precursors [13,21]. The drainage from these islands is higher in THM precursors than the channel waters. Because seawater is very high in bromide, seawater intrusion significantly raises the level of bromide in Delta water [15]. Chlorination of bromide forms bromine, which is more reactive in forming THMs than chlorine [34]. Moreover, bromine-containing THMs weigh more than chlorine-containing THMs, a significant fact given that the MCL is regulated on a weight-based total. Confirming earlier work, the study concluded that a peripheral canal would provide a source water to the SWP with reduced concentrations of THM precursors relative to baseline conditions. The report recommended that a monitoring program be initiated for the measurement of THM formation potential in the Delta and the SWP [33]. Formation potential, an indicator of precursor levels under laboratory conditions, does not measure formation under full-scale water treatment operations.

3.2. Expert Panel Report

Motivated by the findings of its SWP THM study [33], CDWR assembled an independent expert panel in 1982 to evaluate the consequences of using Delta source waters for domestic purposes. Panel findings were documented in a report [35] and are summarized here. The panel concluded that:

> *The current Delta water monitoring program … was developed primarily to monitor quality from an ecological perspective specifically directed towards fishery resources and not to assess human health aspects with respect to drinking water. The program as presently constituted … is not entirely adequate to assess the present or projected suitability of these waters as a source of drinking water supply [35].*

The panel opined that drinking water quality " … should be given a much higher priority in decisions about the Delta." While some of the panel members believed that the traditional public health practice of obtaining drinking water from the best source available should be adhered to (providing support for a peripheral canal), other panel members believed that advanced water treatment could provide adequate public health protection. The panel unanimously agreed that public health should be more broadly considered in decisions about Delta water management.

The expert panel recommended that a monitoring program be initiated to identify the sources of drinking water contaminants in the Delta (e.g., THM precursors, sodium, asbestos, pesticides, and heavy metals), how contaminants are transported through the system, and how contaminant sources impact water quality at drinking water diversions. Furthermore, the panel recommended that information from the monitoring program be incorporated into a comprehensive modeling framework that would support public health decision-making as it related to Delta water management.

3.3. Interagency Delta Health Aspects Monitoring Program

The expert panel recommendation was implemented by CDWR in July 1983 with commencement of the Interagency Delta Health Aspects Monitoring Program (IDHAMP) [36–39]. The program, initially established as an 18-month investigation (and subsequently extended), was conducted in cooperation with the U. S. Bureau of Reclamation, the City of Stockton, East Bay Municipal Utility District, Contra Costa Water District, and several SWP water contractors.

Program participants provided funding as well as technical guidance through a standing committee, with participation by other relevant parties such as the California Department of Public Health and the CSWRCB. The program initially focused on monitoring drinking water quality contaminants identified in the expert panel report [35]. Monthly samples were collected from 15 to 18 stations in areas representing urban drinking water diversions, Delta inflows, in-Delta agricultural drainage, in-Delta channels and sloughs, and the Sacramento–San Joaquin River confluence. As analytical methods became more sophisticated, new and previously unidentified water quality concerns emerged and were monitored by the program, including selenium in the San Joaquin River watershed, rice herbicides in the Sacramento River watershed, and insecticides and waterborne pathogenic protozoa (*Giardia* and *Cryptosporidium*) in both watersheds [25].

3.4. Delta Island Drainage Investigation

Data collected under IDHAMP showed high THM formation potential in Delta island drainage. Delta island drainage refers to return flows collected from lands in the Delta and pumped into surrounding channels. Pumping of drainage is needed because organic-rich soils in the Delta lowlands are subsided relative to water levels in surrounding channels. Motivated by this finding, the Delta Island Drainage Investigation was initiated in 1987 to assess the impacts of Delta island drainage on the quality of drinking water supplies taken from the Delta [39]. Goals of the investigation were to (i) evaluate the quality and quantity of island drainage, (ii) identify processes that affect quality and quantity of island drainage, (iii) determine potential impacts of island drainage on water quality in Delta channels and at drinking water supply intakes, and (iv) explore potential mitigation strategies.

The investigation concluded that Delta island drainage had a higher potential to form THMs than water from Delta channels. While THM formation potential was found to vary from island to island, in general, drainage was found to have four times greater THM formation potential than Delta channel samples. Based on mass balance calculations, island drainage was estimated to contribute 40–50% of the THM formation potential in Delta waters during periods of irrigation and winter leaching [39]. Amy et al. [13], in conjunction with CDWR, corroborated the investigation's conclusion that Delta island drainage has a higher THM formation potential relative to the surrounding Delta channels. Through investigation of the chemical composition of water samples, the authors found distinct differences between Delta island drainage and channel waters, with drainage samples exhibiting higher average molecular weights. Amy et al. [13] also conducted a mass balance analysis and concluded that, on average, island drainage could contribute as much as 20% of the THM formation potential found in the SWP.

3.5. Formation of the MWQI Program

Local water agencies, recognizing that source water quality improvements would have significant consumer cost and public health benefits, commissioned a study to evaluate various Delta water management alternatives for meeting existing and proposed state and federal drinking water quality standards [40]. In addition to placing more stringent requirements on water treatment, the Surface Water Treatment Rule (SWTR) [41] promulgated by USEPA also emphasized watershed protection. Robbins et al. [42] concluded that, based on a national survey of surface water systems, a three-tiered approach that combines watershed controls, reservoir management, and water treatment is often necessary to meet public health objectives for drinking water protection. Arriving at a similar conclusion for the Delta water supplies, Krasner et al. [15] advocated DBP control strategies that included watershed management in addition to water treatment components.

As fundamental understanding of Delta source water quality issues coalesced (based on data made available through the various independent efforts of the 1980s) and regulatory pressures mounted, the need for a unified program to provide water quality data in support of management decisions affecting Delta water supplies was identified. The SWP water contactors, regarding IDHAMP as a model of interagency cooperation and a successful vehicle for public health decision support, requested that CDWR create a standing program that would provide information on known and emerging threats to drinking water quality [25]. Responding to this request, CDWR initiated the MWQI Program in 1990, unifying activities being conducted under IDHAMP and the Delta Island Drainage Investigation [43].

4. Program Evolution

The MWQI program was established as a flexible and pro-active collaboration between CDWR and participating water agencies to address new concerns as they arise [25]. Accordingly, the program has evolved in response to regulatory drivers, technology innovations, and an ever-increasing knowledge base. Here, we chronicle the evolution of the MWQI program by distilling the past 30 years into three eras: the early years (1990–1995), the CALFED years (1996–2005), and the RTDF years (2006–present). The CALFED Bay-Delta Program, a cooperative state-federal planning effort created as part of the 1994 Bay-Delta Accord [8,16], was tasked with developing a 30-year plan that addressed key problem areas in the Delta, including water quality. Although this goal was not achieved under the expired CALFED program, the Delta Stewardship Council (a state agency stemming from the CALFED program and created under the 2009 Delta Reform Act) developed a plan for the long-term management of the Delta's water and environmental resources [4].

4.1. Defining Baseline Conditions: The Early Years (1990–1995)

In 1990, CDWR merged IDHAMP, the Delta Islands Drainage Investigation, and other drinking water quality activities into the MWQI Program. The primary goal of this newly formed program was to assist water agencies in protecting and improving Delta drinking water supplies and to guide research on optimal water treatment processes. In the early years of the MWQI program, nominally between 1990 and 1995, this goal was achieved through continuity and expansion of previous monitoring activities (including monitoring key Delta channel and river stations and agricultural drains for constituents such as pesticides, arsenic, selenium, sodium, and THM formation potential) to characterize baseline conditions in the Delta. Efforts to fully characterize baseline conditions were confounded by persistent drought conditions: California endured a severe six-year drought spanning 1987–1992 with a brief return to drought in 1994. During this period, following a recommendation from the 1982 expert panel report [35], the MWQI program expanded efforts to incorporate information from the monitoring program into a comprehensive modeling framework. Finally, during this period, the program conducted its first SWP watershed sanitary survey. A summary of these early activities is provided below.

4.1.1. Water Treatment Technology and Regulations

During this period, THMs were the only regulated category of DBPs. However, new USEPA regulations were anticipated—i.e., the DBP Rule [44,45]—prompting water agencies to initiate research on advanced water treatment technologies such as ozonation [46] or granular activated carbon (GAC) adsorption [47] and to expand testing for additional DBPs [48] in their finished waters. At the time, chlorine and chloramines were generally the preferred primary and secondary disinfectants, respectively, because of relatively low cost, control of THM formation (significantly reduced during post-chloramination) [49], and high effectiveness in controlling bacterial growth in water distribution systems. Among the water agencies participating in the MWQI program, most were employing these conventional water treatment technologies during the early years of the MWQI program; less than 10 percent of the combined treatment capacity of the agencies employed ozonation (Figure 2).

The DBP Rule, which was promulgated in 1998 [44], lowered the total THM standard from 100 μg/L to 80 μg/L and set limits on other DBPs, including a 60 μg/L limit on the sum of five species of haloacetic acids (HAAs) (another chlorination by-product) and a 10 μg/L limit on bromate, an ozone by-product formed from bromide. The Rule also requires removal of total organic carbon (TOC) via enhanced coagulation or softening. The best available technologies identified for meeting the DBP Rule requirements for THMs and HAAs include enhanced coagulation, enhanced softening, or GAC; ozonation process optimization was identified as the best available technology for meeting the bromate standard. Water agencies anticipated the need for extensive research, retrofitting and upgrading of treatment facilities to meet the DBP Rule. Moreover, the specter of a more stringent Rule magnified the urgency to understand the possible effectiveness of source control in addition to new treatment technologies.

It was clear that water agencies charged with protecting the public health through treating drinking water from the Delta would face serious problems in meeting anticipated regulations [50]. The water agencies recognized that the degree of success they would experience in complying with the new rule would depend, in part, on how well DBP precursors could be reduced in the raw water supply. By reducing the concentration of these precursors in the raw water supply, the formation of known and unknown DBPs (regulated and unregulated DBPs) can be lowered. Moreover, controlling DBPs involves balancing risk/risk tradeoff issues with disinfection requirements. The SWTR had removal requirements for *Giardia* and viruses [41]. An Enhanced SWTR, developed along with the DBP Rule [44,51,52], included removal requirements for *Cryptosporidium*, which can be inactivated with ozone but not with chlorine.

4.1.2. Data Collection & Analysis

The MWQI program has conducted discrete water quality monitoring in the Delta since its inception in 1990 [25]. The spatial and temporal extent of this monitoring has varied in response to program needs and system understanding [25]. Currently (as of 2022), the program conducts routine discrete monitoring at ten locations (see Figure 1) and collects data at other locations in support of special studies. Discrete sampling has typically been conducted at a monthly interval and currently measures organic carbon (total and dissolved), standard minerals (i.e., major anions and cations), bromide, nutrients, and chlorophyll. TOC and dissolved organic carbon (DOC) are quantitative indicators of NOM content. Throughout its history, the MWQI program has evaluated several additional water quality parameters as part of routine monitoring and special studies, including trihalomethane (THM) formation potential, ultraviolet absorbance at 254 nm (UV-254) (a surrogate for the humic portion of the NOM), metals, selenium, pesticides, and herbicides [25].

The discrete water quality monitoring network, the backbone of the MWQI program in its early years, expanded dramatically at program inception. While IDHAMP had maintained a network of 15 to 18 sampling locations [39], in 1990, the MWQI program was collecting water quality samples at about 40 Delta urban intake and channel locations and about 30 agricultural drainage locations [43]. Discrete sampling frequency was typically monthly. Foreshadowing the program's later adoption of real time monitoring, beginning in 1993 the program experimented with the use of autosampler technology to increase sampling frequency in a cost-effective manner. These devices were programmed to collect samples at variable frequencies (i.e., daily and sub-daily), which were subsequently analyzed in a laboratory. Autosamplers were later replaced by auto-analyzers, devices that could analyze selected water quality constituents in near real time (typically within minutes). Highlights of the program's data collection and analysis efforts are summarized below; published program annual reports provide greater detail [43,53–56]. During this period, a parallel effort was undertaken by others [57] to evaluate loadings of key drinking water contaminants to Delta tributaries and to determine if there were source control measures that, if implemented, would improve drinking water quality at Delta intakes.

Measurement of cations and anions in Delta source waters continued under the MWQI program, thereby providing continuity with previous IDHAMP monitoring. The intended purpose of these data was to characterize the major water types in the Delta (e.g., freshwater inflows, seawater intrusion, and agricultural drainage) and their sources. By developing a chemical fingerprint or profile of specific water types, the program planned to assess the movement and degradation of water under specific hydrologic conditions in the Delta. Cation and anion data also assisted modelers and planners in the examination of alternatives to improve the management and distribution of Delta water supplies [43]. A comprehensive analysis of these data, along with related legacy data available on CDWR's Water Data Library website http://www.water.ca.gov/waterdatalibrary/ (accessed on 1 November 2020), was undertaken more recently [58,59].

During this period, several advancements were made in the measurement and understanding of THM precursors and THM formation potential and their relationship to actual THM formation in treated water. Measurements of source water THM precursors, such as bromide, TOC, and DOC became routine. The resulting data showed consistent relationships between TOC and DOC and confirmed that seawater intrusion was the primary source of bromide in Delta waters [15]. Starting in 1990, water samples were also measured for UV-254, another indicator of NOM content, which was found to correlate with DOC in most water samples. UV-254 measurements were conducted for several years, with the intent of providing a quick and inexpensive measurement useful in assessing THM precursor levels in the Delta. In 1992, a modified chemical testing procedure was developed and adopted to improve measurement of THM formation potential in high DOC water samples [53]. This modified procedure was needed because the original THM formation potential assay method was shown to underestimate precursor levels in high DOC samples common in agricultural drainage [60]. Analysis of the program's THM formation potential data revealed a consistent relative distribution of the four THM compounds as a function of bromine incorporation factor [61] and strong correlations were found between precursor concentrations, THM formation potential data, and THM formation in simulated distribution system samples [62,63]. Krasner et al. [21] studied water samples from the Delta and other locations to evaluate research approaches for characterizing NOM, including its source and nature in watersheds, its response to seasonal variations, and its potential to form DBPs.

The relationship between Delta agricultural drainage and Delta source water quality became clearer as the MWQI program maintained and expanded monitoring initiated under the Delta Island Drainage Investigation program. The volume and quality of drainage was found to correlate with seasonal farming activities and regional soils. High drainage volumes were associated with farming activities centered on two periods. In the late fall and early winter, fields are flooded to leach out salt accumulations from the soil, resulting in high drainage volume and high DOC concentrations in the drainage, especially from organic soil areas. The second peak drainage period was observed during summer irrigation. DOC concentrations were found to be lower during the irrigation peak relative to the leaching peak; this difference was thought to be caused by less soil-to-water contact time and a lower water table that resulted in lower soil moisture. High DOC and THM formation potential levels were found to be associated with the organic content of the drained soils. The highest concentrations were typically found in drains located on peat soil areas and the lowest from mineral soil areas [53]. Following statistical analysis of drainage data and soil classifications proposed in CDWR [53], characteristic monthly values of DOC, THM formation potential, and UV-254 were proposed for organic soils (i.e., high-range DOC soils), intermediate organic soils (i.e., mid-range DOC soils), and mineral soils (i.e., low-range DOC soils); these are documented in CDWR [64] and summarized in Figure A2 (see Appendix A).

4.1.3. Modeling

Prior to 1990, efforts to develop a comprehensive modeling framework—following a recommendation from the 1982 expert panel report [35]—was limited to testing available software from USEPA, including a pesticide model (EXAMS) and two water quality transport models (QUAL2E and WASP3) [37]. Absent a transport model, a simple water balance approach was utilized to estimate Delta drainage contributions to THM formation potential in channel waters during the leaching and irrigation periods of water year 1988 [39]. Beginning in 1990, the MWQI program collaborated with the CDWR Delta Modeling Section to develop a framework for modeling THM precursor fate and transport in the Delta. Early model development was documented in a CDWR Division of Planning report [65], the CDWR Delta Modeling Section's "Methodology for Flow and Salinity Estimates in the Sacramento–San Joaquin Delta and Suisun Marsh" series [66–71], and other publications [72–76]. Aspects of this early model development that focused on organic precursor fate and transport provided a foundation for today's MWQI forecasting work.

4.2. Evaluating Potential Baseline Changes: The CALFED Years (1996–2005)

The CALFED years, nominally spanning 1996 through 2005, reflect a period of change from the early years of the MWQI program. Prior to 1996, most of the work done by the program was completed in response to water agency concerns over the contaminants in the source water of the Delta and focused on existing problem definition and possible solutions. With the 1994 Bay-Delta Accord among state and federal agencies with management responsibilities over the Delta, and expansion of the interagency CALFED process [16], the MWQI program expanded its scope to address potential changes in land use, water storage, and conveyance. A summary of program activities during this era is provided below.

4.2.1. Water Treatment Technology and Regulations

During the CALFED years, two clusters of USEPA drinking water regulations were promulgated: the Enhanced SWTRs [44,51,52] and the DBP Rule [44,45]. In response to these rules, the CALFED program established targets for providing safe, reliable, and affordable drinking water as either: (i) average bromide concentrations of 50 µg/L and 3.0 mg/L TOC concentrations at southern and central Delta drinking water intakes or (ii) an equivalent level of public health protection using a cost-effective combination of alternative source waters, source control, and treatment technologies [77]. The CALFED targets were predicated, in part, on data and information developed through the MWQI program.

Because the CALFED targets could not be feasibly achieved in the short time frame required to comply with the promulgated rules, water agencies participating in the MWQI program generally adopted advanced treatment technologies during this era that went beyond the best available technologies outlined in response to the new USEPA drinking water regulations. Typical advanced treatment technologies included GAC adsorption with chlorine as the primary and secondary disinfectant or ozone and chloramines as the primary and secondary disinfectants, respectively. By the end of 2005, more than 70 percent of the combined treatment capacity of the agencies employed ozonation (Figure 2), including Metropolitan's 326 million gallon per day (mgd) Mills plant and 750 mgd Jensen plant, which treat Delta water from the east and west branches of the California Aqueduct, respectively. Nonetheless, continued efforts to develop a Delta "fix" were made to address a possible future tightening of the USEPA regulations and/or to reduce the costs of operating the advanced treatment technologies in place.

4.2.2. Data Collection & Analysis

The MWQI discrete water quality monitoring network was scaled back during the CALFED years, resulting in a network of 11 routine sampling locations by 2005 [78] that was more reflective of the IDHAMP network of the late 1980s. During this time, analysis of organic DBP precursors in channel waters shifted away from formation potential testing and the surrogate UV-254 measurement. THM formation potential testing was discontinued in 1998, reflecting a growing concern that it did not adequately represent formation in drinking water and could be misinterpreted [79]. While UV-254 data continued to be collected, their use for analysis was deemphasized because consistent spatial and temporal relationships with DOC could not be found [79] and was observed to vary from one NOM type to another. For example, agricultural drainage from peat soils (high in humic content) was found to have a UV-254:DOC ratio of 0.044–0.050 $cm^{-1}/mg/L$, whereas agricultural drainage from mineralized soils (lower in humic content) was found to have a ratio of 0.035–0.370 $cm^{-1}/mg/L$ [21]. To place these ranges in context, the Sacramento and San Joaquin rivers are characterized by UV-254:DOC ratios of approximately 0.022 and 0.027 $cm^{-1}/mg/L$, respectively. The Colorado River, outside the study area watershed, represents a source water low in humic content and has a ratio of 0.015 $cm^{-1}/mg/L$. Discrete monitoring of nutrients was resumed in 2002 after being discontinued for several years. Greater emphasis was placed on data analysis, given interest in (i) potential adverse effects of nutrients on finished drinking water quality, (ii) examining their seasonal and spatial trends [20], and (iii) studying their effects on in-channel production of organic carbon [80]. Published program annual reports provide greater detail on MWQI's discrete water quality monitoring during the CALFED years [79–83].

MWQI program efforts to conduct real time monitoring and disseminate the data on the internet were initiated during the CALFED years. The first TOC analyzer was installed along the Sacramento River at Hood in 1999 in support of a pilot study of real time organic carbon monitoring [84]. TOC analyzers were later installed at Banks Pumping Plant in 2001 and along the San Joaquin River at Vernalis in 2005. The capabilities of several instruments were evaluated [85]. Anion analyzers were also installed at these locations to collect continuous data on bromide, chloride, nitrate, and sulfate [86]. These installations are believed to be the first to take continuous measurements of organic carbon, bromide, and other anions in United States waters and to publish the data immediately on the internet [87].

MWQI program staff published several peer-reviewed journal articles on laboratory methods during the CALFED years. While most of these articles related to measurement of THM precursors from organic soils [88–91], one focused on measurement of microbial contamination in natural waters [92]. The latter examined *Giardia* and *Cryptosporidium*, which are pathogens of concern in the Enhanced SWTR.

4.2.3. Modeling

The CALFED Delta Drinking Water Council recommended a multi-faceted assessment program that included preliminary establishment of baseline Delta water quality conditions [93]. A workgroup was formed in 1999 and tasked with developing a model simulation of historical water quality conditions in the Delta [94] to (i) develop confidence in a previously developed hydrodynamic and water quality model of the Delta to establish baseline conditions for salt and organic carbon transport and (ii) establish error bounds for future simulation results. The model in question was the Delta Simulation Model version 2 (referred to as DSM2) [95]. CDWR [96] documents a DSM2 model validation study conducted in support of this CALFED effort. The validation study simulated transport of DOC and UV-254 from October 1990 through December 1997 and employed an approach documented in Jung [94] to characterize Delta island return water quality; this approach built upon earlier work documented in CDWR [64].

CDWR's Delta Modeling Section, acting as a program partner to the MWQI program, continued DSM2 model development for simulating DBP organic precursor transport throughout the CALFED years. Model development activities included formulating planning-level precursor boundary conditions [97] and implementing an algorithm to simulate the transfer of DOC from peat soils to Delta channels (due to leaching and microbial decay) when Delta islands are flooded [98,99]. This algorithm was used for evaluation of the proposed Delta Wetlands project, a project to store water on two Delta islands [100] and a major levee break in 2004 on the Jones Tract in the Delta [101].

Following the MWQI program's early success implementing real time monitoring in the Delta, proof-of-concept efforts were launched to utilize available modeling tools to forecast drinking water quality constituents of concern, notably bromide and DOC [102–104]. To support water quality forecasting capabilities beyond the Delta region, the DSM2 modeling platform was used for initial development of a water quality transport model of the California Aqueduct, South Bay Aqueduct, and Delta-Mendota Canal systems [105,106]. These nascent efforts ushered in a new MWQI program focus on providing decision support for SWP operation, as described in Section 4.3.

4.2.4. Targeted Science Studies

Scaling back of the MWQI discrete water quality monitoring network during the CALFED years was accompanied by a greater program emphasis on targeted science studies related to a broad range of Delta drinking water issues. These studies tended to focus on specific aspects of source waters, contaminant loading, measurement methods and instrumentation, and climate and hydrology. They were conducted to: (i) investigate the origins, fate, transport, and, in some cases, loads of current and emerging contaminants of concern; (ii) investigate seasonal patterns and trends of constituents and examine circulation patterns of contaminants; (iii) refine modeling assumptions; and (iv) assess the impacts of increasing urbanization on levels of water quality constituents of concern. Key studies conducted by the MWQI program, some in collaboration with outside agencies and/or consultants, are listed chronologically in Table A2 (see Appendix A). For example, Jung and Weisser [107] studied flooded peat soil environments, as there were various proposed projects (e.g., the Delta Wetlands project) that were considering using flooded islands for water storage.

4.3. *Providing Decision Support for SWP Operations: The RTDF Years (2006–Present)*

The MWQI program was expanded during 2002–2006 with the incorporation of real time data collection and water quality forecasting. The expansion occurred in phases as new sensors and data dissemination were deployed (e.g., [108]); these new program elements are collectively referred to as Real Time Data and Forecasting (RTDF). For the purpose of this narrative, 2006 is defined as the first year when these elements were in stable operation. The objective of RTDF was to enhance the ability of drinking water agencies to make informed operational decisions based on observed and forecasted changes in Delta water quality [24]. Program activities during this contemporary era were documented in CDWR's Bulletin 132 series beginning in 2012 [5,109–111].

As part of this work, 14 participating water contractors were surveyed to understand their current use of MWQI's real time and forecast data. The survey asked each contractor about the locations where they monitor water quality and how their agency used the water quality data. Responses were then categorized and compiled as shown in Figure A3 (see Appendix A). The survey revealed that use of the RTDF website to check on current conditions was commonplace, especially data reported through real time sensors. The main use of MWQI water quality data is to provide an early warning of changes in source water quality. Contractors also reported special interest in metrics that might affect their water treatment processes, particularly over short-term horizons. Specific examples of contractor uses of RTDF data, obtained in the survey responses, include tracking (i) bromide concentrations to allow adjustment of ozone dosage during treatment, (ii) cyanobacteria levels to address potential taste and odor concerns, (iii) pH and alkalinity, which can affect treatment effectiveness, (iv) organic carbon concentrations (and related measurements) to plan for coagulant dosage, and (v) the contribution of San Luis Reservoir releases, which can affect water quality in the California Aqueduct relative to flows coming directly from the Delta.

To satisfy additional real time water quality data needs by the participating water contractors, monitoring of taste and odor compounds and cyanobacteria was initiated by the MWQI program during this era. Taste and odor compounds are sampled weekly all year throughout the SWP (at selected reservoirs, Delta pump station inlets and at selected locations on the California Aqueduct). Cyanotoxins are collected across the SWP system monthly in May and April, and twice a month from June to October. If toxins are detected, the sampling frequency is increased to weekly. These data have helped inform water contractor operations. Specifically, given that most of the participating water contractors have a defined trigger level for source water taste-and-odor compounds, such timely information can inform decisions related to real time water treatment. The MWQI program is currently evaluating real time sensors for chlorophyll and phycocyanin to possibly serve as early warning detection of harmful algal blooms.

The well-established MWQI program has also provided a unique environment for participating water contractors and CDWR to interact on a regular basis. It has become a forum where the contractors can bring their source water quality concerns to address either a new or proposed regulation or to seek assistance on issues related to the operation of the SWP. Ultimately, these issues may be resolved with other CDWR personnel outside of the MWQI Program, but the MWQI program serves as a resource for locating assistance. For example, water contractors received notification about a new herbicide (Endothall), which would be used to treat aquatic vegetation in the SWP. Due to water quality concerns about the residual herbicide concentration, the contractors and CDWR discussed the issue extensively at MWQI meetings and ultimately developed a monitoring plan to study the fate and transport of the herbicide in the SWP, which was implemented and supported by CDWR.

As another example of interagency coordination, CDWR periodically receives requests from water agencies to transport non-SWP water (typically groundwater) or to transfer water between agencies through the California Aqueduct. These are generally referred to as non-project water turn-ins. The water contractors have raised concerns about such requests in the past; these concerns have related primarily to groundwater quality. To address these concerns, the contractors collaborated with CDWR to establish a facilitation group, which sets an approval process (including minimum water quality criteria) for each turn-in request. An annual report is produced by CDWR to document turn-ins and their impact on downstream water quality [112].

4.3.1. Water Treatment Technology and Regulations

While drinking water regulations have continued to evolve since 2006, they have had varied influence on the treatment of Delta source waters. The shift from chlorination as a primary disinfectant toward ozonation has continued since 2005. By 2020, about 85 percent of the combined treatment capacity of the participating agencies employed ozonation (Figure 2). In the current era, drinking water regulations have played a reduced role in driving MWQI program activities. However, that trend would likely reverse if new regulations were implemented. Examples of new DBPs being considered for regulation include four additional brominated HAAs and nitrosamines (such as N-nitrosodimethylamine or NDMA) [113]. In waters high in bromide, such as Delta waters, the five currently regulated HAAs may only represent around half of the concentration of the nine HAAs [114]. Certain pharmaceuticals discharged from wastewater treatment plants react with chloramines to form NDMA in drinking water [115]. The participating MWQI agencies that use ozone as the primary disinfectant use chloramines as the secondary disinfectant; thus, NDMA is produced at many of their water treatment plants.

4.3.2. Data Collection & Analysis

The MWQI discrete monitoring network has changed little since the CALFED years (Figure 1), with routine monitoring locations sampling for bromide, TOC, DOC, nutrients, standard minerals, and chlorophyll on a monthly basis [78,116]. The total number of discrete monitoring locations may change year to year based on need of special studies. The MWQI program discontinued publication of its annual data reports after water year 2009; however, the program continues to make its data available online through the CDWR

Water Data Library. Representative time series of discrete sample bromide and TOC data from Banks Pumping Plant are shown in Figure 3.

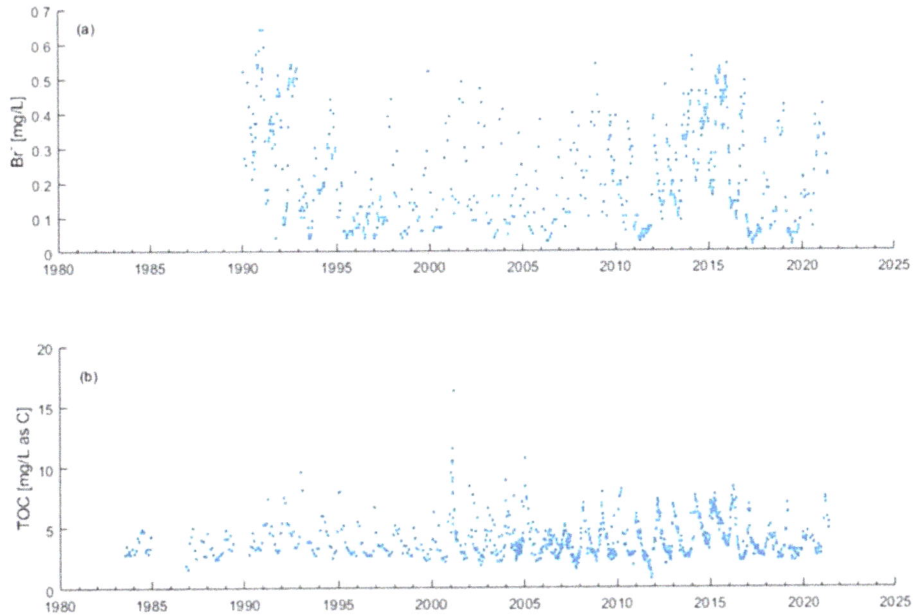

Figure 3. Representative data from the MWQI Program at the Banks Station: (**a**) Bromide (Br^-) and (**b**) TOC. Data are downloaded from the California Water Data Library (WDL Water Quality Data (ca.gov), accessed 21 July 2021).

The MWQI program was expanded during 2002–2006 with the initiation of real time data collection and water quality forecasting. The key elements of this expansion include construction and on-going management of real time monitoring sites, systematic production of water quality forecasts, and management and dissemination of information generated by the program expansion [24]. The objective of this program expansion was to enhance the ability of drinking water treatment plant operators to make informed operational decisions based on observed and forecasted changes in Delta water quality [117].

Five real time monitoring stations are operated and maintained by the MWQI program; four of the stations are located in the Delta and one is located south of the Delta (Figure 1). The Delta stations include Hood (located on the Sacramento River near the town of Hood upstream of most Delta influences—see Figure A4 in Appendix A), Vernalis (located on the San Joaquin River near the town of Vernalis upstream of most Delta influences), Banks Pumping Plant (located at the head of the SWP), and Jones Pumping Plant (located at the head of the Delta-Mendota Canal, which provides water to the CVP). The southern station, Gianelli, is located along the SWP at the Gianelli Pumping-Generating Plant on O'Neill Forebay below San Luis Reservoir. Data, collected from 46 field sensors at the five monitoring stations, measure several key water quality constituents, with organic carbon (both TOC and DOC) and bromide being the primary focus. Of the 46 sensors, 31 are operated by the MWQI program while the remaining sensors are operated by other entities such as the U.S. Geological Survey, the San Luis Delta Mendota Water Authority, and other CDWR programs. Table 1 summarizes MWQI real time station names, California Data Exchange Center (CDEC) identifications, and water quality parameters collected by the MWQI program and other entities; parameters collected by other entities are referred to as "non-program" parameters in the table.

Table 1. MWQI Real Time Stations and Parameters.

Station Name	CDEC ID	Installation Date	MWQI Program Parameters	Non-Program Parameters (Collected by Non-MWQI Programs)
Sacramento River at Hood	SRH	1999	TOC, DOC	Water: Chlorophyll, EC, Dissolved Oxygen, pH, Temperature, Turbidity Atmospheric: Solar Radiation, Temperature, Wind Speed and Direction
San Joaquin River at Vernalis	SJR	2005	TOC, DOC, Bromide, Chloride, Nitrate, Sulfate	Water: Chlorophyll, EC, Dissolved Oxygen, pH, Temperature, Turbidity Atmospheric: Solar Radiation, Temperature, Wind Speed and Direction
Banks Pumping Plant	HRO	2001	TOC, DOC, Bromide, Chloride, Nitrate, Sulfate, EC, Temperature, Dissolved Oxygen, pH, FDOM, Algal Fluorescence	Water: EC, pH, Temperature, Turbidity, Fluorescence, Pump Discharge Atmospheric: Temperature, Wind Speed and Direction
Jones Pumping Plant	TRP	2009	TOC, DOC, Bromide, Chloride, Nitrate, Sulfate, EC, Temperature, Dissolved Oxygen, pH, FDOM, Algal Fluorescence	Water: EC, Temperature, Pump Discharge
Gianelli	ONG	2012	TOC, DOC, Bromide, Chloride, Nitrate, Sulfate, EC, Temperature, Dissolved Oxygen, pH, FDOM, Algal Fluorescence	Water: Pump and Generation Discharge

Note: Abbreviations and definitions: EC = electrical conductivity (surrogate for salinity); TOC = total organic carbon; DOC = dissolved organic carbon; FDOM = Fluorescent Dissolved Organic Matter (refers to the fraction of organic matter that fluoresces). Algal fluorescence refers to the detection of cyanobacterial and algal cells via measurement of fluorescence from specific pigments.

Although the MWQI program discontinued measurement of UV-254 as a surrogate for organic carbon, efforts continue to find a useful real time surrogate. Recent efforts have focused on the use of fluorescence of dissolved organic matter (FDOM) as a real time surrogate for DOC; preliminary results appear promising [118]. The U.S. Geological Survey has been using FDOM sensors in the Delta since 2014 and has found it to be a strong predictor of DOC concentration [119,120]. Real time monitoring data, along with the daily water quality report, are posted on the SWC RTDF web page at http://rtdf.info/ (accessed on 8 October 2022). This web page provides links to CDWR's Water Data Library, CDEC, and the daily MWQI water quality report.

4.3.3. Modeling

Modeling during the early RTDF years focused on developing and enhancing DSM2 forecast simulation capabilities for salinity (specific conductance), bromide, and DOC in the Delta and the downstream aqueduct systems [106,121–123] and conducting tool validation [124]. Most of this work was conducted by CDWR's Delta Modeling Section. As modeling tools matured, CDWR's Division of Operations & Maintenance (O&M) became a program partner and began producing water quality retrospective simulations and 21-day forecasts on a routine basis [125]. The retrospective simulations also characterize Delta water quality through fingerprints that quantify different source contributions (e.g., Sacramento River inflow, San Joaquin River inflow, Delta agricultural drainage) at fixed locations [126,127]. The fingerprint technique has also been used to estimate the fraction of Sacramento and San Joaquin River inflows that are wastewater-derived [22]. The program has also routinely produced "seasonal" water quality forecasts for conditions several months into the future. However, based on several years of experience, the utility of these seasonal forecasts has been questioned (due to large uncertainty) and production of these forecasts was discontinued [128].

Participating water contractors commissioned the development of a DSM2 emulator that employed artificial neural network technology, a common machine learning technique [129]. This model was developed to provide an alternate, easy-to-use tool with greatly reduced run times. However, to date, the tool has not been adopted as part of the MWQI suite of forecasting tools.

Concerns about low alkalinity levels in exported Delta waters have been raised by participating MWQI agencies, prompting the program to develop capabilities to simulate and forecast bicarbonate fate and transport in the Delta. While moderately low alkalinity levels can improve the effectiveness of the coagulation process in water treatment plants, especially when aluminum sulfate is used as a coagulant, extremely low alkalinity levels can depress pH levels and can result in TOC re-stabilization [130]. These new forecasting capabilities were implemented in 2021.

Information from MWQI's routine monitoring and modeling efforts are disseminated in a variety of ways. Discrete and real time data are posted in CDWR's Water Data Library and in CDEC (https://cdec.water.ca.gov/ accessed on 8 October 2022), respectively. A daily water quality report, distributed electronically via email, contains a summary table of real time station data, links to water quality forecasts, and a brief synopsis of current events in and around the Delta. The summary table is updated daily and provides mean daily values, seven-day averages and the percent change over the seven days. Finally, monitoring data and model output, along with the daily water quality report, are posted on the SWC Real Time Data and Forecasting web page at http://rtdf.info/ (accessed on 8 October 2022). This web page provides links to CDEC and the daily MWQI water quality report. Figure 4 provides an example of observed and simulated data at Clifton Court Forebay (located upstream of Banks Pumping Plant) that are routinely updated as part of the MWQI program.

4.3.4. Targeted Science Studies

With an increasing emphasis on providing decision support for SWP operations since 2006, the MWQI program has commensurately placed a decreasing emphasis on targeted science studies in recent years. In spite of this shift in priorities, several important studies have been conducted by the MWQI program (some in collaboration with outside agencies and/or consultants) during the RTDF years; these are listed chronologically in Table A2 along with relevant citations to illustrate the breadth of study. The MWQI program additionally provided support to two key studies that examined wastewater discharges as sources of NDMA and NDMA precursors in the Delta [22,131].

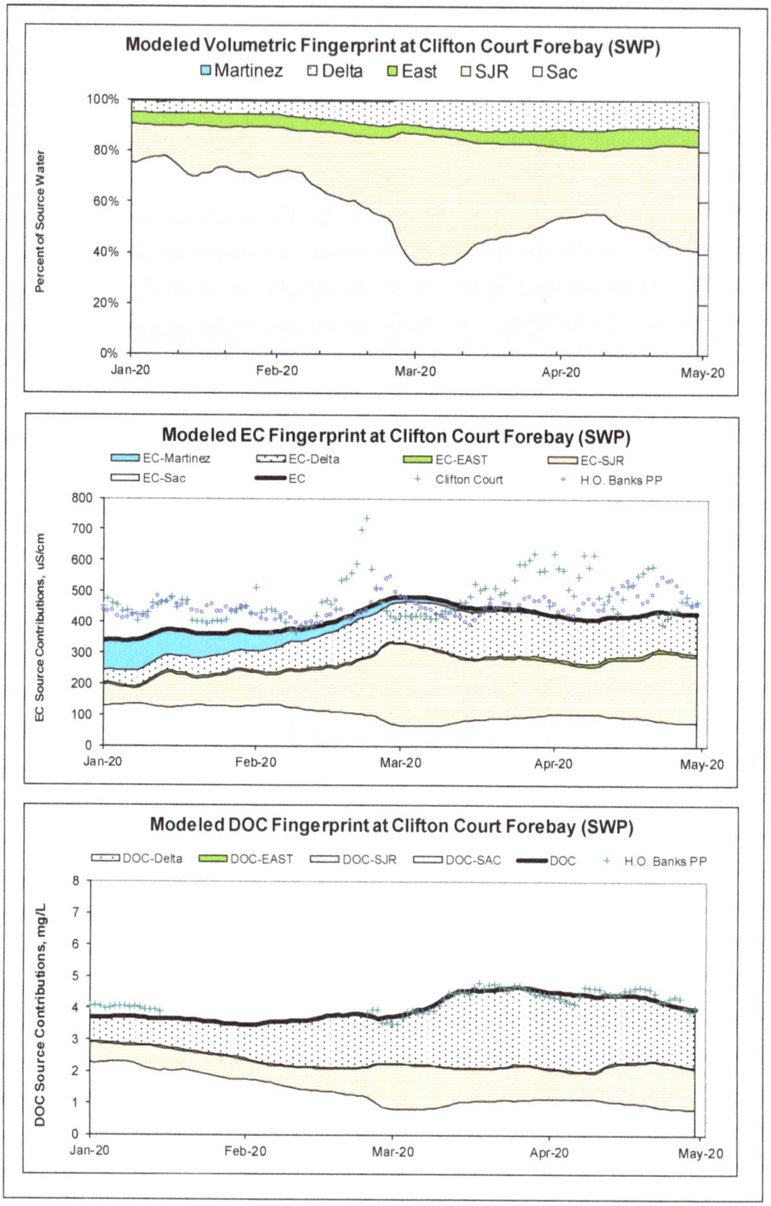

Figure 4. Representative fingerprint charts for Clifton Court Forebay using the DSM2 model. Sources tracked by the model include seawater intrusion (Martinez), in-Delta agricultural return flows (Delta), Cosumnes and Mokelumne Rivers (East), San Joaquin River at Vernalis (SJR), and Sacramento River at Hood (Sac). Symbols show the observed concentrations at these stations.

5. Program Accomplishments & Future Directions

CDWR's MWQI program is the most extensive and cohesive program established to investigate the quality of Delta source water with respect to its suitability for production of drinking water. The program has met this goal over its first three decades of existence by evolving in response to regulatory drivers, technology innovations, and an ever-increasing knowledge base developed through monitoring, targeted scientific studies, and computer modeling. In its successful quest to follow the recommendations set forth by an expert panel in the early 1980s [34], the MWQI program has accumulated several notable accomplishments:

- A monitoring program for constituents of human health significance was instituted, resulting in several long-term, open access discrete water quality data sets of value to scientists and engineers working in the Delta. The monitoring program has successfully balanced the competing needs of maintaining long-term data sets while allowing for on-going adjustments in response to new science and measurement technologies.
- A computer modeling framework for the transport of DBP precursors was developed and, along with data analysis, has been successfully employed to identify sources of key chemical constituents, determine how contaminants from each source are transported through the system, and determine how they affect concentrations at the points where water is diverted from the Delta and along the SWP.
- The MWQI program has provided key information for making decisions on how to manage the Delta and treat its waters to protect public health. For example, real time data on DBP precursors (i.e., DOC and bromide) have provided water agencies with information on how to best operate advanced water treatment plants. This information has supported participating water agencies in their decisions to adopt new water treatment technology; it has also supported the development and update of watershed sanitary surveys of the SWP.
- Real time data collection and water quality forecasting was initiated to enhance the ability of drinking water agencies to make informed operational decisions based on observed and forecasted changes in Delta water quality. Key elements of this operational support include construction and on-going management of five real time monitoring sites, systematic production of water quality forecasts, and management and dissemination of generated information. The MWQI program is believed to be the first in the United States to conduct continuous, real-time monitoring of organic carbon, bromide, and anions and to publish the data immediately on the internet.

The MWQI program, an on-going collaboration between CDWR and participating water agencies, has been designed and managed to be flexible and pro-active in order to address new drinking water quality challenges as they arise. The program's monitoring network, its frequency of data analysis and reporting, and its modeling tools—all of which have evolved over the past three decades—are collectively an effective "finger on the pulse" for Delta and SWP water quality that allow program participants to proactively manage risks by adjusting treatment processes, changing source water blends, or applying other operational tools to obtain optimal source water quality. Given the program's founding principles that drinking water quality concerns will continue to change and that it must adapt to address new challenges, future directions will likely be guided by a host of factors that may trigger changes in treatment plant processes and operations and Delta source water quality, including (but not limited to) identification of emerging contaminants, changes in land and water management practices, Delta island flooding, sea level rise, and climate change [132].

Author Contributions: P.H.H.; S.B.R.; S.W.K. and L.P. have contributed to the writing of this manuscript. All authors have read and agreed to the published version of the manuscript.

Funding: Funding for this work was provided by the State Water Contractors MWQI Specific Projects Committee.

Data Availability Statement: This work does not include primary data; all data referred to are available from their original publications or from the public websites that are cited.

Acknowledgments: The authors wish to acknowledge Bruce Macler (USEPA retired) for sharing his expertise related to the evolution of adopted water treatment technology and regulatory environment. We thank Richard Breuer (CSWRCB, retired), Carol DiGiorgio (CDWR, retired), Kevin Donhoff (MWDSC), Tony Liudzius (MWDSC), Cindy Messer (CDWR), and Shaun Philipart (CDWR) for their review of an early draft of the manuscript. We also thank Greg Gartrell (CCWD retired) for his review of an interim draft of the manuscript and his perspective on the early years of the MWQI program. Finally, we wish to thank members of the MWQI Specific Projects Committee for input on their current use of RTDF data.

Conflicts of Interest: The authors declare no conflict of interest.

Abbreviations

CDEC	California Data Exchange Center
CDWR	California Department of Water Resources
CSWRCB	California State Water Resources Control Board
CVP	Central Valley Project
DBP	Disinfection By-Product
DOC	Dissolved Organic Carbon
DSM2	CDWR's Delta Simulation Model 2
EC	Electrical Conductivity or Specific Conductance
FDOM	Fluorescence of Dissolved Organic Matter
GAC	Granular Activated Carbon
HAA	Haloacetic Acid
IDHAMP	Interagency Delta Health Aspects Monitoring Program
MCL	Maximum Contaminant Level
MWQI	Municipal Water Quality Investigations
NDMA	N-Nitrosodimethylamine
NOM	Natural Organic Matter
QA/QC	Quality Assurance/Quality Control
RTDF	Real Time Data and Forecasting
SWP	State Water Project
SWTR	Surface Water Treatment Rule
THM	Trihalomethane
TOC	Total Organic Carbon
USEPA	United States Environmental Protection Agency
UV-254	Ultraviolet Absorbance at 254 nm
WTP	Water Treatment Plant

Appendix A

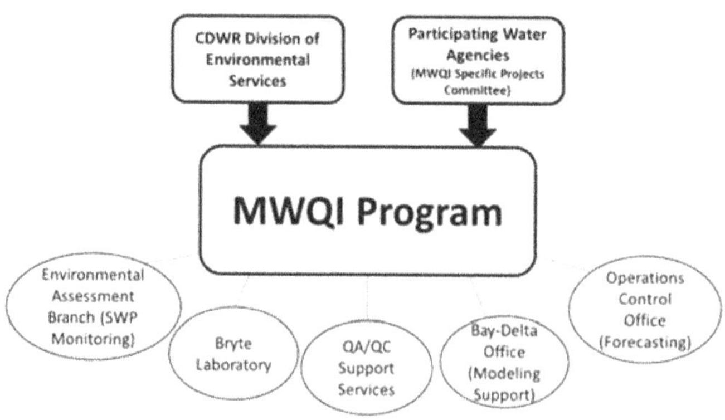

Figure A1. The MWQI program organization structure includes CDWR management, CDWR program participants, and participating urban water agencies.

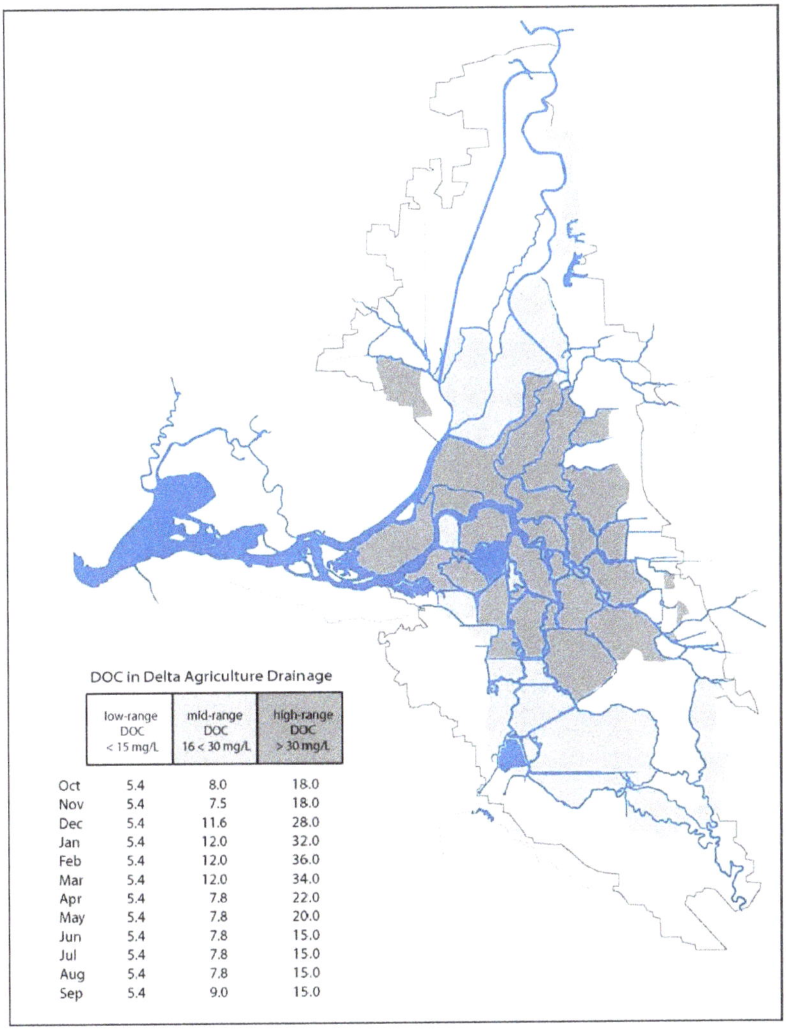

	low-range DOC < 15 mg/L	mid-range DOC 16 < 30 mg/L	high-range DOC > 30 mg/L
Oct	5.4	8.0	18.0
Nov	5.4	7.5	18.0
Dec	5.4	11.6	28.0
Jan	5.4	12.0	32.0
Feb	5.4	12.0	36.0
Mar	5.4	12.0	34.0
Apr	5.4	7.8	22.0
May	5.4	7.8	20.0
Jun	5.4	7.8	15.0
Jul	5.4	7.8	15.0
Aug	5.4	7.8	15.0
Sep	5.4	9.0	15.0

Figure A2. Delta island drainage DOC values used in DSM2 [133]. The islands are divided into three groups, each corresponding to specific ranges in DOC (low, medium, and high).

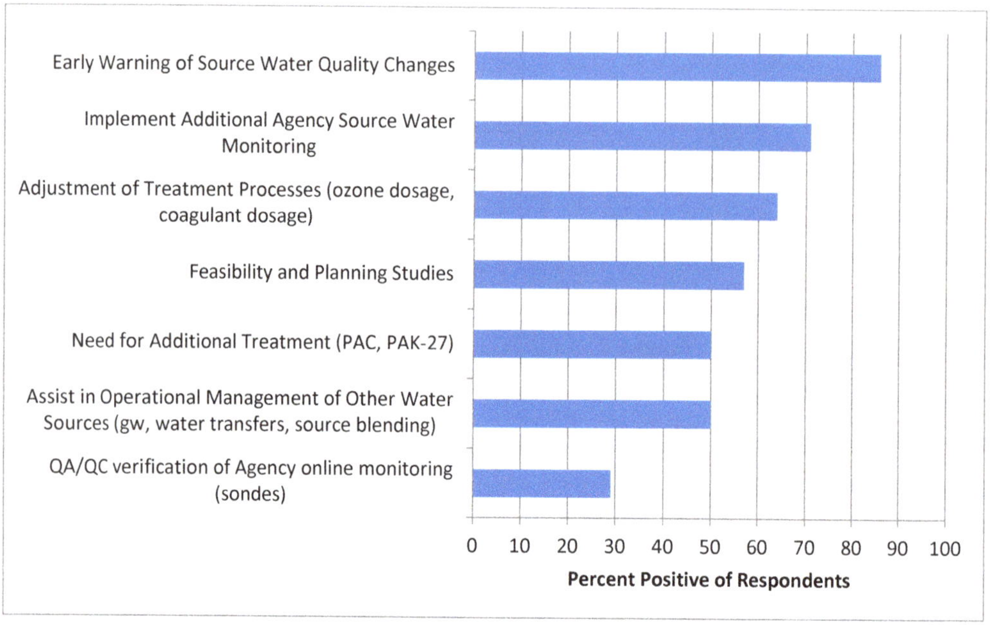

Figure A3. Survey results from 14 water contractors on how they use MWQI real time and forecast data in their regular operations.

Table A1. SWP Watershed Sanitary Surveys.

MWQI Era	Year	Focus	Reference(s)
Early Years	1990	This survey focused on reviewing available water quality data and providing an inventory of contaminant sources in the Sacramento, San Joaquin, and Tulare watersheds and along the aqueducts, with minimal focus on the contaminant sources in the SWP reservoir watersheds. The survey determined that the most significant water quality degradation in the SWP system occurs between the Sacramento River at Greene's Landing near Hood and the south Delta export facilities at Banks and Jones Pumping Plants (see Figure 1). The major sources of this degradation were identified as agricultural drainage from Delta islands, sea water intrusion, inflow from the San Joaquin River, and local discharges in the Stockton area and into Cache Slough.	[134]
CALFED Years	1996	This update focused on the recommendations from the 1990 survey and major changes in the watersheds between 1990 and 1996. The update also provided more details on contaminant sources in several reservoir watersheds (Del Valle, San Luis, Pyramid, Castaic, Silverwood, and Perris), the North Bay Aqueduct Barker Slough watershed, and the open canal section of the Coastal Branch of the California Aqueduct.	[135]
	2001	This update, which provided more details on contaminant sources in the watersheds of the SWP reservoirs and along the aqueducts, contained a detailed analysis of pathogen and indicator micro-organism data from the SWP. A major objective of the update was to provide information needed to comply with the California Department of Public Health's Drinking Water Source Assessment Program requirements.	[136]
RTDF Years	2006	This update, in addition to reviewing significant changes to the watersheds and their impacts on water quality, focused on the Jones Tract levee failure and emergency response procedures and efforts to coordinate pathogen monitoring in response to USEPA's Enhanced SWTR.	[137]
	2011	Similar to the 2006 update, this update concentrated on the key water quality issues that challenge the SWP contractors. This update discussed effects of the 2008 U.S. Fish and Wildlife Service and 2009 National Marine Fishery Service biological opinions, recent droughts and non-SWP aqueduct inflows on water quality, subsidence along the California aqueduct, and monitoring conducted to comply with the Enhanced SWTR. In addition, this update assessed long-term data trends at several locations in the Delta and along the aqueducts.	[138]
	2016	This update, in addition to evaluating key SWP water quality constituents, evaluated water quality impacts associated with grazing and the drought spanning 2012 to 2015. A separate report on contaminants in the San Joaquin River watershed was prepared (CDWR, 2015a).	[139]

Figure A4. Sacramento River at Hood Real-Time Monitoring Station.

Table A2. Cross Section of Applied Science Studies Conducted under the MWQI Program.

MWQI Era	Study Name	Reference
CALFED Years	Characterization of DOC from Delta Island Soils	[81,140]
	Coordinated Pathogen Monitoring Program for SWP	[81,82]
	Treatment of Delta Island Drainage to Reduce TOC Loads	[81]
	Delta Drainage, Surface Withdrawal, and Land Use Data	[141]
	Delta Island Drainage Volume Estimates 1954–1955 verses 1995–1996	[142]
	The North Bay Aqueduct Barker Slough Watershed Water Quality Phase 1 Report	[143]
	Candidate Delta Regions for Treatment to Reduce Organic Carbon Loads	[144]
	Environmental Study of Dredged Materials: Grant Line Canal	[145]
	Seasonal Water Quality Changes in Flooded Peat Soil Environments Due to Peat Soil, Water Depth, and Water Exchange Rate	[107]
	Water Quality Investigations of the Barker Slough Watershed, 1997–2001: North Bay Aqueduct Summary	[146]
	Natomas East Main Drainage Canal Water Quality Investigation	[147,148]
	Movement of Diuron and Hexazinone	[149]
RTDF Years	Staten Island Wildlife-Friendly Farming Demonstration—Water Quality Monitoring	[150]
	Steelhead Creek Water Quality Investigation	[151]
	Jones Tract Flood Water Quality Investigations	[152]
	Identifying Sources of DOC using Radiocarbon Age Dating	[153]
	Source, Fate, and Transport of Endocrine Disruptors, Pharmaceuticals, and Personal Care Products	[131]
	Travel Time and Longitudinal Dispersion Rates in the California State Water Project	[154]
	Lathrop Urban Runoff Study	[155]
	Delta Salinity Constituent Analysis	[58,59]
	Nitrosamine Precursors and Wastewater Indicators in Discharges in the Sacramento–San Joaquin Delta	[22]
	Limnology of the State Water Project Nutrient Budget Study: Nutrients at the Hood Water Quality Station	[156]
	FDOM Final Report: A Two-Year Comparison of Dissolved Organic Carbon to Fluorescence of Dissolved Organic Matter	[118]

References

1. McClusky, D.S.; Elliott, M. *The Estuarine Ecosystem: Ecology, Threats and Management*; OUP Oxford: Oxford, UK, 2004.
2. Adey, W.H.; Loveland, K. *Chapter 22: Estuaries Ecosystem Modeling and Restoration, in Dynamic Aquaria: Building and Restoring Living Ecosystems*; Elsevier: Amsterdam, The Netherlands, 2007.
3. Ross, D.A. *Introduction to Oceanography*; Harper Collins College Publishers: New York, NY, USA, 1995; ISBN 978-0-673-46938-0.
4. Delta Stewardship Council. The Delta Plan. 2013. Available online: https://deltacouncil.ca.gov/delta-plan/ (accessed on 1 November 2020).
5. California Department of Water Resources (CDWR). *Management of the California State Water Project, Bulletin*; CDWR: Sacramento, CA, USA, 2017; p. 132-16.
6. Lund, J.R.; Hanak, E.; Fleenor, W.E.; Bennett, W.A.; Howitt, R.E.; Mount, J.F.; Moyle, P.B. *Comparing futures for the Sacramento–San Joaquin Delta*; University of California Press: Berkeley, CA, USA, 2010.
7. California State Water Resources Control Board (CSWRCB). *Water Rights Decision*; CSWRCB: Sacramento, CA, USA, 2000.
8. Lund, J.; Hanak, E.; Fleenor, W.; Howitt, R.; Mount, J.; Moyle, P. *Envisioning Futures for the Sacramento–San Joaquin Delta*; Public Policy Institute of California: San Francisco, CA, USA, 2007; Available online: http://www.ppic.org/main/publication.asp?i=671 (accessed on 1 November 2020).
9. Gross, E.S.; Hutton, P.H.; Draper, A.J. A Comparison of Outflow and Salt Intrusion in the Pre-development and Contemporary San Francisco Estuary. *San Fr. Estuary Watershed Sci.* **2018**, *16*. [CrossRef]
10. California Department of Public Works (CDPW). Division of Water Resources, Bulletin 27. Variation and Control of Salinity in the Sacramento–San Joaquin Delta and Upper San Francisco Bay. 1931. Available online: http://www.water.ca.gov/waterdatalibrary/docs/historic/Bulletins/Bulletin_27/Bulletin_27__1931.pdf (accessed on 1 November 2020).
11. Edmonston, A.D.; Matthew, R.; Blackie, E.E. Bulletin No. 28, Economic Aspects of a Salt Water Barrier Below Confluence of Sacramento and San Joaquin Rivers. 1931. Available online: https://cawaterlibrary.net/document/bulletin-no-28-economic-aspects-of-a-salt-water-barrier-below-confluence-of-sacramento-and-san-joaquin-rivers/ (accessed on 9 September 2022).
12. California Department of Water Resources (DWR). Bulletin No. 3 The California Water Plan. 1957. Available online: https://cawaterlibrary.net/document/bulletin-no-3-the-california-water-plan/ (accessed on 9 September 2022).
13. Amy, G.L.; Thompson, J.M.; Tan, L.; Davis, M.K.; Krasner, S.W. Evaluation of THM Precursor Contributions from Agricultural Drains. *J. AWWA* **1990**, *82*, 57–64. [CrossRef]
14. Lange, A.L.; Kawczynski, E. Controlling Organics: The Contra Costa County Water District Experience. *J. AWWA* **1978**, *70*, 653–660. [CrossRef]
15. Krasner, S.W.; Schlimenti, M.J.; Means, E.G. Quality Degradation: Implications for DBP Formation. *J. AWWA* **1994**, *86*, 34–47. [CrossRef]
16. Water Education Foundation. *Layperson's Guide to the Delta*; Water Education Foundation: Sacramento, CA, USA, 2007.
17. Najm, I.N.; Krasner, S.W. Effects of Bromide and NOM on By-Product Formation. *J. AWWA* **1995**, *87*, 106. [CrossRef]
18. Wagner, E.D.; Plewa, M.J. CHO Cell Cytotoxicity and Genotoxicity Analyses of Disinfection By-Products: An Updated Review. *J. Environ. Sci.* **2017**, *58*, 64–76. [CrossRef]
19. Kuivila, K.M.; Barnett, H.D.; Edmunds, J.L. Herbicide concentrations in the Sacramento-San Joaquin Delta, California. US Geological Survey Toxic Substances Hydrology Program: Contamination of Hydrologic Systems and Related Ecosystems. 1999; p. 69. Available online: https://ca.water.usgs.gov/archive/reports/wrir994018/CA-0218.pdf (accessed on 9 September 2022).
20. Heidel, K.; Roy, S.; Creager, C.; Chung, C.; Grieb, T. *Conceptual Model for Nutrients in the Central Valley and Sacramento-San Joaquin Delta, Prepared for USEPA Region IX and Central Valley Drinking Water Policy Workgroup*; Tetra Tech: Pasadena, CA, USA, 2006.
21. Krasner, S.W.; Croue, J.; Buffle, J.; Perdue, E.M. Three Approaches for Characterizing NOM. *J. AWWA* **1996**, *88*, 66–79. [CrossRef]
22. Lee, C.T.; Krasner, S.W.; Sclimenti, M.J.; Prescott, M.; Guo, Y.C. Chapter 7: Nitrosamine Precursors and Wastewater Indicators in Discharges in the Sacramento-San Joaquin Delta. In *Recent Advances in Disinfection By-Products*; ACS Symposium Series; American Chemical Society: Washington, DC, USA, 2015. [CrossRef]
23. Schemel, L.E. *Salinity, Alkalinity, and Dissolved and Particulate Organic Carbon in the Sacramento River Water at Rio Vista, California, and at Other Locations in the Sacramento-San Joaquin Delta, 1980*; Water Resources Investigations Report: 83-4059; US Geological Survey: Sacramento, CA, USA, 1984.
24. Richardson, C.M.; Fackrell, J.K.; Kraus, T.E.; Young, M.B.; Paytan, A. Lateral Carbon Exports from Drained Peatlands: An Understudied Carbon Pathway in the Sacramento-San Joaquin Delta, California. *J. Geophys. Res. Biogeosciences* **2020**, *125*, e2020JG005883. [CrossRef]
25. CDWR. Municipal Water Quality Investigations Program History and Studies 1983–2012, Division of Environmental Services. 2013. Available online: www.rtdf.info (accessed on 1 November 2020).
26. CDWR. MWQI Work Plan: January–December 2020, Division of Environmental Services. 2019. Available online: www.rtdf.info (accessed on 9 September 2022).
27. United States. 86 Stat. 816—An Act to Amend the Federal Water Pollution Control Act. 1972. Available online: https://www.govinfo.gov/app/details/STATUTE-86/STATUTE-86-Pg816 (accessed on 9 September 2022).
28. United States. 88 Stat. 1660—Safe Drinking Water Act. 1974. Available online: https://www.govinfo.gov/app/details/STATUTE-88/STATUTE-88-Pg1660-2 (accessed on 1 November 2020).
29. Rook, J.J. Formation of haloforms during chlorination of natural waters. *Water Treat. Exam.* **1974**, *23*, 234–243.

30. Bellar, T.A.; Lichtenberg, J.J.; Kroner, R.C. The Occurrence of Organohalides in Chlorinated Drinking Waters. *J. AWWA* **1974**, *66*, 703–706. [CrossRef]
31. Villanueva, C.M.; Cantor, K.P.; Cordier, S.; Jaakkola, J.J.; King, W.D.; Lynch, C.F.; Porru, S.; Kogevinas, M. Disinfection byproducts and bladder cancer: A pooled analysis. *Epidemiology* **2004**, *15*, 357–367. [CrossRef] [PubMed]
32. U.S. Environmental Protection Agency (USEPA). Federal Register Volume 44, Issue 231 (29 November 1979). Available online: https://www.govinfo.gov/app/details/FR-1979-11-29 (accessed on 9 September 2022).
33. CDWR. State Water Project Trihalomethane Study, Central District. April 1982. Available online: www.rtdf.info (accessed on 1 November 2020).
34. Amy, G.L.; Tan, L.; Davis, M.K. The Effects of Ozonation and Activated Carbon Adsorption on Trihalomethane Speciation. *Water Res.* **1991**, *25*, 191. [CrossRef]
35. McCarty, P.; DeFalco, P.; Doull, J.; Greenberg, A.; Miller, B.J.; O'Connor, D.; Pearl, E. Public Health Aspects of Sacramento-San Joaquin Delta Water Supplies: A Panel Report for the California Department of Water Resources. 31 December 1982. Available online: www.rtdf.info (accessed on 1 November 2020).
36. CDWR. Interagency Delta Health Aspects Monitoring Program Project Report, Central District. May 1985. Available online: www.rtdf.info (accessed on 1 November 2020).
37. CDWR. Interagency Delta Health Aspects Monitoring Program Project Report, Central District. December 1986. Available online: www.rtdf.info (accessed on 1 November 2020).
38. CDWR. The Delta as a Source of Drinking Water: Monitoring Results 1983 to 1987, Interagency Delta Health Aspects Monitoring Program, Central District. August 1989. Available online: www.rtdf.info (accessed on 1 November 2020).
39. CDWR. Project Report of The Interagency Delta Health Aspects Monitoring Program (Municipal Water Quality Investigations Program): Summary of Monitoring Results (January 1988–December 1989), Division of Local Assistance. October 1990. Available online: www.rtdf.info (accessed on 9 September 2022).
40. Brown & Caldwell. Delta Drinking Water Quality Study, Technical Report Prepared for the California Urban Water Agencies. May 1989. Available online: https://static1.squarespace.com/static/5a565e93b07869c78112e2e5/t/5a9a0faa71c10b1a7df4a37c/1520046332824/DeltaDrinkingWaterQuality1989.pdf (accessed on 9 September 2022).
41. USEPA. Federal Register Volume 54, Issue 77 (24 April 1989). Available online: https://www.govinfo.gov/app/details/FR-1989-04-24 (accessed on 9 September 2022).
42. Robbins, R.W.; Glicker, J.L.; Bloem, D.M.; Niss, B.M. Effective Watershed Management for Surface Water Supplies. *J. AWWA* **1991**, *83*, 34–44. [CrossRef]
43. CDWR. Annual Report of The Municipal Water Quality Investigations Program: Summary of Monitoring Results January 1990-December 1990, Division of Local Assistance. February 1993. Available online: www.rtdf.info (accessed on 1 November 2020).
44. USEPA. National Primary Drinking Water Regulations: Disinfectants and Disinfection Byproducts. 1998. Available online: https://www.federalregister.gov/documents/1998/12/16/98-32887/national-primary-drinking-water-regulations-disinfectants-and-disinfection-byproducts (accessed on 9 September 2022).
45. USEPA. National Primary Drinking Water Regulations: Stage 2 Disinfectants and Disinfection Byproducts Rule. 1998. Available online: https://www.federalregister.gov/documents/2006/01/04/06-3/national-primary-drinking-water-regulations-stage-2-disinfectants-and-disinfection-byproducts-rule (accessed on 9 September 2022).
46. Jacangelo, J.G.; Patania, N.L.; Reagan, K.M.; Aieta, E.M.; Krasner, S.W.; McGuire, M.J. Ozonation: Assessing Its Role in the Formation and Control of Disinfection By-Products. *J. AWWA* **1989**, *81*, 74–84. [CrossRef]
47. McGuire, M.J.; Davis, M.K.; Tate, C.H.; Aieta, E.M.; Howe, E.W.; Crittenden, J.C. Evaluating GAC for Trihalomethane Control. *J. Am. Water Work. Assoc.* **1991**, *83*, 38–48. [CrossRef]
48. Krasner, S.W.; McGuire, M.J.; Jacangelo, J.G.; Patania, N.L.; Reagan, K.M.; Aieta, E.M. The Occurrence of Disinfection By-Products in U.S. Drinking Water. *J. AWWA* **1989**, *81*, 41. [CrossRef]
49. Diehl, A.C.; Speitel, G.E., Jr.; Symons, J.M.; Krasner, S.W.; Hwang, C.J.; Barrett, S.E. DBP Formation During Chloramination. *J. AWWA* **2000**, *92*, 76. [CrossRef]
50. State of California. Delta Water Quality: A Report to the Legislature on Trihalomethanes and the Quality of Drinking Water Available from the Sacramento-San Joaquin Delta. October 1991. Available online: https://link.sfpl.org/portal/Delta-water-quality--a-report-to-the-legislature/iD5C56mrWV4/ (accessed on 9 September 2022).
51. USEPA. Perfluoroalkyl Sulfonates; Significant New Use Rule. 2002. Available online: https://www.govinfo.gov/content/pkg/FR-2002-12-09/pdf/02-31011.pdf (accessed on 9 September 2022).
52. USEPA. National Primary Drinking Water Regulations: Stage 2 Disinfectants and Disinfection Byproducts Rule. 2006. Available online: https://www.federalregister.gov/documents/2006/01/04/06-3/national-primary-drinking-water-regulations-stage-2-disinfectants-and-disinfection-byproducts-rule (accessed on 9 September 2022).
53. CDWR. Five-Year Report of The Municipal Water Quality Investigations Program: Summary and Findings During Five Dry Years (January 1987–December 1991), Division of Local Assistance. November 1994. Available online: www.rtdf.info (accessed on 1 November 2020).
54. CDWR. 1992–1993 Biennial Memorandum Report of The Municipal Water Quality Investigations Program, Division of Local Assistance. April 1995. Available online: www.rtdf.info (accessed on 9 September 2022).

55. CDWR. Data Report of The Municipal Water Quality Investigations Program: Summary of Monitoring Results Water Year 1994, Division of Local Assistance. July 1995. Available online: www.rtdf.info (accessed on 1 November 2020).
56. CDWR. Municipal Water Quality Investigations Program Annual Report Water Year 1995, Division of Local Assistance. August 1996. Available online: www.rtdf.info (accessed on 9 September 2022).
57. Study of Drinking Water Quality in Delta Tributaries. 1995. Available online: https://www.cuwa.org/publications-archive/r9wypzpbmbrbhcxtsarfwrg7p3msrk (accessed on 9 September 2022).
58. Denton, R.A. Delta Salinity Constituent Analysis, Prepared for the State Water Project Contractors Authority. February 2015. Available online: https://www.baydeltalive.com/assets/588ee18bdb51ef1619ac6fd28b97f694/application/pdf/Denton_2015_Delta_Salinity_Constituents_Report.pdf (accessed on 9 September 2022).
59. Hutton, P.H.; Sinha, A.; Roy, S.B. Simplified Approach for Estimating Salinity Constituent Concentrations in the San Francisco Estuary & Sacramento-San Joaquin River Delta: A User Guide, Report Prepared for the State Water Contractors. July 2022. Available online: https://rtdf.info/ (accessed on 1 November 2020).
60. Krasner, S.W.; Sclimenti, M.J. Characterization of Natural Organic Matter: Disinfection By-Product Analysis. In Proceedings of the Proc. Workshop on Natural Organic Matter in Drinking Water: Origin, Characterization, and Removal (AWWARF & Lyonnaise des Eaux-Dumez), Chamonix, France, 1 January 1993; pp. 105–113.
61. Hutton, P.H.; Chung, F.I. Bromine Distribution Factors in THM Formation. *J. Water Resour. Plan. Manag. ASCE* **1994**, *120*, 1–16. [CrossRef]
62. Koch, B.; Krasner, S.W.; Sclimenti, M.J.; Schimpff, W.K. Predicting the Formation of DBPs by the Simulated Distribution System. *J. AWWA* **1991**, *83*, 62–70. [CrossRef]
63. Hutton, P.H.; Chung, F.I. Correlating Trihalomethane Data. *J. Environ. Eng. ASCE* **1994**, *120*, 219–241. [CrossRef]
64. CDWR. Representative Delta Island Return Flow Quality for Use in DSM2. Division of Planning, Modeling Support Branch. May 1995. Available online: https://www.sfei.org/sites/default/files/biblio_files/Appendix%206%20Updating%20DSM2%20nutrient%20calibration.pdf (accessed on 9 September 2022).
65. CDWR. Trihalomethane Formation Potential in the Sacramento-San Joaquin Delta: Mathematical Model Development, Division of Planning. October 1991.
66. CDWR. Chapter 3: Trihalomethane Formation Potential Modeling, in Methodology for Flow and Salinity Estimates in the Sacramento-San Joaquin Delta and Suisun Marsh, 12th Annual Progress Report. April 1991. Available online: https://water.ca.gov/Library/Modeling-and-Analysis/Bay-Delta-Region-models-and-tools/Delta-Simulation-Model-II (accessed on 1 November 2020).
67. CDWR. Chapter 9: Trihalomethane Formation Potential Modeling, in Methodology for Flow and Salinity Estimates in the Sacramento-San Joaquin Delta and Suisun Marsh, 13th Annual Progress Report. June 1992. Available online: https://water.ca.gov/Library/Modeling-and-Analysis/Bay-Delta-Region-models-and-tools/Delta-Simulation-Model-II (accessed on 1 November 2020).
68. CDWR. Chapter 7: Trihalomethane Formation Potential Modeling, in Methodology for Flow and Salinity Estimates in the Sacramento-San Joaquin Delta and Suisun Marsh, 14th Annual Progress Report. June 1993. Available online: https://water.ca.gov/Library/Modeling-and-Analysis/Bay-Delta-Region-models-and-tools/Delta-Simulation-Model-II (accessed on 1 November 2020).
69. CDWR. Chapter 5: Disinfection By-Product Formation Modeling, in Methodology for Flow and Salinity Estimates in the Sacramento-San Joaquin Delta and Suisun Marsh, 15th Annual Progress Report. June 1994. Available online: https://water.ca.gov/Library/Modeling-and-Analysis/Bay-Delta-Region-models-and-tools/Delta-Simulation-Model-II (accessed on 1 November 2020).
70. CDWR. Chapter 8: Disinfection By-Product Formation, in Methodology for Flow and Salinity Estimates in the Sacramento-San Joaquin Delta and Suisun Marsh, 16th Annual Progress Report. June 1995. Available online: https://water.ca.gov/Library/Modeling-and-Analysis/Bay-Delta-Region-models-and-tools/Delta-Simulation-Model-II (accessed on 1 November 2020).
71. CDWR. Chapter 3: Impact of Delta Islands on Water Quality, in Methodology for Flow and Salinity Estimates in the Sacramento-San Joaquin Delta and Suisun Marsh, 17th Annual Progress Report. June 1996. Available online: https://water.ca.gov/Library/Modeling-and-Analysis/Bay-Delta-Region-models-and-tools/Delta-Simulation-Model-II (accessed on 1 November 2020).
72. Hutton, P.H.; Chung, F.I. Simulating THM Formation Potential in Sacramento Delta: Part I. *J. Water Resour. Plan. Manag. ASCE* **1992**, *118*, 513–529. [CrossRef]
73. Hutton, P.H.; Chung, F.I. Simulating THM Formation Potential in Sacramento Delta: Part II. *J. Water Resour. Plan. Manag. ASCE* **1992**, *118*, 530–542. [CrossRef]
74. Hutton, P.H.; Enright, C. Simulating THM Precursors Transport with DWRDSM. In Proceedings of the Hydraulics Engineering '93: Proceedings of the 1993 Conference, San Francisco, CA, USA, 25–30 July 1993; pp. 821–826.
75. Hutton, P.H.; Sandhu, N.; Chung, F.I. Predicting THM Formation with Artificial Neural Networks. In Proceedings of the 1996 North American Water and Environment Congress, ASCE, Anaheim, CA, USA, 22–28 June 1996.
76. Hutton, P.H.; Mahadevan, N.; Chung, F.I. Simulating DBP Precursor Transport in Sacramento Delta. In Proceedings of the Proceedings 1996 North American Water and Environment Congress, ASCE, Anaheim, CA, USA, 22–28 June 1996.
77. CALFED Bay-Delta Program, 2000. Programmatic Record of Decision. Available online: https://www.ccwater.com/DocumentCenter/View/3213/CALFED-Programmatic-ROD--Aug-28-2000?bidId= (accessed on 1 November 2021).

78. CDWR. The Municipal Water Quality Investigations Program Summary and Findings of Data Collected from the Sacramento-San Joaquin Delta Region, October 2005 through September 2007, Division of Environmental Services. June 2008. Available online: www.rtdf.info (accessed on 1 November 2020).
79. CDWR. The Municipal Water Quality Investigations Program Summary and Findings from Data Collected August 1998 through September 2001, Division of Environmental Services. July 2003. Available online: www.rtdf.info (accessed on 1 November 2020).
80. CDWR. The Municipal Water Quality Investigations Program Summary and Findings from Data Collected October 2001 through September 2003, Division of Environmental Services. June 2005. Available online: www.rtdf.info (accessed on 1 November 2020).
81. CDWR. Municipal Water Quality Investigations Program Annual Report October 1995-December 1996, Division of Planning and Local Assistance. December 1997. Available online: www.rtdf.info (accessed on 1 November 2020).
82. CDWR. Municipal Water Quality Investigations Program Annual Report 1997–1998, Division of Planning and Local Assistance. October 2000. Available online: www.rtdf.info (accessed on 1 November 2020).
83. CDWR. The Municipal Water Quality Investigations Program Summary and Findings of Data Collected from the Sacramento-San Joaquin Delta Region, October 2003 through September 2005, Division of Environmental Services. December 2006. Available online: www.rtdf.info (accessed on 1 November 2020).
84. CDWR. Real-Time Monitoring of Organic Carbon in the Sacramento River and California State Water Project Using Process Analyzers, Division of Planning & Local Assistance. May 2001. Available online: www.rtdf.info (accessed on 1 November 2020).
85. Ngatia, M.; Pimental, J. Comparisons of Organic Carbon Analyzers and Related Importance to Water Quality Assessments. *San Fr. Estuary Watershed Sci.* **2007**, *5*. [CrossRef]
86. CDWR. Real-Time, Continuous Monitoring of Bromide and Nutrients at Harvey O. Banks Pumping Plant and San Joaquin River near Vernalis, Division of Environmental Services. July 2008. Available online: www.rtdf.info (accessed on 1 November 2020).
87. Sickman, J.; OGonzalez, D.; San Julian, S. Real-time monitoring of organic carbon in the Sacramento River and California State Water Project Using Process Analyzers, prepared for CDWR MWQI program. 11 April 2005. Available online: www.rtdf.info (accessed on 1 November 2020).
88. Chow, A.T.; Guo, F.; Gao, S.; Breuer, R.S. Trihalomethane Formation Potential of Filter Isolates of Electrolyte-Extractable Soil Organic Carbon. *J. Environ. Qual.* **2005**, *34*, 1992–1997. [CrossRef] [PubMed]
89. Chow, A.T.; Guo, F.; Gao, S.; Breuer, R.; Dahlgren, R.A. Filter pore size selection for characterizing dissolved organic carbon and trihalomethane precursors from soils. *Water Res.* **2005**, *39*, 1255–1264. [CrossRef]
90. Chow, A.T.; Guo, F.; Gao, S.; Breuer, R.S. Trihalomethane Reactivity of Water- and Sodium Hydroxide–Extractable Organic Carbon Fractions from Peat Soils. *J. Environ. Qual.* **2006**, *35*, 114–121. [CrossRef]
91. Chow, A.T.; Guo, F.; Gao, S.; Breuer, R.S. Size and XAD fractions of trihalomethane precursors from soils. *Chemosphere* **2006**, *62*, 1636–1646. [CrossRef]
92. DiGiorgio, C.L.; Gonzalez, D.A.; Huitt, C.C. *Cryptosporidium* and *Giardia* Recoveries in Natural Waters by Using Environmental Protection Agency Method 1623. *Appl. Environ. Microbiol.* **2002**, *68*, 5952–5955. [CrossRef]
93. Woodard, R. *Sources and Magnitudes of Water Quality Constituents of Concern in Drinking Water Supplies Taken from the Sacramento-San Joaquin Delta*; Technical Report; CALFED Bay-Delta Program: West Sacramento, CA, USA, 2000.
94. Jung, M. Revision of Representative Delta Island Return Flow Quality for DSM2 and DICU Model Runs, prepared for the CALFED Ad-Hoc Workgroup to Simulate Historical Water Quality Conditions in the Delta, Consultant's Report to the CDWR MWQI Program. December 2000. Available online: www.rtdf.info (accessed on 1 November 2020).
95. CDWR. DSM2: Delta Simulation Model II. 2020. Available online: https://water.ca.gov/Library/Modeling-and-Analysis/Bay-Delta-Region-models-and-tools/Delta-Simulation-Model-II (accessed on 1 November 2020).
96. CDWR. Chapter 3: Simulation of Historical DOC and UVA Conditions in the Delta, in Methodology for Flow and Salinity Estimates in the Sacramento-San Joaquin Delta and Suisun Marsh, 22nd Annual Progress Report. August 2001. Available online: https://water.ca.gov/Library/Modeling-and-Analysis/Bay-Delta-Region-models-and-tools/Delta-Simulation-Model-II (accessed on 1 November 2020).
97. CDWR. Chapter 7: Generating Monthly Dissolved Organic Carbon and UVA at DSM2 Boundaries, in Methodology for Flow and Salinity Estimates in the Sacramento-San Joaquin Delta and Suisun Marsh, 23rd Annual Progress Report. June 2002. Available online: https://water.ca.gov/Library/Modeling-and-Analysis/Bay-Delta-Region-models-and-tools/Delta-Simulation-Model-II (accessed on 1 November 2020).
98. CDWR. Chapter 9: DOC Growth in QUAL, in Methodology for Flow and Salinity Estimates in the Sacramento-San Joaquin Delta and Suisun Marsh, 23rd Annual Progress Report. June 2002. Available online: https://water.ca.gov/Library/Modeling-and-Analysis/Bay-Delta-Region-models-and-tools/Delta-Simulation-Model-II (accessed on 1 November 2020).
99. CDWR. Chapter 7: Implementation of a New DOC Growth Algorithm in DSM2-QUAL, in Methodology for Flow and Salinity Estimates in the Sacramento-San Joaquin Delta and Suisun Marsh, 24th Annual Progress Report. August 2003. Available online: https://water.ca.gov/Library/Modeling-and-Analysis/Bay-Delta-Region-models-and-tools/Delta-Simulation-Model-II (accessed on 1 November 2020).
100. CALFED Bay-Delta Program. In-Delta Storage Program, Draft Summary Report. May 2002. Available online: https://cawaterlibrary.net/wp-content/uploads/2017/05/CALFED-Staff-review-of-Delta-Conveyance-Reports.pdf (accessed on 9 September 2022).

101. CDWR. Chapter 3: Jones Tract 2004 Levee Break DSM2 Simulation, in Methodology for Flow and Salinity Estimates in the Sacramento-San Joaquin Delta and Suisun Marsh, 26th Annual Progress Report. October 2005. Available online: https://water.ca.gov/Library/Modeling-and-Analysis/Bay-Delta-Region-models-and-tools/Delta-Simulation-Model-II (accessed on 1 November 2020).
102. Liudzius, T. California Aqueduct Water Quality Modeling Results for Proof of Concept Water Quality Forecast, MWDSC Memorandum Report. 1 December 2003. Available online: www.rtdf.info (accessed on 1 November 2020).
103. Hutton, P. Seasonal Water Quality Forecast Proof of Concept, MWDSC Memorandum to MWQI RTDF Steering Committee. 22 January 2004. Available online: www.rtdf.info (accessed on 1 November 2020).
104. CDWR. Chapter 8: Real-Time Data and Forecasting Proof of Concept and Development, in Methodology for Flow and Salinity Estimates in the Sacramento-San Joaquin Delta and Suisun Marsh, 25th Annual Progress Report. October 2004. Available online: https://water.ca.gov/Library/Modeling-and-Analysis/Bay-Delta-Region-models-and-tools/Delta-Simulation-Model-II (accessed on 1 November 2020).
105. CH2MHill. DSM2 Extension for the California Aqueduct, South Bay Aqueduct, and Delta-Mendota Canal Phase 2, prepared for State Water Contractors and MWQI Program. June 2005. Available online: www.rtdf.info (accessed on 1 November 2020).
106. CH2MHill. DSM2 Extension for the California Aqueduct, South Bay Aqueduct, and Delta-Mendota Canal Phase 2 Analysis, prepared for State Water Contractors and CDWR' MWQI Program. June 2008. Available online: www.rtdf.info (accessed on 1 November 2020).
107. Jung, M.; Weisser, L. Seasonal Water Quality Changes in Flooded Peat Soil Environments Due to Peat Soil, Water Depth, and Water Exchange Rate, prepared for CDWR Division of Planning & Local Assistance. December 2000. Available online: www.rtdf.info (accessed on 9 September 2022).
108. Breuer, R. *Real Time Data and Forecasting Project Water Quality Weekly Report*; Office of Water Quality, Department of Water Resources: Sacramento, CA, USA, 2004; Volume 1.
109. CDWR. Management of the California State Water Project; Bulletin 132-13. April 2015. Available online: https://ostromworkshop.indiana.edu/pdf/seriespapers/2021spr-colloq/chaudhry.pdf (accessed on 9 September 2022).
110. CDWR. Management of the California State Water Project, Bulletin 132-14. November 2015. Available online: https://water.ca.gov/-/media/DWR-Website/Web-Pages/Programs/State-Water-Project/Management/Bulletin-132/Bulletin-132/Files/Bulletin-132-14-r.pdf (accessed on 9 September 2022).
111. CDWR. Management of the California State Water Project, Bulletin 132-15. July 2016. Available online: https://water.ca.gov/programs/state-water-project/management/bulletin-132 (accessed on 9 September 2022).
112. CDWR. Water Quality Assessment of Non-Project Turn-ins to the California Aqueduct, 2015; Division of Operations & Maintenance. December 2016. Available online: https://www.scwa2.com/wp-content/uploads/2020/01/N21e.2017-SWP-Sanitary-Survey-Update_reduced_version-ID-253435.pdf (accessed on 9 September 2022).
113. USEPA. Contaminant Candidate List 3 (CCL3). 2009. Available online: http://water.epa.gov/scitech/drinkingwater/dws/ccl/ccl3.cfm (accessed on 1 November 2020).
114. Cowman, G.A.; Singer, P.C. Effect of Bromide Ion on Haloacetic Acid Speciation Resulting from Chlorination and Chloramination of Aquatic Humic Substances. *Environ. Sci. Technol.* **1996**, *30*, 16. [CrossRef]
115. Shen, R.; Andrews, S.A. Demonstration of 20 pharmaceuticals and personal care products (PPCPs) as nitrosamine precursors during chloramine disinfection. *Water Res.* **2011**, *45*, 944–952. [CrossRef]
116. CDWR. The Municipal Water Quality Investigations Program Summary and Findings of Data Collected from the Sacramento-San Joaquin Delta Region, October 2007—September 2009, Division of Environmental Services. June 2010. Available online: www.rtdf.info (accessed on 1 November 2020).
117. Bettencourt, M. Get the Data Out: Real Time Data Dissemination from Isolated Monitoring Stations within the Delta. In Proceedings of the AWWA Cal-Nevada Section 2013 Fall Conference, San Diego, CA, USA, 1 October 2013.
118. Del Cid, J. *FDOM Final Report: A Two-Year Comparison of Dissolved Organic Carbon to Fluorescence of Dissolved Organic Matter*; Unpublished Report. MWQI, 2019. Available online: https://rtdf.info/public_docs/Miscellaneous%20RTDF%20Web%20Page%20Information/Other%20MWQP%20and%20DWR%20Publications/FDOM%20Final%20Report_pdf.pdf (accessed on 9 September 2022).
119. Downing, B.D.; Pellerin, B.A.; Bergamaschi, B.A.; Saraceno, J.F.; Kraus, T.E.C. Seeing the light: The effects of particles, dissolved materials, and temperature on in situ measurements of DOM fluorescence in rivers and streams: Effects and compensation for in situ DOM fluorescence, Limnology and Oceanography. *Limnol. Oceanogr. Methods* **2012**, *10*, 767–775. [CrossRef]
120. Downing, B.D.; Bergamaschi, B.A.; Kraus, T.E.C. *Synthesis of Data from High-Frequency Nutrient and Associated Biogeochemical Monitoring for the Sacramento–San Joaquin Delta, Northern California*; Scientific Investigations Report; US Geological Survey: Sacramento, CA, USA, 2017; pp. 2017–5066. 28 p. [CrossRef]
121. CDWR. Chapter 2: Data Quality Associated with EC Values in the DICU Model, in Methodology for Flow and Salinity Estimates in the Sacramento-San Joaquin Delta and Suisun Marsh, 30th Annual Progress Report. June 2009. Available online: https://water.ca.gov/Library/Modeling-and-Analysis/Bay-Delta-Region-models-and-tools/Delta-Simulation-Model-II (accessed on 1 November 2020).

122. CDWR. Chapter 3: DSM2 Dissolved Organic Carbon Boundary Condition Improvement, in Methodology for Flow and Salinity Estimates in the Sacramento-San Joaquin Delta and Suisun Marsh, 32nd Annual Progress Report. August 2011. Available online: https://water.ca.gov/Library/Modeling-and-Analysis/Bay-Delta-Region-models-and-tools/Delta-Simulation-Model-II (accessed on 1 November 2020).
123. CDWR. Chapter 2: Extension of DSM2 for the South Bay and California Aqueducts and Delta Mendota Canal, in Methodology for Flow and Salinity Estimates in the Sacramento-San Joaquin Delta and Suisun Marsh, 34th Annual Progress Report. June 2013. Available online: https://water.ca.gov/Library/Modeling-and-Analysis/Bay-Delta-Region-models-and-tools/Delta-Simulation-Model-II (accessed on 1 November 2020).
124. CDWR. Chapter 9: DOC Validation with DSM2, in Methodology for Flow and Salinity Estimates in the Sacramento-San Joaquin Delta and Suisun Marsh, 32nd Annual Progress Report. August 2011. Available online: https://water.ca.gov/Library/Modeling-and-Analysis/Bay-Delta-Region-models-and-tools/Delta-Simulation-Model-II (accessed on 1 November 2020).
125. Uecker, I. Development of a Multiple Linear Regression Forecast Model for Dissolved Organic Carbon DSM2 Boundary Conditions, internal CDWR memorandum. 2020. Available online: https://link.springer.com/article/10.1007/s40899-020-00377-9 (accessed on 9 September 2022).
126. CDWR. Chapter 14: DSM2 Fingerprinting Methodology, in Methodology for Flow and Salinity Estimates in the Sacramento-San Joaquin Delta and Suisun Marsh, 23rd Annual Progress Report. June 2002. Available online: https://water.ca.gov/Library/Modeling-and-Analysis/Bay-Delta-Region-models-and-tools/Delta-Simulation-Model-II (accessed on 1 November 2020).
127. CDWR. Chapter 6: Fingerprinting: Clarifications and Recent Applications, in Methodology for Flow and Salinity Estimates in the Sacramento-San Joaquin Delta and Suisun Marsh, 26th Annual Progress Report. October 2005. Available online: https://water.ca.gov/Library/Modeling-and-Analysis/Bay-Delta-Region-models-and-tools/Delta-Simulation-Model-II (accessed on 1 November 2020).
128. Hutton, P.; Chen, L.; Roy, S. *Assessment of MWQI Seasonal Water Quality Forecasts*; Final Report; MWQI Specific Project Committee: West Sacramento, CA, USA, 2020.
129. Chen, L.; Roy, S.B.; Hutton, P.H. Emulation of a Process-Based Estuarine Hydrodynamic Model. *Hydrol. Sci. J.* **2018**, *63*, 783–802. [CrossRef]
130. Krasner, S.W.; Amy, G.L. Jar-Test Evaluations of Enhanced Coagulation. *J. AWWA* **1995**, *87*, 93–107. [CrossRef]
131. Guo, Y.C.; Krasner, S.W.; Fitzsimmons, S.; Woodside, G.; Yamachika, N. *Source, Fate, and Transport of Endocrine Disruptors, Pharmaceuticals, and Personal Care Products in Drinking Water Sources in California—Final Project Report*; Water Research Institute: Fountain Valley, CA, USA, 2010.
132. Chen, W.; Haunschild, K.; Lund, J.R.; Fleenor, W.E. Current and Long-Term Effects of Delta Water Quality on Drinking Water Treatment Costs from Disinfection Byproduct Formation. *San Fr. Estuary Watershed Sci.* **2010**, *8*. [CrossRef]
133. Suits, B. DSM2 DOC Modeling Results for MWQI's Study of DOC in Agricultural Drainage Contribution to DOC at Delta Exports, Internal CDWR memorandum. 2003. Available online: https://water.ca.gov/Library/Modeling-and-Analysis/Bay-Delta-Region-models-and-tools/Delta-Simulation-Model-II (accessed on 9 September 2022).
134. Brown & Caldwell. *Sanitary Survey of the State Water Project*; State Water Contractors: Sacramento, CA, USA, 1990.
135. CDWR. Sanitary Survey Update Report 1996, prepared for the State Water Contractors. May 1996. Available online: www.rtdf.info (accessed on 9 September 2022).
136. CDWR. Sanitary Survey Update Report 2001, Prepared for the State Water Contractors. December 2001. Available online: www.rtdf.info (accessed on 9 September 2022).
137. Archibald Consulting, Richard Woodard Water Quality Consultants, and Palencia Consulting Engineers, California State Water Project Watershed Sanitary Survey 2006 Update, Prepared for the SWP Contractors Authority. June 2007. Available online: www.rtdf.info (accessed on 9 September 2022).
138. Archibald Consulting, Palencia Consulting Engineers, and Starr Consulting. California State Water Project Watershed Sanitary Survey 2011 Update, prepared for the SWP Contractors Authority and CDWR. June 2012. Available online: www.rtdf.info (accessed on 1 November 2020).
139. Palencia Consulting Engineers and Starr Consulting. California State Water Project 2016 Watershed Sanitary Survey Update, prepared for the State Water Project Contractors Authority. June 2017. Available online: www.rtdf.info (accessed on 1 November 2020).
140. Fujii, R.; Ranalli, J.A.; Aiken, G.R.; Bergamaschi, B. *Dissolved Organic Carbon Concentrations and Compositions, and Trihalomethane Formation Potentials in Waters from Agricultural Peat Soils, Sacramento-San Joaquin Delta, California: Implications for Drinking-Water Quality*; US Geological Survey: Sacramento, CA, USA, 1998.
141. Templin, W.E.; Cherry, D.E. *Drainage-Return, Surface-Water Withdrawal, and Land-Use Data for the Sacramento-San Joaquin Delta, with Emphasis on Twitchell Island, California*; Open-File Report 97-350; US Geological Survey: Sacramento, CA, USA, 1997.
142. Jung, M.; Tran, Q. Delta Island Drainage Volume Estimates 1954–1955 Verses 1995–1996, Prepared for DWR Division of Planning & Local Assistance. January 1998. Available online: www.rtdf.info (accessed on 1 November 2020).
143. CDWR. The North Bay Aqueduct Barker Slough Watershed Water Quality Phase 1 Report, Division of Planning & Local Assistance, Division of O&M. July 1998. Available online: www.rtdf.info (accessed on 1 November 2020).
144. Jung, M.; Tran, Q. Candidate Delta Regions for Treatment to Reduce Organic Carbon Loads, prepared for DWR's Division of Planning & Local Assistance. January 1999. Available online: www.rtdf.info (accessed on 1 November 2020).

145. CDWR. Environmental Study of Dredged Materials: Grant Line Canal, Office of SWP Planning, Division of Planning and Local Assistance. January 2000. Available online: www.rtdf.info (accessed on 1 November 2020).
146. CDWR. Water Quality Investigations of the Barker Slough Watershed, 1997-2001: North Bay Aqueduct Summary, Division of Environmental Services. May 2002. Available online: www.rtdf.info (accessed on 1 November 2020).
147. CDWR. Natomas East Main Drainage Canal Water Quality Investigation: Initial Technical Report, prepared for the State Water Contractors, Division of Environmental Services. November 2003. Available online: www.rtdf.info (accessed on 1 November 2020).
148. Sickman, J.O.; Zanoli, M.J.; Mann, H.L. Effects of Urbanization on Organic Carbon Loads in the Sacramento River, California. *Water Resour. Res.* **2007**, *43*, W11422. [CrossRef]
149. Prichard, T.; Troiano, J.; Marade, J.; Guo, F.; Canevari, M. Movement of Diuron and Hexazinone in Clay Soil and Infiltrated Pond Water. *J. Environ. Qual.* **2005**, *34*, 2005–2017. [CrossRef]
150. CDWR. Staten Island Wildlife-Friendly Farming Demonstration—Water Quality Monitoring: Final Report, prepared under CALFED Grant #ERP-02-P08, MWQI Unit, 18 June 2007. Available online: www.rtdf.info (accessed on 1 November 2020).
151. CDWR. Steelhead Creek Water Quality Investigation: Final Technical Report, Division of Environmental Services. February 2008. Available online: www.rtdf.info (accessed on 1 November 2020).
152. CDWR. Jones Tract Flood Water Quality Investigations, Division of Environmental Services. July 2009. Available online: www.rtdf.info (accessed on 1 November 2020).
153. Sickman, J.O.; DiGiorgio, C.L.; Davisson, M.L.; Lucero, D.M.; Bergamaschi, B. Identifying sources of dissolved organic carbon in agriculturally dominated rivers using radiocarbon age dating: Sacramento-San Joaquin River Basin, California. *Biodegradation* **2009**, *99*, 79–96. [CrossRef]
154. CDWR. Travel Time and Longitudinal Dispersion Rates in the California State Water Project, Division of Environmental Services. December 2012. Available online: www.rtdf.info (accessed on 1 November 2020).
155. CDWR. Lathrop Urban Runoff Study: Final Technical Report, Division of Environmental Services. February 2015. Available online: www.rtdf.info (accessed on 1 November 2020).
156. CDWR. Limnology of the State Water Project Nutrient Budget Study: Nutrients at the Hood Water Quality Station, Division of Environmental Services. September 2017. Available online: www.rtdf.info (accessed on 1 November 2020).

MDPI
St. Alban-Anlage 66
4052 Basel
Switzerland
Tel. +41 61 683 77 34
Fax +41 61 302 89 18
www.mdpi.com

Water Editorial Office
E-mail: water@mdpi.com
www.mdpi.com/journal/water

www.ingramcontent.com/pod-product-compliance
Lightning Source LLC
Chambersburg PA
CBHW040757250926
51850CB00007BA/1017
* 9 7 8 3 0 3 6 5 7 7 9 4 4 *